Advanced Therapy in Companion Animals

Advanced Therapy in Companion Animals

Guest Editors

Ziyao Zhou
Zhijun Zhong

MDPI

Basel • Beijing • Wuhan • Barcelona • Belgrade • Novi Sad • Cluj • Manchester

Guest Editors

Ziyao Zhou
Department of
Veterinary Surgery
Sichuan Agricultural
University
Chengdu
China

Zhijun Zhong
Department of
Veterinary Surgery
Sichuan Agricultural
University
Chengdu
China

Editorial Office
MDPI AG
Grosspeteranlage 5
4052 Basel, Switzerland

This is a reprint of the Special Issue, published open access by the journal *Veterinary Sciences* (ISSN 2306-7381), freely accessible at: https://www.mdpi.com/journal/vetsci/special_issues/537D79882B.

For citation purposes, cite each article independently as indicated on the article page online and as indicated below:

Lastname, A.A.; Lastname, B.B. Article Title. *Journal Name* **Year**, *Volume Number*, Page Range.

ISBN 978-3-7258-4739-6 (Hbk)
ISBN 978-3-7258-4740-2 (PDF)
https://doi.org/10.3390/books978-3-7258-4740-2

Contents

veterinary sciences

MDPI

Article

Primary Evaluation of Three-Dimensional Printing-Guided Endodontics in the Dog Maxillary

Chengli Zheng [1,2,†], Xiaoxuan Pan [1,†], Jiahui Peng [1,†], Xiaoxiao Zhou [3], Xin Shi [2], Liuqing Yang [2], Yan Luo [1], Haifeng Liu [1], Zhijun Zhong [1], Guangneng Peng [1], Min Yang [1,4,*], Ming Zhang [5,*] and Ziyao Zhou [1,*]

1 College of Veterinary Medicine, Sichuan Agricultural University, Chengdu 611130, China
2 Sichuan Institute of Musk Deer Breeding, Sichuan Institute for Drug Control, Chengdu 611830, China
3 Chengdu Center for Animal Disease Prevention and Control, Chengdu 610041, China
4 Pet Nutrition and Health Research Center, Chengdu Agricultural College, Chengdu 611130, China
5 College of Animal Science and Technology, Sichuan Agricultural University, Chengdu 611130, China
* Correspondence: yang040101@yeah.net (M.Y.); zhangming@sicau.edu.cn (M.Z.); zzhou@sicau.edu.cn (Z.Z.)
† These authors contributed equally to this work.

Simple Summary: Our previous research demonstrated that 3D printing in guided endodontics provides accurate positioning, direction, and length in the mandible region of dogs. However, it may have relative errors in some special teeth, particularly in the case of teeth with multi-roots. This study presents an advanced approach to veterinary endodontic treatments in the maxilla of Beagle dogs, particularly regarding the three-rooted fourth premolar. This study highlights the potential of 3D printing-guided endodontics to improve the precision and success rate of root canal treatments in veterinary dentistry. Limitations include breed-specific anatomical variations and the need for further clinical validation.

Abstract: This study aims to evaluate the feasibility and accuracy of 3D printing-guided endodontics in the maxillary teeth of dogs. CT data from a Beagle dog were processed to create a 3D model of the maxilla, and virtual root canal pathways were established using SOLIDWORKS software (version 29.0.0.5028). Guided endodontic templates were 3D printed and tested in vitro on 20 maxillary teeth (excluding the third molars), with 36 root canals treated using both guided and conventional methods. Results indicated that 3D printing-guided endodontics provided accurate root canal pathways, with minimal deviations in length (average 3.08 ± 1.75%) and angular alignment (average 2.06° ± 0.5°) compared to conventional methods. This research represents a significant step forward in the application of 3D printing technology in veterinary endodontics, offering a promising alternative to traditional methods for treating complex dental conditions in dogs.

Keywords: 3D printing; guided endodontics; maxillary teeth; finite element analysis; veterinary dentistry; root canal treatment

1. Introduction

Our previous research demonstrated that 3D printing in guided endodontics provides accurate positioning, direction, and length in the mandible region of dogs [1]. The benefits of guided endodontics are as follows: (1) finite element analysis (FEA), when merged with customized computed tomography (CT) data, can yield highly accurate representations of individual anatomical conditions and precise three-dimensional information [2–4]; (2) the ability to generate personalized implants, which can decrease the fabrication time and expenses [5]; (3) operative planning might considerably increase the success rate of surgical interventions [6].

However, this situation may have relative errors in some special teeth, particularly in the case of multi-rooted teeth [7]. In the maxilla of dogs, notably, the maxillary fourth premolar and first molar may present a challenge, as the mesiopalatal and buccal roots share a common canal open [8,9]. These complexities pose significant challenges for guided endodontics.

The aim of this study is to address these challenges by simulating dental pulp procedures through the interpretation of CT data, the establishment of mathematical models, precision 3D printing for endodontics, and the assessment of root canal pathways via radiographic analysis. Our study introduces an innovative method to overcome the difficulties encountered in veterinary dental pulp procedures in the maxilla of dogs, particularly regarding the three-rooted fourth premolar.

2. Materials and Methods

2.1. Creation of a 3D Model of the Beagle Maxillary

A Beagle dog without head trauma, scanned in our previous study, was investigated through computer modeling [1]. The dog was provided by the Sichuan Science and Technology Resources Sharing Platform for Beagle Dog Breeding and Experimental Technology Service. The CT scans (Figure 1A) were initially loaded into Mimics Medical (Materialise Co., Ltd., Leuven, Belgium, Version 22.0.306), which is a universal medical computer imaging software program in DICOM format (Figure 1B). A 3D mesh model of the Beagle maxillary was generated using the "New Mask" function. (Figure 1C).

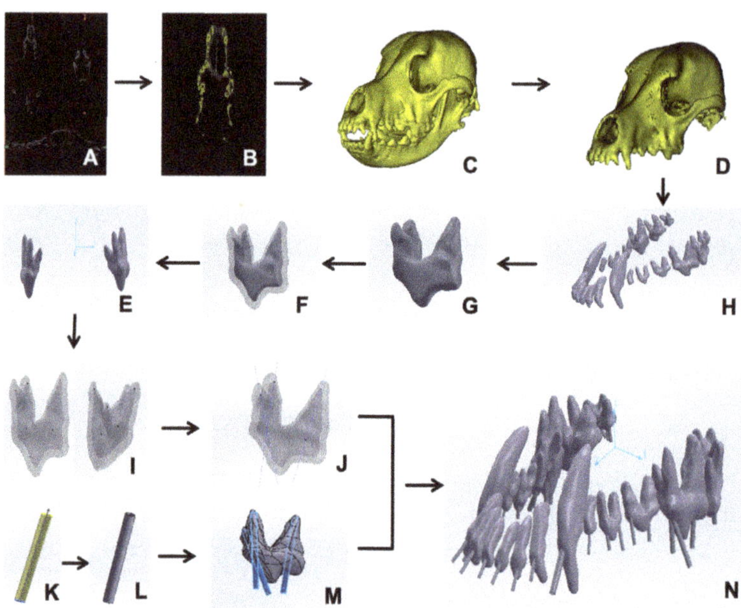

Figure 1. Scheme of the Beagle maxillary 3D model construction. (**A**) CT scanned data. (**B**) Head skull image reconstructed. (**C**) Preliminary 3D head skull model generated. (**D**) Separation of the maxillary skull. (**E**) Teeth segmented. (**F**) Individual tooth segmented. (**G**) Dental pulp reconstructed. (**H**) Dental coordinate reduction. (**I**) Determination of the center and apex of the root canal orifice of the tooth. (**J**) Identification of the root canal orientation. (**K**) Virtual root canal model. (**L**) Accurate root canal model. (**M**) The guided model fit to the pulp. (**N**) All maxillary teeth assembled and positioned.

Next, the stereolithography file was imported into Geomagic Wrap (Geomagic Co., Ltd., Research Triangle Park, NC, USA, version 2021.0.0.3008) for shell smoothing via the "Feature Removal" and "Hole Filling" tools. Subsequently, the "Precise Surface" options were harnessed to finely adjust the surface, permitting targeted modifications and eliminating noise, so that a refined 3D model of the Beagle maxillary bone (Figure 1D) was created. Eventually, the maxillary graphics obtained were saved in STP format and introduced into SOLIDWORKS (Dassault Systèmes S.A. Co., Ltd., Vélizy-Villacoublay, France, version 29.0.0.5028) for the purpose of materialization.

2.2. Extraction of Maxillary Teeth

In SOLIDWORKS, 20 maxillary teeth were identified and virtually extracted using the "Region Growing" function (Figure 1E). Each tooth was reconstructed using the "3D Editing" and "Boolean Operations" functions to eliminate artifacts, resulting in a refined surface polygon for each tooth (Figure 1F) [10]. The pulp chamber and dentin were separately reconstructed (Figure 1G) [10]. Within SOLIDWORKS, the "Coordinate System" feature was neatly employed to generate a 3D coordinate system made up of the XYZ axes, after which they were exported in STL format (Figure 1H).

2.3. Formulation of Root Canal Pathways and Guiding Templates

In SOLIDWORKS, the origin point (0, 0, 0) was designated as the anatomical midpoint of the skull. The "Reference Geometry" feature was used to ascertain the position of the peak point (x1, y1, z1) and of the central point of each dental root canal orifice (x2, y2, z2) for every tooth (Figure 1I). In theory, a direct line should link these two points within the three-dimensional space. Subsequently, the "Curve Through XYZ Points" feature was utilized to determine the linear pathway of the root canal for the maxillary tooth (Figure 1J).

The root canal guide template, with a thickness of 1 mm, was created from the line connecting the root canal to the tooth surface, serving as the guide for the procedure. The maximum diameter of the root canal was measured and the corresponding three-dimensional orifice was produced in a guided model (Figure 1K,L). The guide hole extends to the incisal edge to facilitate surgical maneuvering (Figure 1M). To safeguard the preciseness of each tooth's positioning, the entire skull model was likewise incorporated (Figure 1N). The resulting model was exported as an STL file for 3D printing.

2.4. Root Canal Experiment in Vitro

In vitro root canal experiments were conducted on two distinct Beagle skull models provided by the Sichuan Science and Technology Resources Sharing Platform for Beagle Dog Breeding and Experimental Technology Service. A total of 36 root canals were performed on maxillary teeth, totaling 20 teeth (including 6 incisors, 2 canines, 8 premolars, and 4 molars), using guided endodontic and conventional treatment methods, respectively. The first Beagle maxillary model was utilized for guided endodontic treatment, with initially printed templates for each tooth according to the manufacturer's instructions (SLA550Li, material: photosensitive resin; Ultrust Imaging (Chengdu) Center Co., Ltd., Chengdu, China). The fit of the template to the printed corresponding tooth was verified for accuracy and reproducibility.

This study focused on evaluating the relative error in root canal length measurement in two methods. Therefore, the method did not involve complete root canal treatment. The specific procedures were described according to the guidelines for reporting case reports in endodontics, specifically from the PRICE explanation and elaboration document [11] as follows: A high-speed dental drill (SKU: VetPro 1000, from Midmark, Versailles, OH, USA) was entered into the template to create an access cavity. The pulp chamber was irrigated with sodium hypochlorite to remove residual pulp tissue. Moisture control was achieved using sterile paper points. Standardized endodontic K-files (ISO 3630-1 [12] specifications) (yellow: diameter 0.20 mm, length 21 mm, and taper 0.02; red: diameter 0.25 mm, length 21 mm, and taper 0.02; blue: diameter 0.30 mm, length

21 mm, and taper 0.02) were employed for canal exploration and length determination using a dental X-ray (Ningbo Juxing Peteeth Medical Smart Technology Co., Ltd., Ningbo, China).

Conventional endodontic treatment was performed on another Beagle maxilla, which followed the same procedure as the guided endodontics treatment but without employing the template. In both methods, the root canal length was measured from the center point of the root canal opening to the bottom of the pulp chamber. A comparative assessment of angular deviation was also performed to evaluate the accuracy of the guided root canal treatment versus the conventional techniques.

2.5. Statistical Analysis

Statistical assessments were carried out on the measured root canal lengths and angular discrepancies between the two methods. The comparison of data distributions was executed using the Student's t-test via SPSS v27 (IBM, New York, NY, USA). A p-value below 0.05 was regarded as statistically significant.

3. Results

3.1. Computer-Aided Design of the Root Canal Guided Lines

Based on a three-dimensional coordinate system, a total of 20 guided endodontic templates were designed for 20 maxillary teeth. The coordinates of the apex of the pulp chamber (x1, y1, z1) and the central point of each dental root canal orifice (x2, y2, z2) for each maxillary tooth are listed in Table 1. Taking the left fourth premolar (208) as an illustration, the diameters of the two pulp crowns were 4.70 mm and 4.27 mm, respectively. The coordinates of the mesial palatine root in 3D space were measured as (19.84, 5.76, 1.27) and (22.58, −17.97, 2.63). The coordinates of the mesial buccal root were (15.53, 0.21, 1.64) and (21.37, −8.46, 2.57). The coordinates of the distal root were (23.00, 3.75, −9.03) and (24.10, −9.05, −5.70). Three root canal guide lines can be generated by connecting the lines based on the coordinates.

Table 1. Coordinates of the apex point (x1, y1, z1) and the central point of the dental root canal orifice (x2, y2, z2) for each tooth.

Tooth Number	Root Position	Apex Point (x_1, y_1, z_1)	Central Point of the Dental Root Canal Orifice (x_2, y_2, z_2)	Tooth Number	Root Position	Apex Point (x_1, y_1, z_1)	Central Point of the Dental Root Canal Orifice (x_2, y_2, z_2)
101	/	(−1.84, 5.78, 51.91)	(−0.88, 1.03, 60.62)	201	/	(3.19, 5.84, 51.10)	(4.01, 1.42, 59.73)
102	/	(−4.55, 6.43, 49.65)	(−5.85, 1.17, 59.41)	202	/	(5.74, 7.03, 48.82)	(8.81, 1.21, 58.28)
103	/	(−6.89, 7.03, 46.31)	(−10.34, 0.93, 55.93)	203	/	(6.85, 7.40, 43.00)	(11.84, 2.32, 54.22)
104	/	(−12.30, 13.23, 23.33)	(−13.64, −0.42, 44.52)	204	/	(9.42, 14.41, 25.21)	(13.98, 1.30, 42.73)
105	/	(−13.72, 3.27, 29.12)	(−14.22, −2.82, 33.54)	205	/	(11.87, 3.96, 27.36)	(14.44, −1.42, 31.66)
106	mesial	(−14.52, 1.48, 24.99)	(−14.58, −5.28, 26.86)	206	mesial	(12.36, 3.40, 23.84)	(14.84, −2.93, 25.23)
	distal	(−14.47, 0.42, 19.42)	(−16.16, −5.03, 21.69)		distal	(12.97, 0.95, 18.55)	(15.86, −3.76, 20.54)
107	mesial	(−15.72, 0.36, 13.91)	(−17.09, −6.14, 14.57)	207	mesial	(13.50, 1.32, 13.13)	(16.15, −3.96, 15.00)
	distal	(−18.44, −0.84, 8.17)	(−20.25, −7.27, 9.93)		distal	(16.62, 0.40, 6.68)	(19.04, −5.43, 9.79)
108	mesiobuccal	(−17.27, −0.41, 3.09)	(−22.41, −10.18, 3.83)	208	mesiobuccal	(15.53, 0.21, 1.64)	(21.37, −8.46, 2.57)
	mesiopalatal	(−23.07, 3.62, 3.06)	(−22.41, −10.18, 3.83)		mesiopalatal	(19.84, 5.76, 1.27)	(22.58, −17.97, 2.63)
	distal	(−26.45, 1.46, −6.39)	(−25.10, −10.69, −3.45)		distal	(23.00, 3.75, −9.03)	(24.10, −9.05, −5.70)
109	mesiobuccal	(−27.84, −3.89, 10.04)	(−25.47, −13.04, −11.20)	209	mesiobuccal	(16.34, 0.19, −15.28)	(17.62, −8.05, −14.04)
	mesiopalatal	(−20.62, 2.96, −14.25)	(−19.85, −9.75, −12.74)		mesiopalatal	(24.37, −0.17, −18.00)	(22.86, −8.32, −16.81)
	distobuccal	(−28.12, −3.51, −16.38)	(−25.14, −12.14, −14.89)		distobuccal	(24.83, −1.94, −12.00)	(24.54, −9.57, −13.16)
110	mesiobuccal	(−24.82, 5.03, −19.16)	(−22.23, −8.93, −20.12)	210	mesiobuccal	(16.49, −1.93, −24.18)	(16.80, −6.21, −24.17)
	mesiopalatal	(−18.22, −2.86, 19.99)	(−17.48, −7.54, −20.01)		mesiopalatal	(13.21, −1.88, −21.51)	(13.84, −6.00, −21.97)
	distobuccal	(−21.29, 4.10, −22.51)	(−19.96, −8.32, −21.95)		distobuccal	(20.78, −2.00, −20.96)	(18.96, −7.39, −21.85)

3.2. Three-Dimensional-Printed Maxillary and Guided Endodontic Templates

Three-dimensional-guided endodontic templates were printed based on the created pattern, demonstrating a precise fit between the template and the printed teeth (Figure 2A). Specifically, the left fourth premolar (208) measured 15.35 mm in length, 6.69 mm in width, and 21.30 mm in height. The template dimensions were 15.21 mm × 8.62 mm × 11.29 mm, which were fully compatible with the printed tooth (Figure 2B). The root canal pathway was also fitted to the pulp chamber by using the Beagle maxillary model (Figure 2C,D).

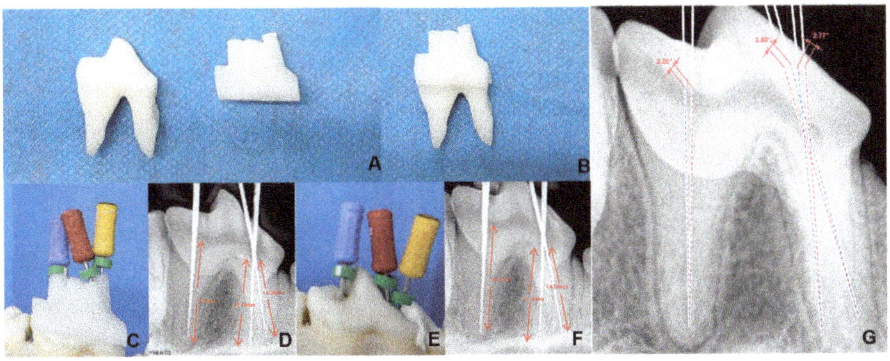

Figure 2. Comparison of treatments by guided endodontics and conventional root canal methods. (**A**) The 3D-printed left fourth premolar tooth and its template. (**B**) Combination of 3D-printed root canal template and the printed tooth for verifying the fits. (**C**) Guided root canal treatment method. (**D**) Radiograph-based guided endodontics. (**E**) Removal of guided molding. (**F**) Radiograph-based conventional root canal method. (**G**) Comparison of root canal angles between the guided (blue line) and conventional (red line) root canal treatment methods.

3.3. Comparison of 3D Printing-Guided Endodontics and Conventional Root Canal Methods

The root canal treatment experiments were performed and evaluated using 36 maxillary pulp cavities of Beagle dogs. The left fourth premolar (208), which possesses a complex root, was selected for X-ray scanning, and the tip of the gum was observed to be straight to the apex without bending, damage, or excessive length. The length of the mesiopalatal root and mesiobuccal root canal was 10.49 mm and 14.26 mm, respectively; meanwhile, the distal root canal measured 14.02 mm (Figure 2D). In contrast, the length of the mesiopalatal and mesiobuccal root canal using the conventional method was 11.19 mm and 14.83 mm, respectively, and the distal root canal measured 13.27 mm (Figure 2F).

The angular deviations of the root canal pathway between the two methods were 2.77° (mesiopalatal root), 2.83° (mesiobuccal root), and 2.20° (distal root), respectively (Figure 2G). To enhance clarity and readability, the root canal orientations of the guided sleeve method and the conventional method were plotted separately in Figure 2G.

Of the 36 root canals, the relative deviations between the guided endodontics and conventional methods ranged from 0.16% (106 distal root) to 6.67% (208 mesiobuccal root), with an average deviation of 3.08% ± 1.75%. The mean angular deviation was 2.06° ± 0.5°, ranging from 1.17° (201 root) to 2.41° (107 mesial root) (Table 2).

Of the maxillary teeth, the premolars exhibited the most considerable variation in root canal length, amounting to 3.55% ± 2.03% of the standard deviation. This variation was significantly higher than that of the incisors (1.44% ± 0.66%, $p = 0.022$) but was similar to that of the molars (3.60% ± 1.03%, $p = 0.932$). However, there was no significant difference in angular deviation among the teeth, with the incisors measuring 1.45° ± 0.35° ($p = 0.6778$ for incisors vs. premolars), premolars 2.35° ± 0.39° ($p = 0.9825$ for premolars vs. molars), and molars 2.09° ± 0.32° ($p = 0.7806$ for incisors vs. molars), respectively.

Table 2. Length measurement of each root canal using guided and conventional techniques.

Tooth Number	Root Position	Guided Endodontics/mm	Classic Method/mm	Deviation Rate	Angular Deviation	Tooth Number	Root Position	Guided Endodontics/mm	Classic Method/mm	Deviation Rate	Angular Deviation
101	/	9.97	9.87	1.00%	2.38°	201	/	9.73	9.5	2.36%	1.17°
102	/	11.16	11.34	1.61%	1.65°	202	/	11.52	11.75	2.00%	2.03°
103	/	11.9	11.98	0.67%	2.08°	203	/	13.29	13.42	0.98%	1.41°
104	/	24.01	24.46	1.87%	1.28°	204	/	23.92	23.82	0.42%	1.34°
105	/	7.55	7.47	1.06%	1.81°	205	/	7.36	7.19	2.31%	2.24°
106	mesial	7.12	7.41	4.07%	2.17°	206	mesial	6.93	6.85	1.15%	1.31°
	distal	6.14	6.13	0.16%	1.45°		distal	5.88	6.13	4.25%	1.64°
107	mesial	6.67	6.54	1.95%	2.41°	207	mesial	6.20	6.56	5.81%	2.4°
	distal	6.90	6.84	0.87%	1.77°		distal	7.04	6.79	3.55%	1.68°
108	mesiobuccal	11.06	11.67	5.52%	2.21°	208	mesiobuccal	10.49	11.19	6.67%	1.83°
	mesiopalatal	13.63	14.31	4.99%	2.27°		mesiopalatal	14.26	14.83	4.00%	1.77°
	distal	12.54	11.91	5.02%	1.72°		distal	14.02	13.27	5.35%	2.2°
109	mesiobuccal	9.53	9.17	3.78%	1.61°	209	mesiobuccal	7.73	7.61	1.55%	1.41°
	palatal	6.99	7.29	4.29%	2.25°		palatal	8.65	8.43	2.54%	1.98°
	distobuccal	9.24	9.57	3.57%	2.07°		distobuccal	8.38	8.75	4.42%	1.36°
110	mesiobuccal	4.77	4.97	4.19%	1.75°	210	mesiobuccal	5.55	5.39	2.88%	2.17°
	palatal	4.94	5.11	3.44%	2.18°		palatal	4.29	4.41	2.80%	1.87°
	distobuccal	5.73	5.47	4.54%	1.94°		distobuccal	4.40	4.63	5.23%	2.28°

4. Discussion

Currently, endodontic treatment in veterinary dentistry mostly relies on the operator's clinical knowledge base and expertise. Several potential causes of failure during pulp access include the following: (1) deviations during entry in the root canal treatment, such as incorrect angles of access to the pulp chamber, incomplete exposure of the chamber, inadequate exposure of the canal orifice, perforations in the pulp chamber floor or canal walls, overlooked canals, or incorrect interpretation of canal anomalies (e.g., S-shaped canal configurations) [13,14]; (2) deviations in canal preparation, which may involve excessive canal enlargement, ledge formation, instrument separation, or incomplete cleaning of the canal. For instance, Lee et al. [15] evaluated the outcomes of endodontic treatments in 204 dogs, noting that multi-root teeth with more complex canal morphologies may result in higher fail rates. Therefore, developing an adjunctive endodontic method with computer-aided design contributes to diminishing the likelihood of complications arising and to enhancing the overall outcome for patients [16,17].

Laboratory and in vitro investigations have demonstrated considerable precision in aligning actual canal trajectories with pre-determined pathways through the application of guided procedures [18,19]. Our previous research indicated that guiding root canal treatment in dogs' mandibular canals is comparable to traditional endodontic techniques employed by dental specialists [1]. This method enables veterinarians to formulate feasible plans for actual clinical procedures, aiming for optimal positioning and angles, ultimately minimizing the removal of valuable dental structures. However, certain complex canal situations remain unknown, particularly in the maxillary fourth premolars, which are among the most commonly fractured and treated teeth in dogs [20]. In this study, our findings suggest that utilizing a 3D-printed guide for root canal treatment can be quite suitable for maxillary teeth, including the frequently encountered complex fourth premolars. However, compared to single-rooted teeth (such as incisors and canines), the use of guided endodontics in multi-rooted teeth indeed results in greater discrepancies. This outcome may be attributed to the more complex fit of the guiding template with the canals, making it easier to introduce angulation errors during clinical procedures. Consequently, it is highly recommended to calibrate the coordinates and dimensions in SOLIDWORKS software, before printing 3D root canal templates, to avoid mismatches between the template and canals after final assembly. This is particularly evident when two or more canals share a common open pulp chamber, significantly affecting the procedural difficulty [21]. However, this issue is a result of the anatomical structure of the teeth rather than our method. In traditional approaches, endodontic treatment of these multi-rooted teeth is also complex and has a high failure rate [22]. We hope our method might serve as a better solution for these intricate canal procedures.

Nonetheless, this study has some inherent limitations, including the following: (1) the guided pulp templates and measured parameters are individualized for Beagles, which means that they may not be applicable to other breeds; (2) even within the same breed, variations in size, sex, age, and dental structure can lead to significant differences in root canal templates produced from dogs of the same breed; (3) the addition of CT scans and the production of 3D guiding templates may extend operation and anesthesia times, which can be particularly impactful for complex dental cases and older animals, posing significant challenges to practitioners' skills and experience [23]. In summary, FEA-guided endodontic treatment represents a safe and direct approach, especially considering the further developmental potential of 3D printing technology [24,25]. However, FEA-guided endodontic treatment may require additional equipment and increase the cost of veterinary dentistry. Additional clinical research might be needed to fully determine the clinical advantages of these new guiding techniques.

Author Contributions: Conceptualization, X.Z. and Z.Z. (Ziyao Zhou); Data curation, C.Z. and Z.Z. (Zhijun Zhong); Investigation, C.Z., X.P. and H.L.; Methodology, J.P., X.S. and L.Y.; Resources, X.Z. and Y.L.; Software, J.P. and X.Z.; Supervision, M.Y. and M.Z.; Writing—original draft, C.Z. and X.P.; Writing—review and editing, G.P., M.Z. and Z.Z. (Ziyao Zhou). All authors have read and agreed to the published version of the manuscript.

Funding: This research was funded by the Natural Science Foundation of Sichuan Province (24NSFSC1281) and the Science and Technology program of the Sichuan Medical Products Administration (2024010).

Institutional Review Board Statement: The animal use protocol for the study has been cleared by the Animal Ethical and Welfare Committee of Sichuan Agricultural University (approval number: 20220202; date: 27 June 2022). The study was conducted in accordance with the local legislation and institutional requirements.

Informed Consent Statement: Not applicable.

Data Availability Statement: The original contributions presented in this study are included in the article. Further inquiries can be directed to the corresponding authors.

Acknowledgments: We would like to acknowledge the support of the Sichuan Science and Technology Resources Sharing Platform for Beagle Dog Breeding and Experimental Technology Service.

Conflicts of Interest: The authors report no conflicts of interest.

References

1. Peng, J.; Yang, J.; Liu, R.; Liu, H.; Zhong, Z.; Peng, G.; Zhang, K.; Zheng, C.; Zhang, M.; Zhou, Z. Evaluating the concept of three-dimensional printing guided endodontics in the dog. *Front. Vet. Sci.* **2024**, *11*, 1481612. [CrossRef] [PubMed]
2. Li, Z.; Zhang, B.; Fang, B.; Gong, H.; Han, Y.; Pei, S.; Zhang, S.; Song, G. Finite element analysis of a three-dimensional cervical spine model with muscles based on CT scan data. *Comput. Methods Biomech. Biomed. Eng.* **2024**, *4*, 1–11. [CrossRef] [PubMed]
3. Liu, R.; Yang, J.; Zhu, Y.; Zhou, X.; Zhou, Q.; Liang, T.; Wang, H.; Luo, Y.; Xie, Y.; Liu, H.; et al. A forecasting model for suitable dental implantation in canine mandibular premolar region based on finite element analysis. *BMC Vet. Res.* **2024**, *20*, 353. [CrossRef]
4. Rodillo, M.; Cunha, R.S.; Bauer, P.A.; Travan, S.; Cavalcanti, B.; McDonald, N. Incidence of missed canals during endodontic treatment of maxillary first and second molars. *Gen. Dent.* **2024**, *72*, 10–14. [PubMed]
5. Prabhuji, V.; Archana, S.; Srirekha, A.; Pai, V.; Champa, C.; Shetty, A. Comparative evaluation of efficacy of guided endodontic technique with and without sleeve for access cavity preparation: An In Vitro study. *J. Oral Biol. Craniofacial Res.* **2024**, *14*, 825–829. [CrossRef]
6. Banka, A.; Patri, G.; Pradhan, P.K.; Lath, H. Comparative evaluation between guided endodontics and conventional techniques for calcific metamorphosis—A systematic review and meta-analysis. *J. Conserv. Dent. Endod.* **2024**, *27*, 891–896. [CrossRef]
7. AlTamimi, E.A.; Agwan, M.A.S.; Ahmad, M.Z. Endodontic Management of Failed Root Canal Treatment in Teeth with Previously Missed Canals: A Report of Two Cases with Rare Root Canal Morphology. *Cureus* **2024**, *16*, e73466. [CrossRef]
8. Hernández, S.Z.; Negro, V.B.; Maresca, B.M. Morphologic Features of the Root Canal System of the Maxillary Fourth Premolar and the Mandibular First Molar in Dogs. *J. Vet. Dent.* **2001**, *18*, 9–13. [CrossRef]
9. Jinesh, A.; Nair, S.J.; Gupta, S.; Chansoria, H.; Rawat, G.; Jinesh, A., Jr.; Chansoria, H.; Rawat, G. Unusual Canal Morphology in Mandibular Premolars with Two Distal and One Mesial Canal: A Case Series. *Cureus* **2024**, *16*, e73832.
10. Maravić, T.; Vasiljević, D.; Kantardžić, I.; Lainović, T.; Lužanin, O.; Blažić, L. Influence of restorative procedures on endodontically treated premolars: Finite element analysis of a CT-scan based three-dimensional model. *Dent. Mater. J.* **2018**, *37*, 493–500. [CrossRef]
11. Nagendrababu, V.; Chong, B.; McCabe, P.; Shah, P.; Priya, E.; Jayaraman, J.; Pulikkotil, S.; Dummer, P. PRICE 2020 guidelines for reporting case reports in Endodontics: Explanation and elaboration. *Int. Endod. J.* **2020**, *53*, 922–947. [CrossRef] [PubMed]
12. *ISO 3630-1:2008*; Dentistry—Root-Canal Instruments—Part 1: General Requirements and Test Methods. International Organization for Standardization (ISO): Geneva, Switzerland, 2008.
13. Armenta, H.B.; Mireles, A.G.R.; Martinez, J.S.; Alvarez, J.P.; Herrera, Y.R.; Maldonado, O.T.; Ojeda, O.A.; Pacheco, J.S.; Sarabia, G.O.; Juarez, F.X.C.; et al. Prevalence and Classification of C-Shaped Canal and Radix in Mandibular Molars Using Cone-Beam Computed Tomography on Mexican Population. *Dent. J.* **2024**, *12*, 212. [CrossRef]

14. Prasad, P.; Galani, M.; Nawal, R.R.; Talwar, S.; Kumar, G. Three-dimensional Assessment of Two-rooted Maxillary Central Incisor with Labiogingival Groove: Endodontic and Periodontal Surgical Management. *Int. J. Clin. Pediatr. Dent.* **2023**, *16*, 528–533. [PubMed]
15. Lee, D.B.; Arzi, B.; Kass, P.H.; Verstraete, F.J. Radiographic outcome of root canal treatment in dogs: 281 teeth in 204 dogs (2001–2018). *J. Am. Vet. Med. Assoc.* **2022**, *260*, 535–542. [CrossRef]
16. Dan, Z.; Weige, X.; Tianguo, L.; Anqi, W.; Li, W.; Wen, K.; Lu, W.; Shiliang, G.; Xuna, T.; Sijing, X. New-designed 3D printed surgical guide promotes the accuracy of endodontic microsurgery: A study of 14 upper anterior teeth. *Sci. Rep.* **2023**, *13*, 15512.
17. Ahn, S.-Y.; Kim, N.-H.; Kim, S.; Karabucak, B.; Kim, E. Computer-aided Design/Computer-aided Manufacturing–guided Endodontic Surgery: Guided Osteotomy and Apex Localization in a Mandibular Molar with a Thick Buccal Bone Plate. *J. Endod.* **2018**, *44*, 665–670. [CrossRef]
18. Torres, A.; Dierickx, M.; Lerut, K.; Bleyen, S.; Shaheen, E.; Coucke, W.; Pedano, M.S.; Lambrechts, P.; Jacobs, R. Clinical outcome of guided endodontics versus freehand drilling: A controlled clinical trial, single arm with external control group. *Int. Endod. J.* **2024**, *58*, 209–224. [CrossRef]
19. Buchgreitz, J.; Buchgreitz, M.; Bjørndal, L. Guided Endodontics Modified for Treating Molars by Using an Intracoronal Guide Technique. *J. Endod.* **2019**, *45*, 818–823. [CrossRef]
20. Tandir, F.; Avdić, R.; Dučić, N.; Džanković, A.; Tandir, R.; Šaljić, E.; Vejzović, A.; Hadžiomerović, N. Position and frequency of lateral canals in carnassial teeth of dogs. *Res. Vet. Sci.* **2024**, *179*, 105384. [CrossRef]
21. Fossati, C. Retreatment of a mandibular second premolar with two canals in two roots. *G. Ital. Endod.* **2018**, *32*, 31–35. [CrossRef]
22. SunMi, K.; Eunsuk, A. Tooth Survival Following Non-Surgical Root Canal Treatment in South Korean Adult Population: A 11-Year Follow-Up Study of a Historical Cohort. *Eur. Endod. J.* **2022**, *7*, 20–26.
23. Fernandes, G.W.; Roberti, G.L.d.F.; Peressoni, V.D.; Antunes, B.E.; Lima, D.L.C.d.; Silveira, T.C.d. Guided Endodontics in Root Canals with Complex Access: Two Case Reports. *Braz. Dent. J.* **2021**, *32*, 115–123.
24. Wei, X.; Du, Y.; Zhou, X.; Yue, L.; Yu, Q.; Hou, B.; Chen, Z.; Liang, J.; Chen, W.; Qiu, L.; et al. Expert consensus on digital guided therapy for endodontic diseases. *Int. J. Oral Sci.* **2023**, *15*, 540–551. [CrossRef]
25. Abdelhafeez, M.M. Applications of Finite Element Analysis in Endodontics: A Systematic Review and Meta-Analysis. *J. Pharm. Bioallied Sci.* **2024**, *16*, S1977–S1980. [CrossRef] [PubMed]

veterinary sciences

MDPI

Article

Telomere Tales: Exploring the Impact of Stress, Sociality, and Exercise on Dogs' Cellular Aging

Luisa Mascarenhas Ladeia Dutra [1,2,*], Flaviane S. Souza [1,3], Angelica Silva Vasconcellos [4], Robert J. Young [1] and Ivana Gabriela Schork [5]

1 School of Science, Engineering & Environment, University of Salford, Manchester M5 4WT, UK; flaviane@ufrb.edu.br (F.S.S.); r.j.young@salford.ac.uk (R.J.Y.)
2 Instituto de Educação Continuada (IEC), Pontifícia Universidade Católica de Minas Gerais, Belo Horizonte 30535-901, Brazil
3 Centro de Ciências da Saúde, Universidade Federal do Recôncavo da Bahia, Cruz das Almas 44380-000, Brazil
4 Programa de Pós-Graduacão em Biodiversidade e Meio Ambiente, Pontifícia Universidade Católica de Minas Gerais, Belo Horizonte 30535-901, Brazil; angelicavasconcellos@pucminas.br
5 Department of Animal and Agriculture, Hartpury University, Gloucester GL19 3BE, UK; ivana.schork@hartpury.ac.uk
* Correspondence: lmldutra@sga.pucminas.br or lulu.mascarenhas@gmail.com

Simple Summary: Dogs, like humans, experience aging, which is influenced by their environment and life experiences. This study explored how different living conditions and life histories affect the aging process in dogs by examining their telomeres—structures located at the ends of chromosomes that preserve the structural integrity of DNA and, therefore, prevent degradation associated with cellular senescence. Shorter telomeres are linked to stress and aging. Through a non-invasive method to collect DNA from dogs, our study investigated the relationship between telomere length and dogs' backgrounds. We found that dogs living in varied environments, such as family homes, kennels, or working as police dogs, showed different rates of telomere shortening. Interestingly, while factors like age and breed did not strongly influence aging, dogs in stressful or less enriching conditions, such as laboratories, had the shortest telomeres. In contrast, working dogs often had longer telomeres, possibly due to structured environments and meaningful activities. These findings help us understand how a dog's life circumstances affect its health and aging, providing valuable insights for improving animal care, health monitoring, and welfare policies. Considering positive and negative experiences, this research could also guide how we support aging in pets and working animals, ensuring healthier lives.

Abstract: Animal welfare is influenced by the cumulative life experiences of an individual. Among these, exposure to chronic stressors has a significant impact on both physical and mental health, contributing to premature aging—a process linked to telomere shortening. Conversely, positive experiences have been shown to mitigate, delay, and sometimes reverse telomere attrition. This suggests that telomere length could be a reliable indicator for assessing animal welfare. This study explored the association between telomere length and characteristics such as life history, environment, and health in domestic dogs. Buccal swabs collected DNA samples from 250 dogs, and telomere length was quantified via qPCR. Our findings revealed that environmental factors significantly influenced telomere length. Dogs housed in kennels or subjected to low physical activity levels exhibited shorter telomeres. Similarly, dogs living in groups of more than five dogs had shorter telomeres, and male dogs were found to have longer telomeres than females. Overall, these results highlight the importance of environmental conditions in influencing telomere length in dogs and the potential to use this biological indicator to evaluate animal welfare.

Vet. Sci. **2025**, *12*, 491 https://doi.org/10.3390/vetsci12050491

Keywords: chronic stress; cumulative experiences; aging; animal welfare

1. Introduction

Dogs (*Canis familiaris*) are not only valued as companions but also for their ability to work closely with humans, a role facilitated by their social nature and trainability [1]. They contribute significantly to various domains such as farming, hunting, guarding, military operations, search and rescue missions, and therapy. However, the demands associated with these roles can negatively impact their welfare [2,3]. For instance, police and military dogs often face acute stress during their duties, whereas laboratory dogs are more prone to chronic stress [4]. Similarly, even pet dogs can experience considerable differences in welfare depending on the household environment, quality of care, and ownership practices [5].

Assessing animal welfare requires tailoring methods to individuals' specific circumstances and environments. However, dogs' diverse lifestyles complicate the evaluation of their well-being. Moreover, many physiological indicators, such as stress hormone levels, reflect stress responses from the past hours or days, offering limited insight into experiences that occurred years earlier but may have significantly influenced the individual's overall quality of life [6].

Telomeres are segments of DNA that protect the ends of chromosomes, gradually shortening with each cellular division. This progressive reduction contributes to replicative senescence or cell death through apoptosis [7]. Factors such as an unhealthy lifestyle (e.g., poor diet) [8] or environmental stressors (e.g., chronic stress) [8] can accelerate telomere shortening. Research on species such as macaques [9], domestic dogs [10], and birds [11] has demonstrated a link between telomere attrition and stress. Evidence suggests that positive experiences may mitigate or even reverse telomere attrition, indicating that telomere length could provide insights into an animal's cumulative life experiences and overall welfare status, whether positive or negative [12–15].

Since stress is a key parameter in animal welfare assessments, telomere shortening is proposed as a potential welfare indicator [12]. Previous studies have explored the relationship between telomere shortening and the most common method of stress assessment, cortisol levels. Some report significant associations between telomere shortening and serum or hair cortisol, while others have found mixed or inconclusive results [12,13]. These discrepancies may be due to species-specific differences, variations in stress measurement methods, or the influence of additional confounding factors.

Research on telomere attrition in dogs has primarily focused on understanding the relationship between telomere length, mortality, lifespan, and the activity of telomerase, an enzyme that preserves telomere integrity in dog tissues [10,14]. Findings suggest that smaller dog breeds have longer telomeres, correlating with their extended lifespans, while larger breeds tend to age faster and have shorter lifespans due to their shorter telomeres [10,14]. Additionally, telomeres play a pivotal role in regulating the lifespan of somatic cells [15]. However, to date, no studies have specifically investigated the connection between telomere attrition and dog welfare.

This study presents the development and validation of a non-invasive method to assess the quality of life in dogs by measuring relative telomere length (rTL) and examining its association with the dogs' background.

2. Materials and Methods

2.1. Ethical Statement

This study was conducted following ethical guidelines and was approved by the relevant ethics committees in Brazil (process no. 0056/2017) and the United Kingdom from University of Salford Ethics Committee (approval number STR1617-22). Biological materials were collected under DEFRA license ITIMP16.1096. No additional ethical approval was required for sample collection at this facility. All institutions and organizations will remain anonymous in accordance with GDPR guidelines and the agreement between the University of Salford and the third parties involved.

2.2. Subjects and Life History Questionnaire

This study included 250 domestic dogs from diverse backgrounds for telomere length assessment. The dogs' DNA samples were collected from various locations in Brazil and Europe (Table 1). The dogs were classified by sex, age, and background, which was divided into five categories: companion animals, shelter dogs, working dogs, laboratory dogs, and rehomed individuals (Table 1).

Table 1. Summary of the characteristics of the different dog groups sampled for relative telomere length assessment.

Background	N	Males	Females	Age (±SD)
Pet	84	45	39	4.05 ± 3.42
Shelter	56	33	21	4.00 ± 2.25
Work (BR)	46	32	14	3.21 ± 2.21
Work (UK)	20	17	3	5.27 ± 4.58
Laboratory	21	8	13	3.50 ± 2.49
Rehomed	23	9	14	1.80 ± 1.33

To limit confounding factors that could compromise the interpretation of our results, we conducted a prescreening selection of dogs. To the best of our knowledge, none of the dogs in the study had long-term chronic conditions or exhibited any recurrent behavioral issues that could affect their overall well-being.

To explore the potential impact of the dogs' environment and life history on their telomere length, a detailed questionnaire was administered to gather information on multiple aspects of each dog's life. Respondents provided details regarding (i) breed, (ii) previous ownership (whether the dog was with its first owner or rehomed), (iii) neutering status, (iv) overall health condition, (v) training background, (vi) dietary habits, (vii) sleeping site, (viii) exercise frequency, and (ix) social interactions with both humans and other animals. Some of these variables were not included in the final analysis due to data completeness constraints and to maintain model clarity. Shelter dogs were group-housed in large enclosures outside, and the duration of stay in the shelter varied individually. Research dogs were group-housed in 900 m^2 enclosures outside. Shelter dogs had a doghouse available in every enclosure that could fit all the pack, keeping them safe from any weather conditions. Laboratory dogs were pair-housed under standardized care conditions, following ethical and welfare guidelines.

The breed of each dog was determined based on the UK Kennel Club criteria [16], with dogs of unknown breed categorized as "Mix". According to these criteria, dogs were grouped into categories: Gundogs, Hounds, Pastorals, Terriers, Toys, Utility, Work, and Mix.

An additional categorization was used to differentiate dogs that fell under the working dog category. For these dogs, animals were grouped based on their work location, either UK or Brazilian working dogs. However, this was not intended solely to reflect the geographical location of the dogs, but rather to highlight the distinct characteristics of this type of activity in each country, such as training and management. The Supplementary Material (Supplementary Tables S1–S3) provides a comprehensive list of individual characteristics for all 250 dogs and the options provided in the questionnaire.

2.3. DNA Sampling and Telomere Length Measurement

Saliva and tissue samples to extract DNA were collected using buccal swabs. The swab was gently rubbed against the inner cheek of the dog's mouth for a few seconds to extract the material. The procedure was carried out by either a veterinarian, the owner, or the caretaker to minimize the animal's stress. Positive reinforcement was used during the procedure to encourage the dog's cooperation.

Genomic DNA was isolated from the buccal cell samples using the Buccalyse DNA Release Kit, following the manufacturer's guidelines. The concentration of the extracted DNA was then quantified with the NanoDrop® ND-1000 UV-Vis Spectrophotometer (Thermo Fisher Scientific, Waltham, MA, USA). We set a target DNA concentration of 100 ng/µL as the standard for inclusion in the reactions. Any samples with concentrations higher than this threshold were diluted to ensure minimal variation across samples, helping maintain consistency in DNA concentration throughout the analysis.

Quantitative PCR (qPCR) was conducted to assess telomere length. Each qPCR run incorporated two components: (i) a tenfold serial dilution derived from a pooled DNA sample of 250 dogs at a concentration of 100.7 ng/µL, which was used to generate the standard curve and optimize the primers; (ii) no-template controls (NTC) to identify potential contamination and primer dimer formation [17]. To minimize plate-to-plate variation, all reactions were performed under identical conditions using the same reagent batches, primer sets, and instrument settings. Although inter-run calibrators (IRCs) were not used in this study, standard curve normalization and technical replicates (each sample run in triplicate) were employed to ensure consistency across runs [18]. The primers employed in this study consisted of telomere-specific primers, telg (5′-ACACTAAGGTTTGGGTTTGGGTTTGGGTTTGGGTTAGTGT-3′) and telc (5′-TGTTAGGTATCCCTATCCCTATCCCTATCCCTATCCCTAACA-3′), as described in [19]. Additionally, 18S primers were used for the reference gene, including 18S sense (5′-GAGGTGAAATTCTTGGACCGG-3′) and 18S antisense (5′-CGAACCTCCGACTTTCGTTCT-3′), from a previously established study [20]. Primer performance in qPCR reactions was assessed using Rotor-Gene Q Series software (version 2.3.1), with efficiency values ranging between 98% and 100%.

In this research, we utilized a monochrome multiplex qPCR assay, adapting the approach described by [18]. The master mix reactions followed the protocol established by [21], and amplifications were conducted on a Rotor-Gene® Q cycler (QIAGEN) equipped with Rotor-Gene Q software (version 2.3.1), using 0.1 mL strip tubes and caps. The Rotor-Gene Q Series software created a standard curve, considering the dilution factors for telomere and reference genes in each sample. The master mix preparation for the qPCR reaction followed the protocol established by [19] and validated for dog samples by [21]. Each reaction mixture contained 10 µL of Power SYBR Green master mix (2×) at a final concentration of 1×, as well as 1 µL of telomere-forward primer (2 mM) and 1 µL of telomere-reverse primer (2 mM), both at a final concentration of 0.1 mM. Additionally, 4 µL of H_2O was included to adjust the reaction volume, along with 4 µL of DNA (5 ng/µL), resulting in a total of 20 ng of DNA per reaction. The qPCR reaction mixture and conditions consisted of an initial denaturation step at 95 °C for 15 min, followed by two cycles of 94 °C for 15 s and 49 °C for 15 s. This was succeeded by 40 amplification cycles at 94 °C for

15 s, 62 °C for 10 s, and 74 °C for 15 s with signal acquisition. Additional steps included 84 °C for 10 s and 88 °C for 15 s, both with signal acquisition. The reaction concluded with a melting curve analysis, incrementally increasing the temperature from 72 °C to 95 °C in 0.5 °C steps every 30 s. Each run incorporated a standard curve generated by the Rotor-Gene Q Series software (version 2.3.1), which accounted for the dilution factors corresponding to the telomere and reference gene quantities for each sample [21].

To measure the relative telomere length, we employed a modified qPCR based on the method described by [22] and utilized a multi-copy gene as the reference instead of a single-copy gene, as validated by [23]. This calculation was performed by determining the ratio of the telomere repeat copy number (T) to the reference gene copy number (S). Each sample was analyzed in duplicate, yielding two T/S values. In cases where discrepancies were observed between the duplicates, a second assay was performed. The final measurement for each sample within a given run was obtained by averaging the duplicate values [22].

2.4. Statistical Analyses

Statistical analyses were conducted using linear discriminant analysis (LDA) to verify whether the grouping of dogs based on their backgrounds (i.e., pet, shelter, work, laboratory, or rehomed dogs) and age was appropriate for subsequent analysis. The LDA required a minimum of two samples per category, leading to the exclusion of a 14-year-old dog from the UK work group, as it did not meet this requirement.

The relationships between distinct characteristics of dogs' lives and telomere length were analyzed using generalized linear models. A normality test indicated that RTL did not meet the normality assumptions (Shapiro–Wilk test result, $W = 0.96327$, and a p-value < 0.001). Additional tests on the skewness data produced a value of 0.68, suggesting a right-skewed distribution. Based on this, a Gamma distribution with a log link was selected as the family for analysis; the log link function helps stabilize variance and leads to more homoscedastic residuals in the model, thereby enhancing the model's performance.

Model selection was based on Akaike's information criterion (AIC), allowing the identification of the most explanatory model. ANOVA was used as a post-hoc test to examine differences in rTL across multiple categories. The initial model included data from 250 sampled dogs to explore how various factors influenced rTL. The variables assessed were sex, age, breed group, origin, neuter status, overall health, training, frequency of food treats, sleep location (housing), exercise, interaction with people, and interaction with other animals.

A subsequent analysis assessed potential differences between police work dogs from Brazil and the UK. Although these dogs shared the same occupation, differences in kenneling conditions and workload could significantly affect rTL. As a result, a second model was designed to assess how the country (Brazil or the UK), work schedule, and job type influenced the rTL of 65 police working dogs. Similar to the first model, rTL was the response variable. Additional explanatory variables included in the full model were sex, breed, category group, neutered status, and health status.

Data normality and the best fit for GLM models were analyzed using RStudio Desktop version 2025.05.0+496 (RStudio Team, 2020) with the "MASS" [24], "car" [25], and "AICcmodavg" [26] packages. A significance level of $p < 0.05$ was used.

3. Results

3.1. Association of Telomere Length and Age

A total of 249 dogs were included in the analysis of relative telomere length. The overall mean rTL was 0.744 (SD = 0.080), with values ranging from 0.550 to 0.990. The median rTL was 0.730, with an interquartile range (IQR) from 0.690 to 0.780 (Table 2).

Table 2. The mean relative telomere length (rTL) found for each age band (in years) assessed using qPCR from the samples of 250 dogs used in the study.

Age	N	Mean	StDev	Variance	CoefVar	Minimum	Median	Maximum	IQR
1	51	0.746	0.087	0.008	11.666	0.62	0.720	0.97	0.110
2	63	0.750	0.083	0.007	11.007	0.55	0.740	0.99	0.095
3	31	0.750	0.079	0.006	10.561	0.62	0.740	0.96	0.080
4	20	0.727	0.063	0.004	8.701	0.64	0.725	0.92	0.075
5	22	0.752	0.072	0.005	9.544	0.61	0.735	0.92	0.090
6	17	0.729	0.070	0.005	9.652	0.61	0.730	0.86	0.090
7	17	0.741	0.070	0.005	9.421	0.58	0.740	0.91	0.050
8	4	0.778	0.116	0.013	14.939	0.62	0.795	0.90	0.078
9	6	0.750	0.052	0.003	6.954	0.70	0.745	0.83	0.073
10	5	0.744	0.069	0.005	9.244	0.68	0.720	0.84	0.100
11	5	0.672	0.094	0.009	14.015	0.55	0.660	0.78	0.130
12	3	0.673	0.093	0.009	13.799	0.57	0.700	0.75	0.090
13	2	0.795	0.205	0.042	25.794	0.65	0.795	0.94	0.145
14	1	0.760	NA	NA	NA	0.76	0.760	0.76	0.000

IQR = interquartile range. N = Number of dogs per age group.

A discriminant analysis was conducted using rTL as a predictor to assess the appropriateness of grouping dogs by age. The results revealed that the accuracy of age group assignment was 25.33%, with younger dogs being more frequently misclassified into incorrect age groups, while older dogs were more often correctly assigned to their actual age category.

When analyzing the mean rTL across the different age groups, it was observed that the mean rTL exhibited less variation in younger dogs than in adults and senior dogs. The standard deviation (SD) in younger dogs was also lower, maintaining this trend (Table 2).

3.2. Model 1: Factors Impacting a Dogs Relative Telomere Length

A generalized linear model (GLM) with a Gamma distribution and log link function was constructed to verify the relationship of several predictors with RT Length. The best-fitting model found that category, sex, sleep location, walking frequency, and contact with other animals all affected telomere length. The model demonstrated a pseudo-R^2 (Cox and Snell) of approximately 0.313, indicating that about 31.3% of the variance in RTL was explained by this model. A significant effect was observed: $F(14, 232) = 4.68$, $p < 0.0001$, $\eta^2 = 0.0082$. A summary of the GLM results is presented in Table 3.

The dogs' category based on their origin was a strong predictor of telomere length, with laboratory dogs ($\beta = -0.1737 \pm 0.0448$, $p = 0.0001$), followed by pet dogs ($\beta = -0.1277 \pm 0.0426$, $p = 0.0030$), having, on average, shorter telomeres than other groups (Figure 1).

Male dogs had significantly higher rTL than females ($\beta = 0.02 \pm 0.0124$, $p = 0.03$) (Figure 2).

In relation to different management practices, dogs that were exercised every other day showed a significant increase in telomere length ($\beta = 0.13 \pm 0.02$, $p \leq 0.01$) compared to dogs that received different types of exercise. Dogs that slept outside in a kennel had a significantly shorter rTL than those sleeping elsewhere ($\beta = -0.09 \pm 0.03$, $p = 0.01$). Likewise, contact with other animals was also negatively associated with rTL, with dogs having contact with one to two animals ($\beta = -0.06 \pm 0.02$, $p \leq 0.01$) and those in contact with morethan five animals ($\beta = -0.07 \pm 0.0294$, $p = 0.01$) showing a significant reduction in telomere length. Other factors, such as sleeping in the owner's bed, contact with three to four animals, and certain categories (rehomed, shelter, UK work), were not significant predictors of rTL ($p > 0.05$).

Table 3. GLM results for the optimal model describing the effect of group category, sex, sleep site, exercise, and interaction with other animals on the relative telomere length of domestic dogs [1].

Parameters	Estimate ± SD	t Value	*p*
Intercept	−0.1888 ± 0.0526	−3.588	0.0004 ***
Sex male [a]	0.0270 ± 0.0124	2.170	0.0311 *
Sleep kennel outside [b]	−0.0918 ± 0.0358	−2.565	0.0109 *
Sleep owner's bed	0.0062 ± 0.0215	0.288	0.7732
Walk every other day [c]	0.1351 ± 0.0289	4.683	<0.0001 ***
Walk not walked	0.0695 ± 0.0295	2.358	0.0192 *
Walk once a week	−0.0350 ± 0.0344	−1.016	0.3108
Contact 1–2 [d]	−0.0635 ± 0.0232	−2.735	0.0067 **
Contact 3–4	−0.0392 ± 0.0244	−1.605	0.1098
Contact 5+	−0.0749 ± 0.0294	−2.545	0.0116 *
Category lab [e]	−0.1737 ± 0.0448	−3.879	0.0001 ***
Category pet	−0.1277 ± 0.0426	−3.000	0.0030 **
Category rehomed	−0.0122 ± 0.0373	−0.326	0.7444
Category shelter	−0.0330 ± 0.0362	−0.911	0.3632
Category UK work	0.0529 ± 0.0309	1.711	0.0884

[1] Best model fit AICc = −619.84. Non-significant (NS) variables were removed from the model. The remaining NS factors could not be discarded without compromising the model efficiency. [a] Sex: Female is the reference group. [b] Sleeps outside, inside, shares owner's bed: Inside the house is the reference group. [c] Frequency of exercise—Walks every day, once a week, not walked, or every other day: Walks every day is the reference group. [d] Social interactions—No contact (kept alone), 1–2 animals, 3–4 animals, more than 5 animals: No contact is the reference group. [e] Background: Lab dogs, pet dogs, rehomed dogs, shelter dogs, UK working dogs, BR working dogs: BR work is the reference group. *** $p \leq 0.001$, ** $p \leq 0.01$, * $p \leq 0.05$.

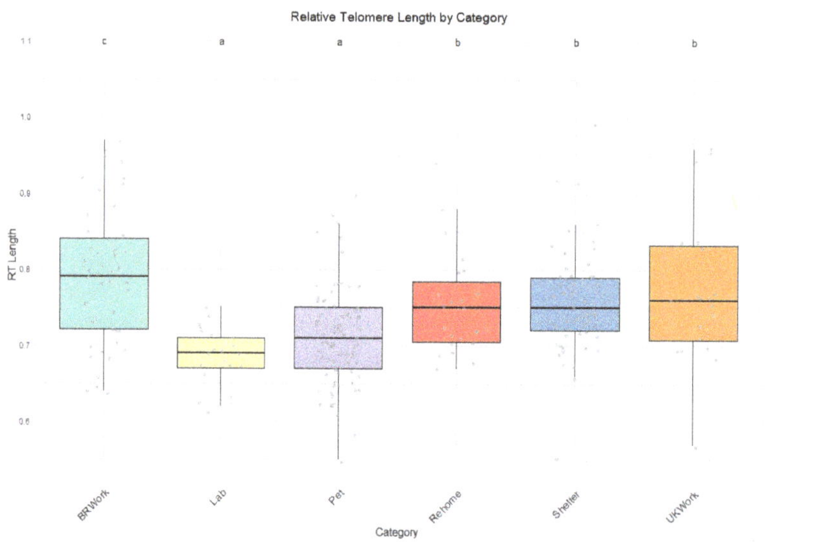

Figure 1. Relative telomere length in the sampled dogs (N = 250) based on their background category. Different colors represent different dogs' groups (green = BR work, yellow = laboratory, purple = pet, red= rehomed, blue = shelter, orange = UK work). Gray circles indicate the dispersion of data points. Boxes represent interquartile ranges, tick lines represent the median, and whiskers show maximum and minimum values. Different letters indicate statistical significance at $p < 0.05$.

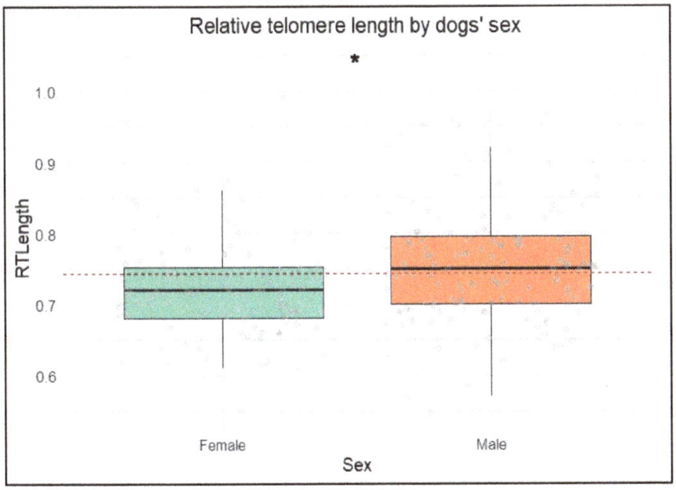

Figure 2. Differences in relative telomere length between male and female dogs. Different colors represent different sexes (green = female, orange = male). Gray circles indicate the dispersion of data points. Boxes represent interquartile ranges, tick lines represent the median, and whiskers show maximum and minimum values. The star indicates statistical significance at $p < 0.05$.

3.3. Model 2: Relative Telomere Length in Relation to Dogs' Work Type

A total of 65 dogs were included in this analysis. The overall mean rTL was 0.784 (SD = 0.093, range = 0.570–0.970). The interquartile range (IQR) was 0.120, and the coefficient of variation (CV) was 11.9%, indicating moderate relative dispersion in rTL values (Table 4).

Table 4. Distribution of the mean relative telomere length (rTL) for 65 working dogs included in the study based on the type of work performed.

Work Type	N	Mean	StDev	Variance	Min	Median	Max	IQR	CoefVar
Breeder	2	0.705	0.092	0.008	0.64	0.705	0.77	0.065	13.039
Drug/gun	14	0.828	0.084	0.007	0.7	0.825	0.96	0.135	10.164
Education	6	0.707	0.052	0.003	0.64	0.695	0.79	0.047	7.362
Explosives	10	0.764	0.057	0.003	0.69	0.765	0.84	0.085	7.435
Retired	6	0.788	0.133	0.018	0.57	0.78	0.94	0.13	16.817
Search	7	0.693	0.042	0.002	0.64	0.69	0.77	0.035	6.107
Tracking	20	0.826	0.086	0.007	0.68	0.815	0.97	0.122	10.395

IQR = interquartile range. N = number of dogs per work type.

When comparing geographic groups, individuals from Brazil (N = 45) had a higher mean rTL (0.790 ± 0.091, CV = 11.57%) than those from the UK (N = 20, mean 0.772 ± 0.098, CV = 12.72%). RTLength also varied by working type; the drug/gun category (N = 14) exhibited the highest mean RTLength (0.828 ± 0.084, IQR = 0.135, CV = 10.16%), whereas search and rescue (N = 7) had the lowest mean RTLength (0.693 ± 0.042, CV = 6.11%) (Table 4).

A subsequent GLM test was conducted to investigate the background factors that may affect relative telomere length (rTL) in police working dogs from Brazil and the UK. Although both groups shared the same occupation, they had differences in their environments and routines. These included variations in the walk routine, sleeping site, food reinforcement (treats), and training style, which were consistent within each group but differed between the two. The GLM test returned a significant effect ($F(8, 56) = 4.16$,

$p < 0.0001$, $\eta^2 = 0.0062$), and the overall model demonstrated a pseudo-R^2 (Cox and Snell) of approximately 0.1963, indicating that nearly 20% of the variance in the relative telomere length was explained by the model. The results for this model, including the effects of these factors on rTL, are summarized in Table 5.

Table 5. GLM results for the optimal model examining the impact of group category, sex, and work type on the relative telomere length of working dogs [1].

Parameters	Estimate ± SD	t Value	p
Intercept	0.6400 ± 0.0784	8.156	<0.0001 ***
Group UK work[a]	−0.0511 ± 0.0262	−1.950	0.0561
Sex male [c]	0.0381 ± 0.0234	1.627	0.1093
Working type: capture [d]	0.1730 ± 0.0833	2.076	0.0425 *
Working type: drugs and guns	0.1606 ± 0.0829	1.936	0.0579
Working type: educational	0.0412 ± 0.0861	0.478	0.6343
Working type: explosives	0.1220 ± 0.0850	1.435	0.1569
Working type: retired	0.1506 ± 0.0883	1.704	0.0939
Working type: search and rescue	0.0342 ± 0.0855	0.400	0.6905

[1] Best models for relative telomere length AICc = −644.40. Non-significant (NS) variables were removed from the model. The remaining NS factors could not be discarded without compromising the model efficiency. [a] UK work or BR work: BR work is the reference group. [c] Sex: Female is the reference group. [d] Type of work: Breeder, capture, drugs and guns, educational, explosives, retired, search and rescue: Breeder is the reference group. * $p \leq 0.05$, *** $p \leq 0.0001$.

The results showed marginally significant differences between UK and Brazilian dogs ($\beta = -0.0511 \pm 0.0262$, $p = 0.0561$), suggesting that work-type location does have a weak effect on telomere length.

Additionally, the type of work performed by the dogs was associated with differences in rTL. Dogs engaged in tracking ($\beta = 0.1730 \pm 0.0833$, $p = 0.0425$) and those working with drugs and guns ($\beta = 0.1606 \pm 0.0829$, $p = 0.0579$) had longer relative telomere lengths than those involved in other types of work (Figure 3). These findings suggest that geographic location and the specific nature of a dog's work can significantly impact their rTL.

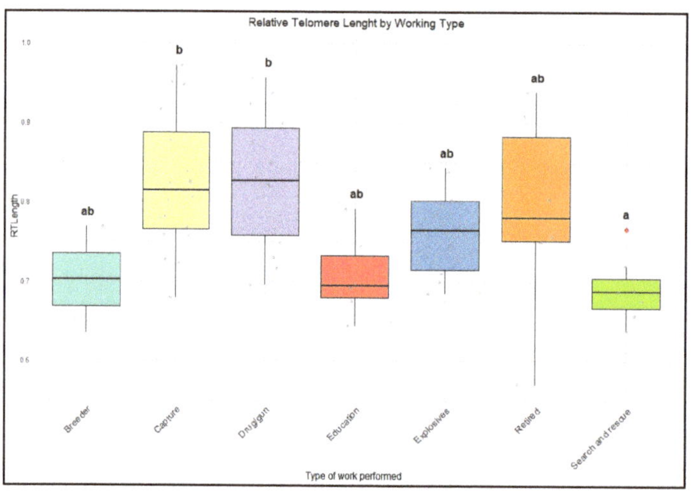

Figure 3. Relative telomere length (rTL) differences between police dogs based on the type of work the individuals perform. Different colors represent different types of work (emerald green = breeder,

yellow = capture, purple = drugs and guns, red= education, blue = explosives, orange = retired, and lime green = search and rescue). The gray circles indicate the dispersion of data points, and the red circles indicate outliers. Boxes represent interquartile ranges, tick lines represent the median, and whiskers show maximum and minimum values. Different letters indicate statistical significance at $p < 0.05$.

4. Discussion

This research has highlighted the complex interactions among dogs' backgrounds, environments, and husbandry practices that influence their relative telomere length (rTL). While previous studies have primarily focused on individual factors, such as age, social isolation, or reproduction, this study takes a broader approach by examining multiple factors and their cumulative effects on rTL.

While age is generally expected to correlate with telomere length, our results did not meet these assumptions. Our findings reveal that dogs with similar routines and social interactions tend to exhibit comparable rates of cellular senescence, as indicated by their rTL. However, the discriminant analysis still identified some inconsistencies among dogs' classification, which suggests that other characteristics may play a significant role in determining telomere attrition, reinforcing that the dog's life experiences may impact their biological aging process.

Based on this premise, our study found that laboratory dogs exhibited the lowest relative telomere lengths (rTLs) compared to the other groups. This finding supports previous research, which highlights that dogs in laboratory environments often experience a poorer quality of life due to suboptimal housing conditions, inadequate handling, and social isolation [27]. These adverse conditions are associated with higher stress responses, as reflected in increased cortisol levels, and often lead to the development of stereotypic behaviors [28,29]. Previous studies have reported mixed findings regarding the relationship between telomere shortening and cortisol levels, with some identifying significant associations [30,31], while others remain inconclusive [30]. Early life experiences, including shelter histories, may significantly influence the behavior and biological aging of dogs, as prior studies have shown that such factors can impact welfare and development [32]. Our results add to the growing body of evidence suggesting that environmental and social stressors in laboratory settings can accelerate biological aging, as indicated by the shortened telomeres in these dogs.

Pet dogs also have shorter telomeres than other groups. Among different dog groups, pets are perceived as having a better quality of life, largely due to their unique relationship with humans and the dedicated healthcare they receive. However, pet dogs may face significant challenges due to their living conditions, such as a lack of physical activity, overfeeding, or poor socialization management [32,33], which can lead to chronic stress [32,34] and even impact their lifespan [35]. For instance, our results suggest that increased exercise has a positive impact on telomere length. In contrast, contact with other animals may affect relative telomere length, further supporting the evidence for the association of lifestyle factors in pet dogs and their impact on health and welfare. Furthermore, other factors, such as mental stimulation through training, have been shown to influence telomere attrition [36,37] and have also demonstrated variability in telomere lengths across dog breeds, correlating with differences in lifespan. If pet ownership trends favor specific traits and management styles, this could directly contribute to the observed reduction in telomere length among pet dogs.

Despite our results not finding a specific association with breed and training, this is most likely due to the heterogeneous nature of the sample and the grouping of breeds by broader types. Therefore, further studies could focus on exploring the various factors present in pets' lives to better understand the impact that ownership choices may have on dogs' aging.

Across the groups, males exhibited longer relative telomere lengths (rTLs) than females. This finding contrasts with studies conducted in rats and humans, which have typically found that females have longer telomeres than males. In mammals, the longer lifespan of females is often attributed to having two X chromosomes, which enhances DNA repair mechanisms [38–40]. The unexpected result in dogs may suggest that the factors influencing rTL in dogs are different from those observed in other mammals, potentially reflecting species-specific biological mechanisms or the influence of environmental and management aspects [10,41]. Furthermore, we did not find any significant association between rTL and reproductive status, despite the existing body of evidence on the relationship between hormone levels and neutering, which could potentially interfere with dogs' health. More than 50% of our sample consisted of entire dogs, which could explain the result; however, further research could support understanding whether changes in the hormone profile after sterilization contribute to telomere attrition.

In our study, the location where pet dogs slept was associated with telomere length, with dogs that slept in kennels having shorter telomeres than those allowed inside the house or on their owner's bed. While research on dog sleeping arrangements is scarce, some studies have suggested that co-sleeping with owners may contribute to better animal welfare. Previous research [42] has found that co-sleeping could reduce anxiety in dogs, although they noted that more comprehensive data across a dog's lifetime is necessary to understand its effects fully.

Outdoor kenneled dogs may face negative environmental factors, such as harsh weather, loud noises, and discomfort. These conditions can induce physiological stress, leading to increased oxidative stress. Studies in other species have linked environmental metrics with elevated stress responses and telomere attrition (see review [43]).

Moreover, sleep quality and consistency may be compromised in kennel environments, further exacerbating stress responses. Research with laboratory dogs has highlighted the association between sleep, stress, and environmental metrics, such as noise and temperature [44]. Given that sleep quality is associated with telomere length maintenance [45], outdoor sleeping conditions may indirectly contribute to accelerated cellular aging.

In this study, dogs that were walked every day had shorter relative telomere lengths (rTLs) than dogs that were never walked or walked every other day. This finding was initially unexpected, as walking is often considered a key element of responsible dog ownership and necessary to ensure a dog's welfare [46]. However, a possible explanation for this result lies in the walks' dynamics.

For instance, walking equipment, such as leashes and harnesses, might cause discomfort or restrict the dog's ability to move freely. This restriction could generate stress, impacting telomere length [47]. Furthermore, daily walks may be brief, with the primary purpose being functional (e.g., to fulfill the dog's physiological needs) rather than providing opportunities for more enriching experiences, such as exercise, socialization, or mental stimulation. Researchers [48] suggested that some dog owners might take their dogs on short walks to avoid feeling guilty about not providing exercise without considering that longer walks could offer additional benefits. Short walks, particularly when they do not meet the dog's need for physical and mental stimulation, can lead to frustration and stress, factors commonly associated with higher cortisol levels. Over time, this chronic stress could accelerate telomere attrition.

Additionally, walking is often used to mitigate behavioral issues in dogs. However, if the owner is not adequately advised by a professional, daily walks could expose the dog to stressors that, when chronic, may contribute to adverse health outcomes, including impacts on rTL [45,47,49].

Since this study focused on background factors rather than specific activity patterns, follow-up questions regarding the duration of daily walks, the dog's emotions during the walk, or the type of walking equipment used were not included in the questionnaire. In dogs that are reactive to other dogs on walks, this activity tends to have a more negative effect because it stresses them, and their owners are also stressed—a double hit on dog welfare [50]. Thus, dog walks may include both negative and positive aspects, and for some dogs, the negatives might outweigh the positives; therefore, further investigations are necessary to better understand the relationship between walking habits and telomere length.

Our study found that dogs in contact with other animals had shorter relative telomere lengths than those with no contact. Although socialization is a well-established factor in the healthy development of dogs, social dynamics can pose additional challenges to these animals.

Animals kept in crowded environments may experience an increased likelihood of social confrontations and heightened stress responses, particularly when spatial restrictions and the inability to avoid other individuals are present, as group living can lead to greater competition for resources [45]. Studies conducted on mice and hyenas demonstrate that crowded groups and social hierarchies impact telomere attrition. Similarly, a study on African grey parrots found that social isolation resulted in shorter telomeres. These findings help illustrate the complexity of the relationship between social settings and telomere dynamics.

The relationships found in our results do not establish a clear parameter for types of interaction that may affect telomere length, as both dogs with only one to two contacts and those with five or more contacts had shorter telomeres. However, certain characteristics of the surveyed groups may help further understand what might underlie these associations. In our study, many dogs had contact with others; still, they were not necessarily housed together, or they may have been housed with individuals without the possibility of avoidance, such as in shared kennels. Furthermore, the presence of another individual does not necessarily equate to companionship, especially if dogs live in a mixed-species household. All these factors may contribute to the variation observed in the results.

As our questionnaire did not provide further detail beyond the number of animal contacts to better understand these findings, future studies could manipulate social settings more deliberately to explore how varying social environments impact telomere dynamics in dogs.

Previous studies have established a relationship between telomere shortening and dogs' breed-predicted lifespan [10]; however, our study did not find any significant effects of age or breed on relative telomere length (rTL). This discrepancy could be attributed to the heterogeneous nature of our sample, which included a wide range of breeds, ages, and sizes. The variability in telomere lengths within the different breeds and age groups may have masked any potential relationships, highlighting the complexity of factors influencing rTL in dogs. Future research with a more homogenous sample or larger sample size could provide further insights into how age and breed specifically impact telomere dynamics.

Finally, despite differences in management and housing conditions, as well as the distribution of rTL, varying among locations, there was no significant difference between the two groups of police dogs (from Brazil and the UK). Only the type of work the dogs were engaged in showed a substantial correlation with telomere length.

Based on the type of work, dogs commonly used in street patrols, known as capture dogs, showed increased telomere length compared to other groups. Police dogs are generally trained for specific duties, such as security, patrol, or detection; sometimes, they are cross-trained for multiple purposes [44]. Overall, research in this area is still limited, but previous studies indicate that the nature of a dog's work may increase stress responses and influence their reproductive performance and life quality [48].

Due to the inherent characteristics of the type of work, patrol dogs may work in less demanding conditions and be less exposed to risk than dogs in other groups, such as those involved in search and rescue and explosive detection [51]. Previous studies have shown that these dogs experience significant psychological and physiological stress due to frustration and high-intensity tasks, which can affect their performance and welfare [51,52]. Moreover, specific training protocols, such as exposure to harmful chemicals in drug detection training, could also potentially influence telomere dynamics, further highlighting the complex relationship between training-related stress and telomere attrition.

Training experience and social interactions with handlers may also play a role. Dogs with higher trainability and less exposure to acute stressors tend to maintain better telomere dynamics, as observed in studies linking behavioral factors to telomere length [37]. Studies dedicated to evaluating the stress dynamics between dogs and handlers have highlighted that older and experienced dogs [53] and also experienced handlers [54] can contribute to better performance and alleviate the stress of working demands.

Lastly, differences in stress responses are also found among dog breeds used in police work, influenced by genetic, physiological, and behavioral [55] factors, all factors that can independently or jointly influence cell aging. Common breeds used for patrolling and other breeds selected for cooperative tasks, such as German Shepherds, may handle stress better than independent worker breeds due to their genetic predisposition for focused attention and human interaction [56].

We propose that future research incorporate location-specific analyses to better understand how regional factors—such as job responsibilities, environmental stressors, and work intensity—affect telomere length in working dogs.

Overall, our study identified several background factors that may impact a dog's telomere length; to our knowledge, no previous research involving non-human species has examined the influence of these multiple background factors on telomere attrition dynamics.

Our findings align with the framework proposed by [50], highlighting telomere attrition as a key hallmark of aging in domestic dogs. This study emphasizes that telomere shortening is critical to biological aging and is influenced by various environmental and behavioral factors. Specifically, our research found that housing conditions, exercise, and social interactions had a significant impact on dogs' relative telomere length (rTL). These factors suggest that lifestyle choices can accelerate or delay aging through telomere attrition. This supports the notion that life experiences, such as sleeping sites and daily walks, are integral to understanding biological aging in dogs.

Despite our considerable sample size, which included dogs from diverse backgrounds, sexes, and age groups, we diluted the overall sample size within specific categories. This dilution may limit our ability to identify additional factors influencing relative telomere length. Moreover, another factor that could hinder comparisons among dogs is the significant number of variables that define a dog's household environment. Therefore, future research should focus on uniform sampling to effectively identify factors contributing to premature aging in dogs, using methodologies similar to those employed in epidemiological studies of human aging and health [49]. Such an approach would enable the precise identification of environmental factors that contribute to biological aging, thus providing valuable insights into the dynamics of aging in non-human species.

5. Conclusions

Our findings contribute to a holistic perspective, advocating that environmental and behavioral experiences should be considered when evaluating canine health and welfare. This aligns with the suggestion that telomere length may serve as a valuable biomarker for assessing the impact of these factors on aging and overall welfare. In contrast to many animal welfare assessments that focus solely on one polarity, this approach provides a more comprehensive understanding of how various beneficial and harmful life experiences contribute to a dog's overall well-being and aging process. By considering the full spectrum of experiences, we can better evaluate and address the needs of dogs throughout their lives.

Our results demonstrated that dogs from different backgrounds experience varying stressors throughout their lives, which can contribute to shorter telomere lengths. Interestingly, while individual characteristics such as age and breed may not directly lead to accelerated aging, dogs with similar traits living in comparable environments tend to show similar patterns of telomere shortening. Laboratory dogs were found to have the shortest relative telomere lengths among the groups studied, supporting the assumption that dogs in such conditions generally experience lower welfare than those in other conditions. On the other hand, police dogs used for patrolling and capture exhibit longer telomeres than their counterparts. Given the challenging nature of their work and the diverse roles that police dogs can undertake, understanding the impacts of various activities may play a vital role in their day-to-day activities and overall health and welfare.

The variation in telomere length among pet dogs was primarily associated with their sleeping arrangements, suggesting that housing conditions significantly elicit stress responses that could accelerate telomere attrition. Factors such as exercise and social interactions were also found to influence telomere length across different dog groups. However, further research is needed to understand the thresholds between these variables better—whether they help prevent telomere attrition or contribute to stress responses that may accelerate aging. Understanding these dynamics is crucial for improving the welfare and longevity of dogs.

While this study provides valuable insights into telomere length as a potential welfare biomarker, several limitations should be acknowledged. First, chronic stress was not directly assessed through additional physiological or behavioral markers, limiting our ability to correlate telomere attrition with specific stress responses. Second, the absence of inter-run calibrators (IRCs) in qPCR may introduce minor plate-to-plate variation, though standard curve normalization and technical replicates were employed to mitigate this. Third, factors such as dietary habits and social interactions, although recorded, were excluded from the final analysis due to data completeness constraints. Additionally, variation in housing conditions across groups, particularly among shelter and laboratory dogs, may have influenced stress exposure differently, and the impact of shelter duration remains an open question. Future studies should integrate complementary stress biomarkers, such as hair cortisol and oxidative stress indicators, alongside telomere length analysis to provide a more comprehensive assessment of canine welfare.

Early identification of accelerated aging is vital, as it helps detect health issues at an early stage, making treatment more straightforward, affordable, and likely more effective [57]. This strategy could prolong the service life of working dogs while safeguarding their well-being. As a result, the findings of this study may have significant economic benefits for both pet owners and organizations employing working dogs, potentially reducing veterinary costs, extending the active service period of the dogs, and enhancing their overall health and quality of life.

Despite the limitations in sample sizes across the groups, we believe this initial approach to evaluating the effect of the environment on relative telomere length holds significant potential as a tool for measuring animal welfare through aging. The findings suggest that the cumulative experiences of animals can influence biological aging, making telomere length a promising indicator. Further studies are needed to explore the relationship between life experiences and aging, providing a deeper understanding of how environmental and social factors contribute to animals' overall health and well-being.

Supplementary Materials: The following supporting information can be downloaded at https://www.mdpi.com/article/10.3390/vetsci12050491/s1, Table S1: Information from the 250 dogs (*Canis familiaris*) sampled for relative telomere length (rTL) analysis used in the current study; Table S2: Information regarding the background of the dogs (*Canis familiaris*) that were sampled for this study to explore the impact of stress, sociality, and exercise on dogs' cellular aging; and Table S3: Categories of police working dogs, as grouped for exploring the impact of stress, sociality, and exercise on dogs' cellular aging. Figure S1: Informed consent statement.

Author Contributions: Conceptualization, L.M.L.D. and R.J.Y.; methodology, L.M.L.D. and R.J.Y.; software, L.M.L.D.; validation L.M.L.D. and F.S.S.; formal analysis, L.M.L.D.; investigation, L.M.L.D.; resources, L.M.L.D. and I.G.S.; data curation, L.M.L.D., I.G.S. and F.S.S.; writing—original draft preparation, L.M.L.D. and I.G.S.; writing—review and editing, L.M.L.D. and I.G.S.; visualization, L.M.L.D., I.G.S., F.S.S., R.J.Y. and A.S.V.; supervision, R.J.Y. and A.S.V.; project administration, L.M.L.D.; funding acquisition, L.M.L.D. All authors have read and agreed to the published version of the manuscript.

Funding: The Minas Gerais Research Foundation—FAPEMIG provided funding to A.S.V. for this study (scholarship 309124/2022-0), and a Science Without Borders scholarship was provided by the Conselho Nacional de Desenvolvimento Científico e Tecnológico (CNPq) to L.M.L.D. (scholarship 206418/2014-0), I.G.S. (scholarship 202351/2015-7, and F.S.S. (scholarship 208427/2017-1).

Institutional Review Board Statement: Data for this study were collected with ethical approval from the University of Salford Ethics Committee (approval number STR1617-22 on 25 January 2017) and the Animal Ethics Committee of the Universidade Federal de Ouro Preto (process no. 0056/2017 on 18 January 2017).

Informed Consent Statement: Informed consent was obtained from all subjects involved in the study.

Data Availability Statement: The dataset is available upon request from the authors.

Acknowledgments: The authors would like to express their gratitude to all the dogs' tutors and staff members, veterinarians, and institutions whose invaluable help made this work possible.

Conflicts of Interest: The authors declare no conflicts of interest.

References

1. Dotson, M.J.; Hyatt, E.M. Understanding dog–human companionship. *J. Bus. Res.* **2008**, *61*, 457–466. [CrossRef]
2. Kim, M.J.; Oh, H.J.; Hwang, S.Y.; Hur, T.Y.; Lee, B.C. Health and temperaments of cloned working dogs. *J. Vet. Sci.* **2018**, *19*, 585–591. [CrossRef]
3. Svartberg, K. Breed-typical behaviour in dogs—Historical remnants or recent constructs? *Appl. Anim. Behav. Sci.* **2006**, *96*, 293–313. [CrossRef]
4. Polgár, Z.; Blackwell, E.J.; Rooney, N.J. Assessing the welfare of kennelled dogs—A review of animal-based measures. *Appl. Anim. Behav. Sci.* **2019**, *213*, 1–13. [CrossRef]
5. Malkani, R.; Paramasivam, S.; Wolfensohn, S. Preliminary validation of a novel tool to assess dog welfare: The Animal Welfare Assessment Grid. *Front. Vet. Sci.* **2022**, *9*, 940017. [CrossRef]
6. Palme, R. Non-invasive measurement of glucocorticoids: Advances and problems. *Physiol. Behav.* **2019**, *199*, 229–243. [CrossRef]
7. Aubert, G.; Lansdorp, P.M. Telomeres and Aging. *Physiol. Rev.* **2008**, *88*, 557–579. [CrossRef]
8. Cai, N.; Chang, S.; Li, Y.; Li, Q.; Hu, J.; Liang, J.; Song, L.; Kretzschmar, W.; Gan, X.; Nicod, J.; et al. Molecular signatures of major depression. *Curr. Biol.* **2015**, *25*, 1146–1156. [CrossRef]
9. Maestripieri, D.; Hoffman, C.L. Chronic stress, allostatic load, and aging in nonhuman primates. *Dev. Psychopathol.* **2011**, *23*, 1187–1195. [CrossRef]

10. Fick, L.J.; Fick, G.H.; Li, Z.; Cao, E.; Bao, B.; Heffelfinger, D.; Parker, H.G.; Ostrander, E.A.; Riabowol, K. Telomere Length Correlates with Life Span of Dog Breeds. *Cell Rep.* **2012**, *2*, 1530–1536. [CrossRef]
11. Sohn, S.H.; Subramani, V.K. Dynamics of telomere length in the chicken. *World's Poult. Sci. J.* **2014**, *70*, 721–736. [CrossRef]
12. Bateson, M. Cumulative stress in research animals: Telomere attrition as a biomarker in a welfare context? *BioEssays* **2016**, *38*, 201–212. [CrossRef]
13. Epel, E.S.; Blackburn, E.H.; Lin, J.; Dhabhar, F.S.; Adler, N.E.; Morrow, J.D.; Cawthon, R.M. Accelerated telomere shortening in response to life stress. *Proc. Natl. Acad. Sci. USA* **2004**, *101*, 17312–17315. [CrossRef]
14. Kraus, C.; Pavard, S.; Promislow, D.E.L. The size-life span trade-off decomposed: Why large dogs die young. *Am. Nat.* **2013**, *181*, 492–505. [CrossRef]
15. Nasir, L.; Devlin, P.; Mckevitt, T.; Rutteman, G.; Argyle, D.J. Telomere Lengths and Telomerase Activity in Dog Tissues: A Potential Model System to Study Human Telomere and Telomerase Biology. *Neoplasia* **2001**, *3*, 351–359. [CrossRef]
16. The Kennel Club. (n.d.). Breed Standards. Available online: https://www.thekennelclub.org.uk/breed-standards/ (accessed on 8 May 2025).
17. Olsen, M.T.; Bérubé, M.; Robbins, J.; Palsbøll, P.J. Empirical evaluation of humpback whale telomere length estimates; Quality control and factors causing variability in the singleplex and multiplex qPCR methods. *BMC Genet.* **2012**, *13*, 77. [CrossRef]
18. Cawthon, R.M. Telomere length measurement by a novel monochrome multiplex quantitative PCR method. *Nucleic Acids Res.* **2009**, *37*, e21. [CrossRef]
19. Plot, V.; Criscuolo, F.; Zahn, S.; Georges, J.Y. Telomeres, age and reproduction in a long-lived reptile. *PLoS ONE* **2012**, *7*, e40855. [CrossRef]
20. Dutra, L.; Souza, F.; Friberg, I.; Araújo, M.; Vasconcellos, A.; Young, R. Validating the use of oral swabs for telomere length assessment in dogs. *J. Vet. Behav.* **2020**, *40*, 16–20. [CrossRef]
21. O'Callaghan, N.J.; Fenech, M. A quantitative PCR method for measuring absolute telomere length. *Biol. Proced. Online* **2011**, *13*, 3. [CrossRef]
22. Cawthon, R.M. Telomere measurement by quantitative PCR. *Nucleic Acids Res.* **2002**, *30*, e47. [CrossRef] [PubMed]
23. Wang, F.; Pan, X.; Kalmbach, K.; Seth-smith, M.L.; Ye, X.; Antumes, D.M.F. Robust measurement of telomere length in single cells. *Proc. Natl. Acad. Sci. USA* **2013**, *110*, E1906–E1912. [CrossRef]
24. Venables, W.N.; Ripley, B.D. *Modern Applied Statistics with S*, 4th ed.; Springer: Berlin/Heidelberg, Germany, 2002. Available online: https://www.stats.ox.ac.uk/pub/MASS4/ (accessed on 30 January 2018).
25. Fox, J.; Weisberg, S. *An R Companion to Applied Regression*, 3rd ed.; SAGE Publications: Thousand Oaks, CA, USA, 2019. Available online: https://www.john-fox.ca/Companion/ (accessed on 30 January 2018).
26. Mezerolle, M. AICcmodavg: Model Selection and Multimodel Inference Based on (Q)AIC(c) [R Package Version]. 2020. CRAN. Available online: https://cran.r-project.org/package=AICcmodavg (accessed on 30 January 2018).
27. Schipper, L.L.; Vinke, C.M.; Schilder, M.B.H.; Spruijt, B.M. The effect of feeding enrichment toys on the behaviour of kenneled dogs (*Canis familiaris*). *Appl. Anim. Behav. Sci.* **2008**, *114*, 182–195. [CrossRef]
28. Beerda, B.; Schilder, M.B.; Bernadina, W.; Van Hooff, J.A.; De Vries, H.W.; Mol, J.A. Chronic stress in dogs subjected to social and spatial restriction. II. Hormonal and immunological responses. *Physiol. Behav.* **1999**, *66*, 243–254. [CrossRef]
29. Wojciechowska, J.I.; Hewson, C.J. Quality-of-life assessment in pet dogs. *J. Am. Vet. Med. Assoc.* **2005**, *226*, 722–728. [CrossRef]
30. Blasco, M.A. Telomeres and human disease: Ageing, cancer and beyond. *Nat. Rev. Genet.* **2005**, *6*, 611–622. [CrossRef]
31. Heimbürge, S.; Kanitz, E.; Otten, W. The use of hair cortisol for the assessment of stress in animals. *Gen. Comp. Endocrinol.* **2019**, *270*, 10–17. [CrossRef] [PubMed]
32. Buttner, A.P.; Awalt, S.L.; Strasser, R. Early life adversity in dogs produces altered physiological and behavioral responses during a social stress-buffering paradigm. *J. Exp. Anal. Behav.* **2023**, *120*, 6–20. [CrossRef]
33. Beerda, B.; Schilder, M.B.H.; Van Hooff, J.A.R.A.M.; De Vries, H.W. Manifestations of chronic and acute stress in dogs. *Appl. Anim. Behav. Sci.* **1997**, *52*, 307–319. [CrossRef]
34. Sonntag, Q.; Overall, K.L. Key determinants of dog and cat welfare: Behaviour, breeding and household lifestyle. *OIE Rev. Sci. Tech.* **2014**, *33*, 213–220. [CrossRef]
35. Broom, D.M.; Fraser, A.F. *Domestic Animal Behaviour and Welfare*, 5th ed.; CABI International: Wallingford, UK, 2015.
36. Dreschel, N.A. The effects of fear and anxiety on health and lifespan in pet dogs. *Appl. Anim. Behav. Sci.* **2010**, *125*, 157–162. [CrossRef]
37. Weixlbraun, J.; Chapagain, D.; Cornils, J.S.; Smith, S.; Schwarzenberger, F.; Hoelzl, F. Impact of trainability on telomere dynamics of pet dogs (*Canis lupus* familiaris): An explorative study in aging dogs. *PLoS ONE* **2025**, *20*, e0317332. [CrossRef]
38. Cherif, H.; Tarry, J.L.; Ozanne, S.E.; Hales, C.N. Ageing and telomeres: A study into organ- and gender-specific telomere shortening. *Nucleic Acids Res.* **2003**, *31*, 1576–1583. [CrossRef]

39. Lemaître, J.-F.; Ronget, V.; Tidière, M.; Allainé, D.; Berger, V.; Cohas, A.; Colchero, F.; Conde, D.A.; Garratt, M.; Liker, A.; et al. Sex differences in adult lifespan and aging rates of mortality across wild mammals. *Proc. Natl. Acad. Sci. USA* **2020**, *117*, 8546–8553. [CrossRef]
40. Mayer, S.; Brüderlein, S.; Perner, S.; Waibel, I.; Holdenried, A.; Ciloglu, N.; Hasel, C.; Mattfeldt, T.; Nielsen, K.V.; Möller, P. Sex-specific telomere length profiles and age-dependent erosion dynamics of individual chromosome arms in humans. *Cytogenet. Genome Res.* **2006**, *112*, 194–201. [CrossRef]
41. Wilbourn, R.V.; Moatt, J.P.; Froy, H.; Walling, C.A.; Nussey, D.H.; Boonekamp, J.J. The relationship between telomere length and mortality risk in non-model vertebrate systems: A meta-analysis. *Philos. Trans. R. Soc. B Biol. Sci.* **2018**, *373*, 20160447. [CrossRef] [PubMed]
42. Cannas, S.; Talamonti, Z.; Mazzola, S.; Minero, M.; Picciolini, A.; Palestrini, C. Factors associated with dog behavioral problems referred to a behavior clinic. *J. Vet. Behav.* **2018**, *24*, 42–47. [CrossRef]
43. Chatelain, M.; Drobniak, S.M.; Szulkin, M. The association between stressors and telomeres in non-human vertebrates: A meta-analysis. *Ecol. Lett.* **2020**, *23*, 381–398. [CrossRef]
44. Schork, I.G.; Manzo, I.A.; de Oliveira, M.R.B.; da Costa, F.V.; Young, R.J.; de Azevedo, C.S. The cyclic interaction between daytime behavior and the sleep behavior of laboratory dogs. *Sci. Rep.* **2022**, *12*, 478. [CrossRef]
45. Sabot, D.; Lovegrove, R.; Stapleton, P. The association between sleep quality and telomere length: A systematic literature review. *Brain Behav. Immun. Health* **2023**, *28*, 100577. [CrossRef] [PubMed]
46. Westgarth, C.; Christley, R.M.; Marvin, G.; Perkins, E. I walk my dog because it makes me happy: A qualitative study to understand why dogs motivate walking and improved health. *Int. J. Environ. Res. Public Health* **2017**, *14*, 936. [CrossRef]
47. Lafuente, M.P.; Provis, L.; Schmalz, E.A. Effects of restrictive and non-restrictive harnesses on shoulder extension in dogs at walk and trot. *Vet. Rec.* **2019**, *184*, 64. [CrossRef]
48. Westgarth, C.; Christley, R.M.; Marvin, G.; Perkins, E. Functional and recreational dog walking practices in the UK. *Health Promot. Int.* **2021**, *36*, 109–119. [CrossRef] [PubMed]
49. Tomiyama, A.J.; Donovan, A.O.; Lin, J.; Puterman, E.; Lazaro, A.; Chan, J.; Dhabhar, F.S.; Wolkowitz, O.; Kirschbaum, C.; Blackburn, E.; et al. Does cellular aging relate to patterns of allostasis? An examination of basal and stress reactive HPA axis activity and telomere length. *Physiol. Behav.* **2012**, *106*, 40–45. [CrossRef]
50. Carr, D.; Friedmann, E.; Gee, N.R.; Gilchrist, C.; Sachs-Ericsson, N.; Koodaly, L. Dog walking and the social impact of the COVID-19 pandemic on loneliness in older adults. *Animals* **2021**, *11*, 1852. [CrossRef] [PubMed]
51. Bartolomé, E.; Sánchez-Guerrero, M.J.; Perdomo-González, D.I.; Valera, M. Stress and behavior assessment in police dogs due to challenging situations: Differences due to training objectives. *Dog Behav.* **2023**, *9*. [CrossRef]
52. Dickinson, S.; Feuerbacher, E.N. Frustration and its impact on search and rescue canines. *Front. Vet. Sci.* **2025**, *12*, 1546412. [CrossRef]
53. Horváth, Z.; Igyártó, B.Z.; Magyar, A.; Miklósi, Á. Three different coping styles in police dogs exposed to a short-term challenge. *Horm. Behav.* **2007**, *52*, 621–630. [CrossRef]
54. Wojtaś, J.; Zieliński, D.; Karpiński, M. Does the sex of the search and rescue (SAR) dog handler affect the work of the rescue team? *Dog Behav.* **2021**, *7*, 23–32. [CrossRef]
55. Maclean, E.L.; Gesquiere, L.R.; Gee, N.R.; Levy, K.; Martin, W.L.; Carter, C.S.; Oliva, J.L. Effects of affiliative human–animal interaction on dog salivary and plasma oxytocin and vasopressin. *Front. Psychol.* **2017**, *8*, 1606. [CrossRef] [PubMed]
56. Gácsi, M.; Topál, J.; Miklósi, Á.; Dóka, A.; Csányi, V. Attachment behavior of adult dogs (*Canis familiaris*) living at rescue centers: Forming new bonds. *J. Comp. Psychol.* **2001**, *115*, 423–431. [CrossRef]
57. Stolz, M.; Gottardi, R.; Raiteri, R.; Miot, S.; Martin, I.; Baschong, W.; Daniels, A.U.; Staufer, U.; Raducanu, A.; Du, M. Early detection of aging cartilage and osteoarthritis in mice and patient samples using atomic force microscopy. *Nat. Nanotechnol.* **2009**, *4*, 186–192. [CrossRef] [PubMed]

veterinary sciences MDPI

Article

A Nutritional Supplement Containing Curcumin C3 Complex, Glucosamine, and Chondroitin Alleviates Osteoarthritis in Mice and Canines

Enpei Zheng [1], Ting Cen [1], Ye Ma [1], Ziyuan Weng [1], Chuanheng Jiang [1], Luxi Hou [2], Jun Leng [2] and Changmin Hu [1,3,*]

1 Department of Clinical Veterinary Medicine, College of Veterinary Medicine, Huazhong Agricultural University, Wuhan 430070, China; zep2023@webmail.hzau.edu.cn (E.Z.); cen15225336226@163.com (T.C.); mayeczj@webmail.hzau.edu.cn (Y.M.); guiqideyouxiang@163.com (Z.W.); 15680039817@163.com (C.J.)
2 Shenzhen Redray Biotechnology Co., Ltd., Shenzhen 518100, China; 15884401207@163.com (L.H.); lenjun@reddog.com.cn (J.L.)
3 The Veterinary Teaching Hospital, College of Veterinary Medicine, Huazhong Agricultural University, Wuhan 430070, China
* Correspondence: hcm@mail.hzau.edu.cn

Simple Summary: This study evaluates the use of Curcumin C3 Complex-glucosamine-chondroitin (C3GC) to alleviate osteoarthritis (OA) symptoms in dogs and mice. OA was induced in mice through medial meniscus destabilization, and the subjects received 8 weeks of dietary supplementation with glucosamine-chondroitin (GC) alone or in combination with Curcumin C3 Complex (C3C). The findings showed that C3GC offers greater bone and cartilage protection than the use of GC alone. Supplementation in OA canines significantly decreased pain scores, serum MMP-3, and TNF-α, reflecting anti-inflammatory activity. The findings support the use of C3GC for alleviating OA in mice and dogs, though further clinical evidence is needed.

Abstract: Osteoarthritis (OA) is a chronically progressive degenerative arthropathy characterized by the loss of cartilage, changes in subchondral architecture, and ongoing inflammation resulting in reduced mobility and pain. This study assessed the treatment potential of a combination of chondroitin and glucosamine enriched with Curcumin C3 Complex (C3GC) in modulating the pathophysiological features in mouse models with surgically induced OA and in dogs with naturally occurring OA. A cohort of 24 male C57BL/6 mice aged 3 months old were surgically destabilized with medial meniscus (DMM) to cause osteoarthritis. These animals underwent a nutritional intervention with C3GC or with GC over a course of 8 weeks. In order to evaluate cartilage health and subchondral bone structure, we carried out a combination of behavioral tests, micro-computed tomography (micro-CT), and histopathological examinations. In addition, a cohort of 12 OA-diagnosed retired police dogs were administered C3GC supplements or conventional care over a course of 30 days, with pain measurement and serum concentrations of MMP-3 and TNF-α determined before and after treatment. According to our findings, the administration of C3GC was determined to preserve subchondral microarchitectural structure integrity ($p < 0.05$) and resulted in better motor function in comparison with GC. In animals taking nutritional supplements, the OARSI scores of joint tissue sections were reduced, with the medial tibial plateau OARSI score being particularly low in the C3GC group ($p < 0.0001$). In dogs, treatment with C3GC resulted in a 24.5% reduction in serum MMP-3 levels ($p < 0.01$), and there was also a 20.8% decrease in serum TNF-α levels ($p < 0.05$), along with a decrease in subjective pain assessment. The results are in support of the chondroprotective, anti-inflammatory, and analgesic properties of C3GC and justify future research on the potential utility of C3GC in treating osteoarthritis.

Keywords: osteoarthritis; Curcumin C3 Complex; chondroprotection; anti-inflammatory; preclinical models

1. Introduction

Osteoarthritis (OA) is a common disease in older dogs, and the pain and inflammation it causes can affect the quality of life of the animals. Joint instability, incoordination, uneven weight bearing, and damage caused by developmental disorders, trauma, or iatrogenic factors can lead to abnormal pressure on the articular cartilage and chronic inflammation, thereby inducing osteoarthritis in animals [1].

Different studies have reported varying incidence rates of canine OA. In the UK, the incidence rate reported in primary veterinary clinics is 6.6%, while statistics from referral hospitals indicate 20%. Referral data in North America suggest that the incidence of OA in dogs over 1 year old is 20% and 80% in dogs over 8 years old. Overall, the incidence increases with age, and the commonly affected joints in animals are the knee joint, hip joint, and elbow joint [2]. Additionally, osteoarthritis in pet dogs can impose economic and psychological burdens on their owners. In 2003, the expenditure on cruciate ligament diseases in dogs alone in the United States was USD 1.32 billion [3].

The treatment options for osteoarthritis are relatively scarce; moreover, non-surgical therapies typically only alleviate symptoms, while surgical methods are considered a last resort. Clinical non-surgical methods primarily involve multimodal therapies, including pharmacological pain relief, non-pharmacological pain relief, appropriate exercise, weight management, and supplementation with joint nutrients. Although non-steroidal anti-inflammatory drugs (NSAIDs) have a beneficial effect on alleviating osteoarthritis, the risk of organ damage from long-term use of NSAIDs cannot be ignored [4]. Curcumin has demonstrated chondroprotective activity through the inhibition of NF-κB signaling, inhibition of matrix metalloproteinases (MMPs), and decreased oxidative stress in models of osteoarthritis [5]. Studies of human osteoarthritis patients have demonstrated a decrease in pain scores similar to those achieved with non-steroidal anti-inflammatory drugs (NSAIDs) when curcumin is administered in doses of more than 500 mg daily [6]. Its bioavailability continues to be a significant issue, and hence formulations that increase absorption, such as C3GC, are required. The present study evaluates the therapeutic efficacy of C3GC in osteoarthritic police dogs, a population with excessive joint loading due to physical exercise. Current clinical research extensively investigates curcumin's therapeutic potential in osteoarthritis treatment and nutritional supplementation [7]. Mechanistic studies have progressively elucidated its key pharmacological actions, particularly focusing on anti-inflammatory and antioxidant properties. These investigations further demonstrate curcumin's chondroprotective effects through both cellular preservation and apoptosis inhibition in chondrocytes [8]. Additionally, curcumin has shown therapeutic or delaying effects on diseases such as Alzheimer's disease, rheumatoid arthritis, liver diseases, and diabetes in related studies involving humans or rats. In dogs, they have demonstrated effects in the treatment of osteoarthritis and degenerative myelopathy, as well as liver regeneration, antibacterial and antiviral properties, wound healing, and anti-inflammatory and anti-itch effects [9,10].

Currently, the main therapeutic targets for osteoarthritis treatment focus on modulating chondrocyte autophagy, growth factor signaling pathways (e.g., TGF-β/MMP-13), inflammatory responses, and harmful molecular signaling, with additional strategies targeting autophagy inhibition and reactive oxygen species (ROS) reduction [11]. The destabilization of the medial meniscus mouse model, induced by surgical transection of the medial meniscotibial ligament, is widely utilized to mimic post-traumatic OA and evaluate therapeutic interventions [12]. This model is characterized by mild cartilage lesions at 2 weeks post-surgery, followed by rapid subchondral bone lesions, sclerosis, and osteophyte formation [13], while studies have demonstrated its utility in exploring OA-related mechanisms such as autophagy dysregulation, ROS accumulation, and growth factor signaling imbalances [14,15]. The knee joint anatomy across terrestrial mammals is conserved, including bone morphology, ligament/meniscus organization, and cartilage histology [16]. Furthermore, canine orthopedic models have generated significant ethical debates [17]. Meanwhile, the DMM-induced pathologies, including cartilage degeneration, synovial hyperplasia, bone sclerosis, and osteophyte formation, closely parallel naturally occurring OA in canines, which is diagnosable via biomarkers [18,19]. Building on these interspecies parallels, this study evaluates the analgesic and anti-inflammatory efficacy of compound C3GC (800 mg/kgbw/day in mice; 112.5 mg/kgbw/day in dogs) using both DMM-induced OA in mice and naturally occurring OA in retired police dogs. The dosing regimen for the active components was established through a systematic review of the existing clinical and preclinical literature [20,21].

2. Materials and Methods

2.1. Experimental Products

C3 curcumin glucosamine-chondroitin was sourced from Shenzhen Redray Biotechnology Co., Ltd. (Shenzhen, China), containing 10% Curcumin C3 Complex (from Sabinsa Corporation, Nanjing, China), 20% glucosamine hydrochloride, and 10% chondroitin sulfate. In addition, it also contains dimethyl sulfoxide (DMSO), sodium hyaluronate, undenatured type II collagen, manganese, vitamin E, and other excipients.

2.2. Induction of Osteoarthritis in Mice and Treatment

C57BL/6 mice (male, 8–10 weeks old) were obtained from the Animal Center of Huazhong Agricultural University (ID Number: HZAUMO-2023-0328) and were allowed to acclimate with free access to water and food, starting the experiment at 3 months of age. Except for the sham surgery group, all groups underwent arthritis modeling using the destabilization of the medial meniscus (DMM) method. The mice were anesthetized with isoflurane inhalation, initially exposed to 4% isoflurane for 2–3 min via an induction chamber, followed by maintenance anesthesia at 1–2% isoflurane delivered through a face mask with an oxygen flow rate of 1 L/min. The modeling was performed on the right hind limb of the mice, with an incision made near the patellar ligament to open the joint capsule, exposing the medial meniscus ligament, which was then severed to destabilize the meniscus. In the sham surgery group (SHAM), the joint capsule was also opened, and the ligament was visible, but no transverse cutting was performed. In the post-operative phase, the mice received a heated recovery pad aimed at offering warmth and relaxation and were monitored closely until the regaining of ambulation, an indicator of a return to a state of normal mobility. Moreover, all of the animals received a single dose of antibiotics post-surgery.

After one week of recovery, a total of 24 male mice were randomly divided into a SHAM group, a DMM group, a C3GC group, and a GC group, with similar initial body weights. In the C3GC group, C3 curcumin glucosamine-chondroitin was added to the diet at a ratio of 0.5% (0.08% Curcumin + 0.42% other components). In comparison, the GC group was fed a diet containing 0.42% glucosamine-chondroitin and other main components. The

SHAM and DMM groups were fed standard feed without any nutritional additives. All of the experimental diets were prepared by Wuhan Wanqian Jiaxing Biotechnology Co., Ltd. (Wuhan, China), using a standardized mixing protocol to ensure homogeneity. The basic components of mouse feed are shown in Supplementary Table S1. Fresh feed was replaced every 72 h. Daily feed intake was recorded per cage, and the average curcumin consumption was estimated as 128 mg/kgbw/day based on a mean daily intake of 4.0 g feed per mouse (body weight 25–30 g). The mice were maintained for 8 weeks post-surgery.

2.3. Behavioral Tests Related to Osteoarthritis

The rotarod test apparatus was provided by Anhui Zhenghua Biologic Apparatus Facilities (Huaibei, Anhui, China). The rotarod induces forced exercise to assess functional parameters by measuring the time or endurance of cycling. Each animal is trained twice within 48 h prior to the experiment. Each animal undergoes three rotarod tests, with a 30 min interval, accelerating from 4 rotations per minute to 40 rotations per minute, maintaining the test for a total of 3 min. Each test ends when the animal falls off the rotarod. The time taken for the animal to fall off the rotarod is recorded and used in subsequent analyses, with each animal repeating the test three times.

The time taken for the animal to fall from the top of the rod is measured through the pole test, allowing for the assessment of the animal's grip strength and agility. The duration of the descent is recorded during the experiment. During the test, the mice were placed with their heads facing upwards on top of the rod, and the animals would typically orient themselves downwards and descend continuously along the rod to return to their housing cage. We recorded the latency period of the animals' automatic turning (Tturn) and the total time taken to descend to the bottom of the rod (Ttotal), with each animal being tested three times.

2.4. Von Frey Test

Mechanical pain was assessed in all mice using an electronic von Frey aesthesiometer (ZS-CITY Beijing Zhongshi-Dichuang Science and Technology Development Co., Ltd., Beijing, China). The instrument was configured with a probe for mice, and the mice were acclimated in a single chamber on top of a metal wire mesh platform for 10 min prior to the test. The von Frey filaments were used to stimulate the plantar surface of the hind paw with increasing force to determine tactile sensitivity. A positive response is indicated by the hind paw withdrawal reflex when stimulation is applied, and the instrument analyzes and records the maximum force applied during stimulation. After each mechanical pain measurement, the mice were allowed to rest for 5 min, and each animal was tested five times.

2.5. Micro-CT

At Week 8, after anesthetizing and euthanizing the mice, blood was collected and the right knee joint was immediately harvested, including part of the femur and tibia, with skin and excess tissue removed. After fixation in 4% formaldehyde for 24 h, four samples were randomly selected from each group and scanned using a micro-computed tomography scanner (Skyscan 1276 Micro-CT, manufactured by Bruker Magnetic Resonance, Ettlingen, Germany). The area below the joint surface of the tibial plateau was obtained for analysis. The parameters included trabecular separation (Tb.Sp, mm) and bone mineral density (BMD).

2.6. Histological Evaluation of Mouse Joint Tissues

After euthanizing the mice, the affected limbs' joints from each group were collected. The cartilage tissues from each group were fixed in 4% formaldehyde for 24 h and decalcified with 5% formic acid for 1 week. Subsequently, the cartilage from each group was dehydrated, embedded in paraffin, and sectioned into 5 μm tissue slices. The degeneration of each knee joint's articular cartilage was assessed using Safranin O-fast green staining and hematoxylin–eosin staining. The degenerative changes were evaluated and analyzed using the Osteoarthritis Research Society International (OARSI) scoring system [22,23].

2.7. Assessment of Antioxidant and Anti-Inflammatory Effects

The oxidative stress levels in serum were assessed using a superoxide dismutase (SOD) typing test kit (hydroxylamine method), and the serum inflammatory factor levels were evaluated using a mouse tumor necrosis factor-alpha (TNF-α) enzyme-linked immunosorbent assay kit. The kits were sourced from NanJing JianCheng Bioengineering Institute (Nanjing, China).

Complete blood counts were performed before the experiment began and at the time of euthanasia, measuring the total white blood cell count (WBC), lymphocyte count (LYM), red blood cell count (RBC), and hematocrit (HCT). Additionally, biochemical blood tests were conducted to measure serum creatinine (Crea), blood urea nitrogen (BUN), alanine aminotransferase (ALT), aspartate aminotransferase (AST), and alkaline phosphatase (ALP). Three samples from each group were randomly selected for testing. The results are included in the Supplementary Materials (Tables S2 and S3).

2.8. Study of the Anti-Inflammatory and Analgesic Effects of C3GC on Dogs with Osteoarthritis

A cohort of 12 retired police dogs (8 males, 4 females; age 6–9 years; body weight 19.64 ± 6.811 kg) with osteoarthritis was randomly selected from a Police Dog Facility in Nanjing, China. All of the dogs were evaluated by a veterinarian and diagnosed with joint pain. The canine cohort comprised three distinct working breeds, as follows: Springer Spaniels ($n = 6$), German Shepherds ($n = 5$), and Chinese Kunmin ($n = 1$). The dogs were randomly assigned to the C3GC treatment group ($n = 6$) or the control group ($n = 6$), ensuring equivalent distributions of body weight, gender, and breed. The specific statistics of the experimental dogs after grouping are shown in Supplementary Table S4. Inflammatory marker levels and levels of pain were assessed by blind observers. Throughout the study period, all dogs received a standardized maintenance diet provided by Nanjing Police Canine Nutrition Center (Nanjing, China). The feed formula is presented in Supplementary Table S5. C3GC was administered at a daily dose of 112.5 mg/kgbw added to the diet of the dogs. The trial lasted for one month, during which the base's caretakers administered oral tablets to the experimental dogs. A modified subjective pain score test (referenced HCPI, Supplementary Figure S1) and serum MMP3 and TNF-α testing were conducted at the beginning and end of the trial. The MMP3 kit was sourced from Jiangsu Enzyme Immuno Industrial Co., Ltd. (Yancheng, Jiangsu, China).

2.9. Statistical Analysis

The data in this experiment are presented as the mean ± standard deviation (Mean ± SD). Statistical analyses were performed with GraphPad Prism 9.0 (GraphPad Software, San Diego, CA, USA). For multigroup comparisons, one-way ANOVA was used to assess the significance of intergroup parameters. Comparisons between the two groups were conducted using an unpaired t-test. $p < 0.05$ is considered statistically significant, and $p < 0.01$ is considered highly significant.

The dog trial data were analyzed using a two-way ANOVA for within-group comparisons. Prior to the analysis, Grubbs' test was conducted to identify and exclude outlier data.

3. Results

3.1. C3GC Improves Motor Functions and Helps in Restoring Functions in OA Mice

The experiment tested the motor coordination function of mice to examine the effects of C3GC on the symptoms of osteoarthritis and the behavior of affected animals. In the rotarod test, the duration of rotarod performance in DMM mice was significantly decreased compared to that of the SHAM group ($p < 0.05$), while the results for the mice fed C3GC and GC showed a statistically insignificant increase compared to those of the DMM group. In the pole test, there was a significant difference between the SHAM group and the DMM group ($p < 0.05$) in the eighth week, indicating a certain decline in the motor ability of the animals after surgery. Meanwhile, the completion time of the surgical mice fed with C3GC was significantly better than that of the surgical mice not fed with C3GC, thus demonstrating that C3GC has a notable enhancement effect on the motor ability of osteoarthritic mice (Figure 1).

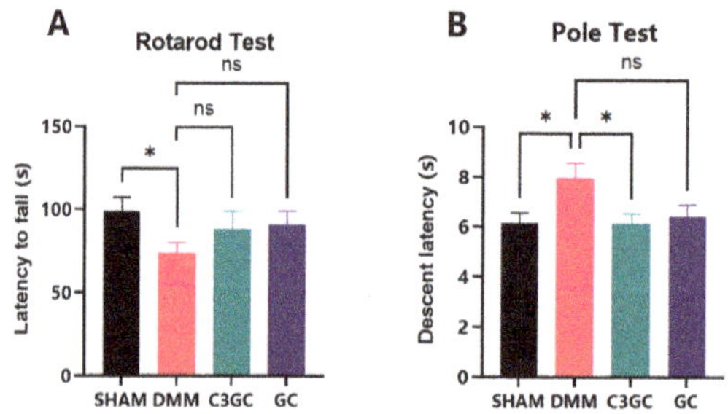

Figure 1. C3GC supplementation mitigates motor impairment in osteoarthritic mice at eight weeks post-surgery. (**A**) Latency-to-fall during rotarod testing quantifies forced exercise endurance in mice. The DMM group showed a significant reduction in motor tolerance ($p < 0.05$ vs. SHAM), while C3GC-treated mice showed a degree of restoration, but without statistical significance ($p = 0.2493$ vs. DMM). (**B**) Pole test descent latency evaluates murine motor function and movement efficiency. C3GC improved latency in descending in comparison with the DMM group ($p < 0.05$), implying an improvement in motor coordination. Data presented as Mean ± SD, $n = 6$, * $p < 0.05$, ns = not significant.

3.2. C3GC Alleviates Mechano-Hyperalgesia in Osteoarthritis

Von Frey tests were conducted to quantify the mechanical sensitivity observed in mice after the experiment. There were no significant differences between the groups prior to the experiment. At Week 8, the mechanical pain threshold of the DMM mice was significantly lower compared to that of the SHAM group, while the C3GC group and the GC group showed mean increases of 18.58% and 21.56%, respectively, compared to the DMM group (Figure 2).

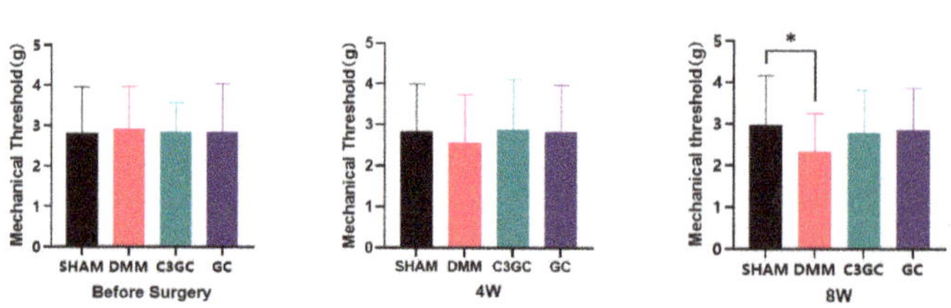

Figure 2. C3GC regulates the mechanosensitivity in osteoarthritic mice. The DMM group showed a significant reduction in the threshold of withdrawal ($p < 0.05$ vs. SHAM), suggesting enhanced pain sensitivity. C3GC supplementation partially restored thresholds, but the differences were not statistically significant ($p > 0.05$ vs. DMM). Data are presented in Mean ± SD, $n = 6$, * $p < 0.05$.

3.3. C3GC Protects Subchondral Bone Integrity and Prevents Degenerating Cartilage in Osteoarthritis

3.3.1. Subchondral Bone Remodeling and Degeneration Association

Surgical joint samples were collected from mice with post-operative osteoarthritis to investigate the effects of C3GC on bone spur formation and osteoporosis. Three-dimensional visualization revealed calcified tissue manifestations, including meniscal enlargement and ectopic mineralization nodules of various sizes. At eight weeks of treatment, bone spurs or meniscal calcification developed on the surgical joint surfaces. The mice treated with C3GC showed a significant reduction in Tb.Sp in the tibial plateau region ($p < 0.01$), indicating altered bone structure, while a notable reduction ($p < 0.05$) was observed in the GC-treated group, as compared to the DMM group of mice. Mice that were simultaneously administered C3GC and chondroitin showed an increase in BMD, but there was no significant difference compared to the DMM group (Figure 3).

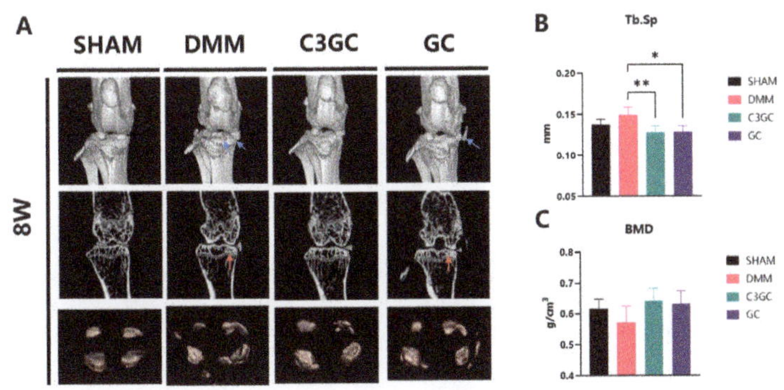

Figure 3. Micro-CT images of the affected limb joints of mice and ROI area analysis results at eight weeks post-surgery. (**A**) Micro-CT 3D reconstruction or 2D representative images of each group, with blue arrows indicating meniscus calcification and joint bone spurs and red arrows indicating bone sclerosis of the medial plateau of the tibia. The increased mineralized tissue was visualized in 3D (distal view). (**B**,**C**) Analysis of trabecular separation (Tb.Sp) and bone mineral density (BMD) in the ROI area of mouse samples at the eighth week after feeding through micro-CT. All data are presented as Mean ± SD, $n = 4$, * $p < 0.05$, ** $p < 0.01$.

3.3.2. C3GC Preserves Cartilage and Reduces the Severity of Osteoarthritis

The OARSI score was implemented through S-O staining and HE staining, and the structural integrity of the knee joint cartilage in mice was examined under a microscope after 8 weeks of feeding. The results showed that the DMM group exhibited severe erosion of the cartilage layer and bone remodeling, while the C3GC group and the GC group only showed a mild loss of surface lamina or erosion to the calcified cartilage. The OARSI score results showed significant osteoarthritis in the medial femoral condyle of the operated mice ($p < 0.0001$). However, the degree of osteoarthritis in the mice treated with C3GC and GC decreased significantly compared to the DMM mice ($p < 0.0001$). Significant osteoarthritis changes were also observed in the medial tibial plateau of the operated mice ($p < 0.0001$), while the mice in the C3GC group showed significant improvement ($p < 0.0001$). There was no significant difference in the GC group when compared to DMM mice (Figure 4).

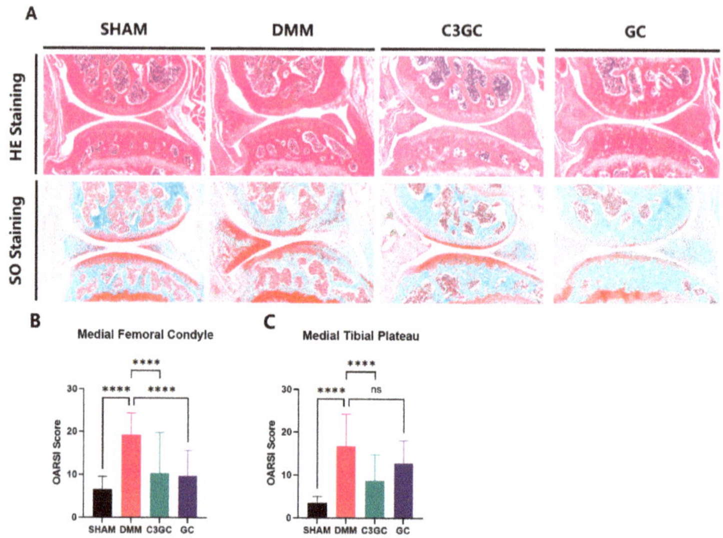

Figure 4. C3GC protects cartilage integrity and attenuates osteoarthritic degeneration in mice. (**A**) Representative SO- and HE-stained sections revealing cartilage integrity (4×). DMM mice revealed severe surface fibrillation, subchondral bone remodeling, and proteoglycan depletion, whereas C3GC-treated mice showed a preserved cartilage structure with reduced matrix degradation. (**B**,**C**) The OARSI score of the medial femoral condyle and medial tibial plateau revealed a significant cartilage loss in DMM mice ($p < 0.0001$ vs. SHAM). C3GC reduced the OARSI score substantially ($p < 0.0001$ vs. DMM), proposing chondroprotective benefits over treatment with glucocorticoids alone. Data are presented as Mean ± SD, $n = 4$, **** $p < 0.0001$, ns = non-significance.

3.4. Curcumin Exerts Anti-Inflammatory Effects in OA Mice Models

To examine the effects of oral C3GC on serum anti-inflammatory and antioxidant factors, SOD and TNF-α levels were estimated in the serum of mice through ELISA. By Week 8, C3GC was able to decrease the level of TNF-α by 31.4% ($p = 0.0702$), while GC only decreased by 14.3% ($p = 0.5151$). The serum level of SOD in C3GC revealed an increasing trend. The SOD level in the C3GC group exhibited an increase of 8.16% when compared to the DMM group ($p > 0.05$) and a rise of 28.44% in comparison to the GC group (Figure 5).

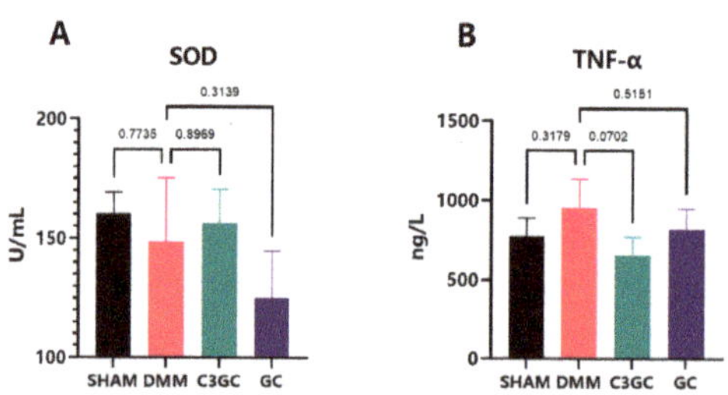

Figure 5. C3GC alleviates systemic inflammation in osteoarthritis. (**A**,**B**) C3GC reduced serum TNF-α levels ($p = 0.0702$ vs. DMM), revealing anti-inflammatory effects. SOD showed an upward trend, suggesting potential oxidative stress modulation ($p > 0.05$). Data are presented as Mean ± SD, p value displayed on comparison lines, $n = 3$.

3.5. The Analgesic and Anti-Inflammatory Effects of C3 Curcumin in Dogs with Osteoarthritis

Subjective pain scoring and serum inflammatory factor testing were conducted on recruited dogs before (Day 0) and after (Day 30) the 1-month dietary intervention. The subjective pain score decreased by 53.3% after one month of C3GC feeding, although there was no significant difference. In addition to this, the serum levels of MMP-3 significantly dropped by 24.5% compared to pre-treatment ($p < 0.01$), and the TNF-α levels also significantly decreased by 20.8% ($p < 0.05$). No significant changes were observed in the control group between pre-treatment and post-treatment measurements for either subjective pain scores or serum inflammatory markers (Figure 6).

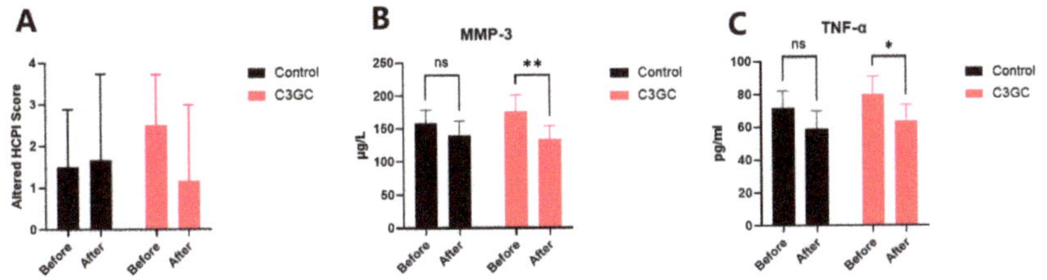

Figure 6. Supplementation with C3GC relieves osteoarthritic dogs from pain and inflammation. (**A**) Subjective pain scores decreased in C3GC-treated dogs on Day 30. (**B**) Serum MMP-3 was notably lower in the C3GC group on Day 30 ($p < 0.01$). (**C**) Serum TNF-α was notably lower in the C3GC group on Day 30 ($p < 0.05$), suggesting chondroprotective and anti-inflammatory effects. Values are expressed as Mean ± SD, * $p < 0.05$, ** $p < 0.01$, ns = non-significance.

4. Discussion

This study demonstrates the role of C3GC in delaying the progression of osteoarthritis in mice and dogs, supported by histopathological, behavioral, and anti-inflammatory evidence. Curcumin is an antioxidant that alleviates pain in osteoarthritis patients and promotes function [6,24]. The combined use of glucosamine and chondroitin exhibits complementary anti-catabolic and anti-inflammatory effects, and its therapeutic efficacy is comparable to that of celecoxib [25]. Other studies have found that chondroitin can alleviate pain symptoms and improve joint function, while glucosamine significantly improves joint stiffness [26]. This study suggests that the use of supplements based on glucosamine and chondroitin can delay cartilage degeneration, preserve the microstructure of subchondral trabecular bone, and promote motor coordination. The incorporation of curcumin leads to improved outcomes, including improved climbing performance, reduced OARSI scores, and decreased levels of TNF-α. Other components in the formulation, such as DMSO and undenatured type II collagen, may provide complementary support. DMSO has been reported to reduce synovial inflammation in OA models [27], while undenatured collagen supplementation was associated with improved mobility in canine osteoarthritis without adverse effects [28].

The symptoms of osteoarthritis in mice involve a complex interaction between pain and functional impairment. Motor functional impairment and indirect pain can be assessed through spontaneous activities (e.g., gait analysis, open-field exploration), whereas forced exercise paradigms, such as rotarod testing, are used to quantify pain-related movement avoidance, endurance, and motor learning [29]. This experiment assessed motor abilities through rotarod testing and found that mice in the C3GC and GC groups exhibited longer latency to fall, which suggested a potential trend of improved motor performance compared to that of the DMM group. However, these differences did not reach statistical significance. The pole test, a validated method for evaluating motor retardation [30], further corroborated the functional decline in DMM mice. It has been widely used in Parkinson's models because it can evaluate the animal's motor coordination [31]. We analyzed the flexibility and coordination of animals through the pole test, finding that the DMM group exhibited a significant decline in rod climbing ability, and that C3GC had a significant positive effect on the motor coordination and flexibility of osteoarthritic mice, outperforming the GC group.

Pain assessment in animal models of osteoarthritis can be conducted through behavioral experiments [32], and the experimental results verified that C3GC could alleviate functional impairment, suggesting an indirect capacity to mitigate pain caused by osteoarthritis. Additionally, some studies have found that curcumin oil can reverse the prolonged rod climbing time in a Parkinson's disease mouse model [33].

The von Frey test is commonly used to assess pain in osteoarthritic mice. In traumatic osteoarthritis, the occurrence of inflammation-induced mechanical sensitization leads to a decrease in the mechanical stimulation threshold in animals for both the affected joint and remote areas [34,35]. In the experiment, mechanical pain assessments were conducted on mice, revealing that the pain hypersensitivity threshold decreased after surgery. However, there was no significant difference in mechanical pain responses between the surgical mice treated with C3GC and the sham surgery group, indicating that the addition is beneficial for alleviating osteoarthritis pain. Research has found that topical curcumin nanoparticles can improve mechanical pain performance and alleviate OA-related pain, however, the analgesic effect is not significant when taken orally [5]. Therefore, the control effect of curcumin on osteoarthritis symptoms may be related to the formulation. Curcumin treatment in chikungunya-infected mice can increase their mechanical pain threshold levels [36]. Additionally, studies in rats have found that lipid-particle-loaded curcumin (30 mg/kg) has a protective effect against joint pain hypersensitivity comparable to naproxen (25 mg/kg) [37].

Subchondral bone sclerosis and progressive cartilage degeneration are widely regarded as hallmarks of OA. During this process, abnormal bone remodeling leads to insufficient subchondral bone mineralization, while pathological changes such as microdamage, bone marrow edema-like lesions, and bone cysts occur in the subchondral bone [38]. Therefore, we studied the microstructure of subchondral trabecular bone using micro-CT. We found that, after 8 weeks of OA induction, Tb.Sp increased and BMD decreased, indicating trabecular damage and a decline in bone mineral density. These findings suggest that supplementation with C3GC and GC preserved the microstructure of subchondral trabecular bone in osteoarthritic mice and prevented bone loss. Compared to the GC group, the differences between the C3GC group and the DMM group were more significant, possibly indicating that C3GC had a better therapeutic effect on osteoarthritis. Similar results were observed in mice with knee osteoarthritis. In the fourth week of the experiment, the Tb.Sp level of DMM mice injected intraperitoneally with curcumin was significantly higher than that of the surgery group [39]. The micro-CT results showed that curcumin treatment significantly reduced Tb.Sp and inhibited structural changes in the skeleton of rats that had developed osteoporosis [40].

The pathological features of osteoarthritis include the loss of proteoglycan staining, cartilage fibrosis, and degeneration, damage, and erosion of the cartilage matrix. In our histological analysis, Safranin O-fast green staining was used to assess the extent of cartilage damage in mice. The OARSI semi-quantitative scoring system [22,23] was used to evaluate the histological changes in the knee joints of osteoarthritic mice. In our results, the scores of the DMM group mice significantly increased, demonstrating that osteoarthritis changes occurred in the mice after surgery. Feeding C3GC and GC to mice resulted in a decrease in the OARSI score related to the affected cartilage layer of the joints, especially in the mice that received C3GC. The experimental results indicate that C3GC protects cartilage and slows down cartilage degeneration in osteoarthritis model mice, with effects superior to those of GC. Other studies have demonstrated that both the oral administration of curcumin [5] and the injection of curcumenol [41] can significantly reduce OARSI scores in DMM-induced osteoarthritic mice. Additionally, there is a wealth of literature confirming that chondroitin sulfate and glucosamine exhibit protective effects on cartilage in experimental animal models of osteoarthritis and in companion animals [42]. In osteoarthritis progression, initial cartilage surface erosion is subsequently exacerbated by progressive matrix degradation [43]. These results indicate that C3GC mitigates osteoarthritis through cartilage tissue preservation.

The inflammatory response and oxidative stress in chondrocytes and other joint tissues are associated with the progression and severity of osteoarthritis, making them ideal targets for treatment. Polyphenolic compounds such as curcumin exert their anti-inflammatory and antioxidant activities in osteoarthritis models, inhibiting the inflammatory response and ROS generation during the OA process [44]. The pro-inflammatory factors IL-1β and TNF-α are elevated in the synovial fluid, cartilage, synovium, and subchondral bone of osteoarthritic joints [44]. In this study, the serum levels of TNF-α were elevated in surgically operated mice and decreased after feeding with C3GC, with a greater reduction than that observed in the GC group. In a mouse model of traumatic osteoarthritis, both curcumin and curcumin-encapsulated nanoparticles inhibited the mRNA expression of pro-inflammatory factors such as IL-1β and TNF-α [5]. A study by Wang et al. [40] demonstrated that curcumin can inhibit IL-1β, IL-6, and TNF-α in surgical OA models. They also showed that curcumin alleviates osteoarthritis through the p38MAPK pathway. Superoxide dismutase (SOD) is an important antioxidant enzyme. In this experiment, there is an upward trend in the SOD levels within the C3GC group in comparison to the DMM group and the GC group. In other animal models, formulations containing curcumin have been shown to enhance the levels of SOD, CAT, and GPX, playing a regulatory role in antioxidant enzymes during the progression of OA [45]. A study using curcumin and chondroitin sulfate in rats showed that both the individual use and combined administration of the two can increase serum SOD levels [46].

TNF-α and IL-1β can regulate the expression of MMP-3 in OA through the NF-κb, MAPK, and PI3K/Akt pathways. During the pathogenesis of OA, increased mechanical load and degradation of the cartilage matrix lead to elevated levels of MMP-3, which can promote angiogenesis in the cartilage, the accumulation of inflammatory cells, and the differentiation of mesenchymal cells into chondrocytes [47]. This study detected the levels of MMP-3 and TNF-α in canine serum and found that C3GC can effectively reduce the levels of MMP-3 and TNF-α in osteoarthritic animals, alleviating their pain. Research on OA rats showed that curcumin could significantly downregulate the levels of MMP-3 in their serum [45]. A nutritional supplement obtained by co-micronizing palmitoyl amino glucosamine with curcumin can also reduce the levels of serum MMP-3 and inflammatory factors TNF-a and IL-1β induced by sodium iodoacetate [48]. These studies indicate that curcumin and its extract formulations can improve the inflammatory response in osteoarthritis to varying degrees.

This study has certain limitations. To mitigate these issues, several methodological refinements were implemented to address potential confounding factors. In the murine studies, daily supplement intake was quantified via feed weight monitoring to compensate for dosing uncertainties, as C3GC was incorporated into the feed under ad libitum conditions. By comparing the C3GC group with the GC control group, the improved effects specifically attributed to the incorporation of C3 in the C3GC composition were highlighted. In the canine trials, although lacking a placebo group, single-blinded assessment protocols coupled with serum inflammatory marker quantification provided objective efficacy evaluation.

5. Conclusions

This study investigated the effect of C3GC by using murine behavior tests, stifle micro-CT, cartilage pathological section, cytokines levels, and subjective pain scores in canines.

The treatment with C3GC and GC—especially C3GC—can relieve pain and symptoms, preserve cartilage, and protect the microstructure of the subchondral trabecular bone in the DMM surgical mouse model. C3GC can also significantly reduce levels of MMP-3 and TNF-α in dogs with osteoarthritis. Moreover, further research is needed to identify the optimal biological effect dosage and establish a foundation for clinical translation and application.

Supplementary Materials: The following supporting information can be downloaded at: https://www.mdpi.com/article/10.3390/vetsci12050462/s1, Table S1: Nutritional components of mouse food; Table S2: Biochemical results of mice post-feeding; Table S3: CBC results of mice after feeding; Table S4: Statistical data of experimental dogs after grouping; Table S5: Nutritional components of dog food; Figure S1: Altered HCPI scoring.

Author Contributions: Conceptualization, E.Z. and T.C.; methodology, E.Z.; validation, E.Z., T.C., Y.M., Z.W. and C.J.; formal analysis, E.Z., T.C. and Y.M.; investigation, E.Z., T.C. and Y.M.; resources, C.H.; data curation, E.Z. and T.C.; writing—original draft preparation, E.Z. and T.C.; writing—review and editing, Z.W., C.J. and C.H.; visualization, E.Z.; supervision, L.H. and J.L.; project administration, L.H. and J.L.; funding acquisition, L.H. All authors have read and agreed to the published version of the manuscript.

Funding: This research was funded by the Distinguished Teacher Workshop of Huazhong Agricultural University (2023), and the Teaching Veterinary Hospital of Huazhong Agricultural University (2024).

Institutional Review Board Statement: The animal study was approved by the Huazhong Agricultural University Animal Ethical and Welfare Committee with the approval number HZAUMO-2023-0328, HZAUDO-2024-0007.

Informed Consent Statement: All owners of the dogs included in this study gave informed consent for the use of the clinical data gathered during the examinations.

Data Availability Statement: The data presented in this study are available upon request from the corresponding author.

Acknowledgments: We would like to thank for the technical support of Yanting Du, Lei Kuang, Xuan Wu, Ruiling Yin, Xun Zhong, Zeqing Yan, Shaheen Majeed, and Md. F. Kulyar for their technical support.

Conflicts of Interest: Authors Luxi Hou and Jun Leng were employed by the company Shenzhen Redray Biotechnology Co., Ltd. The remaining authors declare that the research was conducted in the absence of any commercial or financial relationships that could be construed as a potential conflict of interest.

References

1. Ray, A.; Ray, B.K. An inflammation-responsive transcription factor in the pathophysiology of osteoarthritis. *Biorheology* **2008**, *45*, 3–4. [CrossRef]
2. Anderson, K.L.; O'Neill, D.G.; Brodbelt, D.C.; Church, D.B.; Meeson, R.L.; Sargan, D.; Summers, J.F.; Zulch, H.; Collins, L.M. Prevalence, duration and risk factors for appendicular osteoarthritis in a UK dog population under primary veterinary care. *Sci. Rep.* **2018**, *8*, 5641. [CrossRef]
3. Wilke, V.L.; Robinson, D.A.; Evans, R.B.; Rothschild, M.F.; Conzenius, M.G. Estimate of the annual economic impact of treatment of cranial cruciate ligament injury in dogs in the United States. *J. Am. Vet. Med. Assoc.* **2005**, *227*, 1604–1607. [CrossRef] [PubMed]
4. Bindu, S.; Mazumder, S.; Bandyopadhyay, U. Non-steroidal anti-inflammatory drugs (NSAIDs) and organ damage: A current perspective. *Biochem. Pharmacol.* **2020**, *180*, 114147. [CrossRef]
5. Zhang, Z.; Leong, D.J.; Xu, L.; He, Z.; Wang, A.; Navati, M.; Kim, S.J.; Hirsh, D.M.; Hardin, J.A.; Cobelli, N.J.; et al. Curcumin slows osteoarthritis progression and relieves osteoarthritis-associated pain symptoms in a post-traumatic osteoarthritis mouse model. *Arthritis Res. Ther.* **2016**, *18*, 128. [CrossRef]

6. Feng, J.; Li, Z.; Tian, L.; Mu, P.; Hu, Y.; Xiong, F.; Ma, X. Efficacy and safety of curcuminoids alone in alleviating pain and dysfunction for knee osteoarthritis: A systematic review and meta-analysis of randomized controlled trials. *BMC Complement. Med. Ther.* **2022**, *22*, 276. [CrossRef]
7. Zeng, L.; Yu, G.; Hao, W.; Yang, K.; Chen, H. The efficacy and safety of Curcuma longa extract and curcumin supplements on osteoarthritis: A systematic review and meta-analysis. *Biosci. Rep.* **2021**, *41*, BSR20210817. [CrossRef] [PubMed]
8. Chin, K.Y. The spice for joint inflammation: Anti-inflammatory role of curcumin in treating osteoarthritis. *Drug Des. Dev. Ther.* **2016**, *10*, 3029–3042. [CrossRef]
9. Amalraj, A.; Pius, A.; Gopi, S.; Gopi, S. Biological activities of curcuminoids, other biomolecules from turmeric and their derivatives—A review. *J. Tradit. Complement. Med.* **2016**, *7*, 205–233. [CrossRef]
10. Kepinska-Pacelik, J.; Biel, W. Turmeric and Curcumin-Health-Promoting Properties in Humans versus Dogs. *Int. J. Mol. Sci.* **2023**, *24*, 14561. [CrossRef]
11. Sun, M.M.; Beier, F.; Pest, M.A. Recent developments in emerging therapeutic targets of osteoarthritis. *Curr. Opin. Rheumatol.* **2017**, *29*, 96–102. [CrossRef] [PubMed]
12. Rios, J.L.; Sapède, D.; Djouad, F.; Rapp, A.E.; Lang, A.; Larkin, J.; Ladel, C.; Mobasheri, A. Animal Models of Osteoarthritis Part 1-Preclinical Small Animal Models: Challenges and Opportunities for Drug Development. *Curr. Protoc.* **2022**, *2*, e596. [CrossRef]
13. Fang, H.; Huang, L.; Welch, I.; Norley, C.; Holdsworth, D.W.; Beier, F.; Cai, D. Early Changes of Articular Cartilage and Subchondral Bone in The DMM Mouse Model of Osteoarthritis. *Sci. Rep.* **2018**, *8*, 2855. [CrossRef]
14. Li, Y.; Ding, Z.; Liu, F.; Li, S.; Huang, W.; Zhou, S.; Han, Y.; Liu, L.; Li, Y.; Yin, Z. Luteolin regulating synthesis and catabolism of osteoarthritis chondrocytes via activating autophagy. *Heliyon* **2024**, *10*, e31028. [CrossRef]
15. Wang, J.; Yang, J.; Fang, Y.; Lou, C.; Yu, H.; Li, Y.; Lv, J.; Chen, H.; Cai, L.; Zheng, W. Vinpocetine protects against osteoarthritis by inhibiting ferroptosis and extracellular matrix degradation via activation of the Nrf2/GPX4 pathway. *Phytomedicine* **2024**, *135*, 156115. [CrossRef] [PubMed]
16. Mancini, I.A.D.; Rieppo, L.; Pouran, B.; Afara, I.O.; Bragança, F.M.S.; van Rijen, M.H.P.; Kik, M.; Weinans, H.; Toyras, J.; van Weeren, P.R.; et al. Effects of body mass on microstructural features of the osteochondral unit: A comparative analysis of 37 mammalian species. *Bone* **2019**, *127*, 664–673. [CrossRef]
17. Oláh, T.; Cai, X.; Michaelis, J.C.; Madry, H. Comparative anatomy and morphology of the knee in translational models for articular cartilage disorders. Part I Large Animals. *Ann. Anat.* **2021**, *235*, 151680. [CrossRef] [PubMed]
18. Hong, J.I.; Park, I.Y.; Kim, H.A. Understanding the Molecular Mechanisms Underlying the Pathogenesis of Arthritis Pain Using Animal Models. *Int. J. Mol. Sci.* **2020**, *21*, 533. [CrossRef]
19. Jones, G.M.C.; Pitsillides, A.A.; Meeson, R.L. Moving Beyond the Limits of Detection: The Past, the Present, and the Future of Diagnostic Imaging in Canine Osteoarthritis. *Front. Vet. Sci.* **2022**, *9*, 789898. [CrossRef]
20. Bhathal, A.; Spryszak, M.; Louizos, C.; Frankel, G. Glucosamine and chondroitin use in canines for osteoarthritis: A review. *Open Vet. J.* **2017**, *7*, 36–49. [CrossRef]
21. Koroljević, Z.D.; Jordan, K.; Ivković, J.; Bender, D.V.; Perić, P. Curcuma as an anti-inflammatory component in treating osteoarthritis. *Rheumatol. Int.* **2023**, *43*, 589–616. [CrossRef] [PubMed]
22. Glasson, S.S.; Chambers, M.G.; Van Den Berg, W.B.; Little, C.B. The OARSI histopathology initiative-recommendations for histological assessments of osteoarthritis in the mouse. *Osteoarthr. Cartil.* **2010**, *18*, S17–S23. [CrossRef] [PubMed]
23. Pritzker, K.P.; Gay, S.; Jimenez, S.A.; Ostergaard, K.; Pelletier, J.P.; Revell, P.A.; Salter, D.; van den Berg, W.B. Osteoarthritis cartilage histopathology: Grading and staging. *Osteoarthr. Cartil.* **2006**, *14*, 13–29. [CrossRef]
24. Lopresti, A.L.; Smith, S.J.; Jackson-Michel, S.; Fairchild, T. An Investigation into the Effects of a Curcumin Extract (Curcugen®) on Osteoarthritis Pain of the Knee: A Randomised, Double-Blind, Placebo-Controlled Study. *Nutrients* **2021**, *14*, 41. [CrossRef]
25. Zeng, C.; Wei, J.; Li, H.; Wang, Y.L.; Xie, D.X.; Yang, T.; Gao, S.G.; Li, Y.S.; Luo, W.; Lei, G.H. Effectiveness and safety of Glucosamine, chondroitin, the two in combination, or celecoxib in the treatment of osteoarthritis of the knee. *Sci. Rep.* **2015**, *5*, 16827. [CrossRef]
26. Zhu, X.; Sang, L.; Wu, D.; Rong, J.; Jiang, L. Effectiveness and safety of glucosamine and chondroitin for the treatment of osteoarthritis: A meta-analysis of randomized controlled trials. *J. Orthop. Surg. Res.* **2018**, *13*, 170. [CrossRef]
27. Dalle, S.; Koppo, K. Is inflammatory signaling involved in disease-related muscle wasting? Evidence from osteoarthritis, chronic obstructive pulmonary disease and type II diabetes. *Exp. Gerontol.* **2020**, *137*, 110964. [CrossRef]
28. Gencoglu, H.; Orhan, C.; Sahin, E.; Sahin, K. Undenatured Type II Collagen (UC-II) in Joint Health and Disease: A Review on the Current Knowledge of Companion Animals. *Animals* **2020**, *10*, 697. [CrossRef] [PubMed]
29. Drevet, S.; Favier, B.; Brun, E.; Gavazzi, G.; Lardy, B. Mouse Models of Osteoarthritis: A Summary of Models and Outcomes Assessment. *Comp. Med.* **2022**, *72*, 3–13. [CrossRef]
30. Karl, T.; Pabst, R.; von Hörsten, S. Behavioral phenotyping of mice in pharmacological and toxicological research. *Exp. Toxicol. Pathol.* **2003**, *55*, 69–83. [CrossRef]

31. Taylor, T.N.; Greene, J.G.; Miller, G.W. Behavioral phenotyping of mouse models of Parkinson's disease. *Behav. Brain Res.* **2010**, *211*, 1–10. [CrossRef] [PubMed]
32. Piel, M.J.; Kroin, J.S.; van Wijnen, A.J.; Kc, R.; Im, H.J. Pain assessment in animal models of osteoarthritis. *Gene* **2014**, *537*, 184–188. [CrossRef] [PubMed]
33. Geng, X.; Zhang, H.; Hu, M.; Liu, X.; Han, M.; Xie, J.; Li, Z.; Zhao, F.; Liu, W.; Wei, S. A novel curcumin oil solution can better alleviate the motor activity defects and neuropathological damage of a Parkinson's disease mouse model. *Front. Aging Neurosci.* **2022**, *14*, 984895. [CrossRef]
34. Suokas, A.K.; Walsh, D.A.; McWilliams, D.F.; Condon, L.; Moreton, B.; Wylde, V.; Arendt-Nielsen, L.; Zhang, W. Quantitative sensory testing in painful osteoarthritis: A systematic review and meta-analysis. *Osteoarthr. Cartil.* **2012**, *20*, 1075–1085. [CrossRef]
35. Miller, R.E.; Malfait, A.M. Osteoarthritis pain: What are we learning from animal models? *Best Pract. Res. Clin. Rheumatol.* **2017**, *31*, 676–687. [CrossRef]
36. Sengupta, S.; Tripathi, A. Evaluation of analgesic and prophylactic activity of curcumin against chikungunya-infected acute/chronic arthralgic mice. *J. Med. Virol.* **2023**, *95*, e28661. [CrossRef]
37. Arora, R.; Kuhad, A.; Kaur, I.P.; Chopra, K. Curcumin loaded solid lipid nanoparticles ameliorate adjuvant-induced arthritis in rats. *Eur. J. Pain* **2015**, *19*, 940–952. [CrossRef]
38. Li, G.Y.; Yin, J.M.; Gao, J.J.; Cheng, T.S.; Pavlos, N.J.; Zhang, C.Q.; Zheng, M.H. Subchondral bone in osteoarthritis: Insight into risk factors and microstructural changes. *Arthritis Res. Ther.* **2013**, *15*, 223. [CrossRef]
39. Ding, D.; Liu, G.; Yan, J.; Zhang, Q.; Meng, F.; Wang, L. Curcumin alleviates osteoarthritis in mice by suppressing osteoclastogenesis in subchondral bone via inhibiting NF-kB/JNK signaling pathway. *PLoS ONE* **2024**, *19*, e0309807. [CrossRef]
40. Wang, X.; Yu, H.; Zhang, Y.; Chang, X.; Liu, C.; Wen, X.; Tian, F.; Li, Y. Curcumin Alleviates Osteoarthritis Through the p38MAPK Pathway: Network Pharmacological Prediction and Experimental Confirmation. *J. Inflamm. Res.* **2024**, *17*, 5039–5056. [CrossRef]
41. Chen, W.; Xiao, J.; Zhou, Y.; Liu, W.; Jian, J.; Yang, J.; Chen, B.; Ye, Z.; Liu, J.; Xu, X.; et al. Curcumenol regulates Histone H3K27me3 demethylases KDM6B affecting Succinic acid metabolism to alleviate cartilage degeneration in knee osteoarthritis. *Phytomedicine* **2024**, *133*, 155922. [CrossRef] [PubMed]
42. Fernandez-Martin, S.; Gonzalez-Cantalapiedra, A.; Muñoz, F.; Garcia-Gonzalez, M.; Permuy, M.; Lopez-Pena, M. Glucosamine and Chondroitin Sulfate: Is There Any Scientific Evidence for Their Effectiveness as Disease-Modifying Drugs in Knee Osteoarthritis Preclinical Studies?—A Systematic Review from 2000 to 2021. *Animals* **2021**, *11*, 1608. [CrossRef] [PubMed]
43. Yunus, M.H.M.; Nordin, A.; Kamal, H. Pathophysiological Perspective of Osteoarthritis. *Medicina* **2020**, *56*, 614. [CrossRef] [PubMed]
44. Ansari, M.Y.; Ahmad, N.; Haqqi, T.M. Oxidative stress and inflammation in osteoarthritis pathogenesis: Role of polyphenols. *Biomed. Pharmacother.* **2020**, *129*, 110452. [CrossRef]
45. Yabas, M.; Orhan, C.; Er, B.; Tuzcu, M.; Durmus, A.S.; Ozercan, I.H.; Sahin, N.; Bhanuse, P.; Morde, A.A.; Padigaru, M.; et al. A Next Generation Formulation of Curcumin Ameliorates Experimentally Induced Osteoarthritis in Rats via Regulation of Inflammatory Mediators. *Front. Immunol.* **2021**, *12*, 609629. [CrossRef]
46. Guan, T.; Ding, L.G.; Lu, B.Y.; Guo, J.Y.; Wu, M.Y.; Tan, Z.Q.; Hou, S.Z. Combined Administration of Curcumin and Chondroitin Sulfate Alleviates Cartilage Injury and Inflammation via NF-kB Pathway in Knee Osteoarthritis Rats. *Front. Pharmacol.* **2022**, *13*, 882304. [CrossRef]
47. Wan, J.; Zhang, G.; Li, X.; Qiu, X.; Ouyang, J.; Dai, J.; Min, S. Matrix Metalloproteinase 3: A Promoting and Destabilizing Factor in the Pathogenesis of Disease and Cell Differentiation. *Front. Physiol.* **2021**, *12*, 663978. [CrossRef]
48. Gugliandolo, E.; Peritore, A.F.; Impellizzeri, D.; Cordaro, M.; Siracusa, R.; Fusco, R.; D'Amico, R.; Paola, R.D.; Schievano, C.; Cuzzocrea, S.; et al. Dietary Supplementation with Palmitoyl-Glucosamine Co-Micronized with Curcumin Relieves Osteoarthritis Pain and Benefits Joint Mobility. *Animals* **2020**, *10*, 1827. [CrossRef]

veterinary sciences

MDPI

Review

Clinical Trials of Cancer Immunogene Therapies in Companion Animals: An Update (2017–2024)

Gerardo C. Glikin * and Liliana M. E. Finocchiaro *

Unidad de Transferencia Genética, Área Investigación, Instituto de Oncología "Ángel H. Roffo", Universidad de Buenos Aires, Av. San Martín 5481, Buenos Aires 1417, Argentina

* Correspondence: gglikin1@gmail.com (G.C.G.); finolili@hotmail.com (L.M.E.F.)

Simple Summary: Through the transfer of genetic information, immunogene therapy aims to increase tumor rejection in cancer patients by increasing the immunogenicity of certain tumor antigens as well as decreasing cancer tolerance mechanisms. Treating cancer-bearing veterinary patients can greatly accelerate translational research, providing benefits to both veterinary and human medicine. Recent findings have highlighted the safety and effectiveness of various immunogene therapy strategies. Expanding on our previous reviews, which explored advancements from 1996 to 2016, this updated review focuses specifically on veterinary cancer immunogene therapy, summarizing published studies in the field from 2017 to 2024.

Abstract: This review summarizes the findings of veterinary clinical trials on immunogene therapy published between 2017 and 2024. Various tumor types, including melanoma (canine and feline), mastocytoma (canine), mammary adenocarcinoma (canine), osteosarcoma (canine), and sarcoid (equine), were treated using diverse strategies. Non-viral vectors were predominantly used to deliver genes encoding tumor-associated antigens, cytokines, or suicide enzymes. Among these non-viral methods, electrotransfer was the most commonly employed technique for introducing therapeutic genes into cells. Generally, these procedures resulted in minimal or no adverse side effects, and treated animals often showed significant improvements, such as enhanced quality of life, delayed or suppressed tumor recurrence or metastasis, and increased survival times. Some of these innovative approaches hold great potential as adjunct therapies to standard treatments. The promising outcomes from immunogene therapy studies in companion animals strongly support their application in veterinary oncology and provide valuable preclinical data (including safety assessments and proof-of-concept studies) for analogous human clinical trials.

Keywords: cancer; gene therapy; immunotherapy; companion animals; comparative oncology; clinical trials

1. Introduction

Nearly thirty years have passed since Quintin-Colonna and colleagues published their groundbreaking 1996 study [1] on ex vivo interleukin-2 immunogene therapy for feline fibrosarcoma and canine melanoma. In two previous reviews [2,3], we examined advancements in this field from its inception through to the end of 2016. The present review covers the developments published from January 2017 to December 2024. Since the beginning, most of the veterinary cancer gene therapy trials carried out have aimed to enhance the immune response against tumor cells.

The primary goal of immunotherapy is to enhance the body's immune system to eliminate tumor cells and establish a long-lasting antitumor immune response, avoiding local invasion and the spread of metastatic disease. Among the different immunotherapeutic strategies, immunogene therapy strengthens the antitumor immune response by using genetic engineering techniques to introduce genes that usually encode tumor-associated antigens, cytokines, co-stimulatory molecules, and/or chemokines, into different kinds of patient's cells in vivo or ex vivo (depending on the chosen approach) [4,5].

This review focuses on studies that have implemented strategies to actively stimulate a systemic antitumor immune response through the expression of transgenes, which, in the field of companion animal oncology, mostly encode either tumor antigens or cytokines.

2. Immunogene Therapy in Veterinary Oncology

What makes clinical trials of cancer immunogene therapy in companion animals significant? Despite progress in tumor treatment for both companion animals and human patients, there is still a critical need for novel therapeutic strategies to combat various malignancies, as current treatments often have limited effectiveness and cause significant side effects. Spontaneous tumors in companion animals serve as valuable translational models because pets exhibit relatively large body sizes, diverse genetic backgrounds, and intact immune systems. These models provide robust proof-of-concept evidence in a context comparable to human patient populations while simultaneously offering insights into toxicity and long-term efficacy. Currently, there is renewed interest in utilizing canine clinical trials to drive progress in cancer therapies [6–8].

Different gene therapy approaches were used in clinical studies against various types of tumors: from non-viral vectors such as "naked" plasmids, lipoplexes (cationic lipids/DNA complexes), non-replicative, and conditionally replicative viral vectors such as adenovirus, herpesvirus, poxvirus [2,3]. In some cases, physical methods such as jet injection, microseeding, and electrotransfer were also used to improve plasmid-mediated gene transfer.

As in our earlier reviews [2,3], we have focused exclusively on clinical data gathered from client-owned animals with naturally occurring tumors.

Table 1 displays twenty-six veterinary clinical trials reported from 2017 to 2024. They mainly involved non-viral gene transfer (19/26), sometimes enhanced by cationic lipids (2/26) and often by physical methods (15/26: 12 electrotransfer and 3 jet injection). In the remaining trials, viral vectors were used (7/26: 2 adenovirus, 2 vesicular stomatitis virus, 1 herpes simplex virus, 1 influenza virus, 1 poxvirus). Of the 26 veterinary clinical trials seen in Table 1, 23 correspond to canine patients, 2 to equines, and 1 to felines. Except for the feasibility/safety studies that covered different types of tumors (3 trials), those that focused on specific types include, in descending order of number of trials, canine melanoma (10), canine mastocytoma (4), canine mammary carcinoma (2), canine osteosarcoma (2), equine sarcoid (2), canine brain tumors (1), canine lymphoma (1), and feline melanoma (1). Regarding the types of trials, 9 were pilot studies, 5 were retrospective studies (of which 2 were controlled), and 12 were prospective studies (of which 2 were controlled and 4 compared with historical controls).

Early reports about feasibility and safety of autologous ex vivo immunogene therapies such as genetically modified dendritic cells [9] and chimeric antigen recombinant T cells (CAR-T) [10] or tumor infiltrating CAR-T cells (CAR-TILs) [11] were not included in this review.

3. Antigen Expression-Based Immunogene Therapy

3.1. *Xenogeneic Tyrosinase*

Spontaneous canine mucosal melanoma is a highly aggressive tumor of the oral cavity, digit/footpad, and muco-cutaneous junctions; it is chemo- and radio-resistant, does not respond well to treatment with conventional biological response modifiers, and shares similar metastatic phenotypes and site selectivity [2]. At the time of diagnosis, the disease is often metastatic and has an extremely poor prognosis because of rapid invasion of surrounding normal tissue and high likelihood of regional and distant metastasis early in the course of the disease. Its high metastatic rate and the inefficacy of current therapies warranted the investigation into novel therapies [12].

Preclinical and clinical studies have shown that xenogeneic DNA vaccination with tyrosinase gene family members can produce immune responses against melanoma, resulting in tumor rejection or protection and prolongation of survival. Conversely, syngeneic vaccination with orthologous DNA does not induce immune responses. Although tyrosinase may not seem to be a preferred target in amelanotic canine melanoma when determined by immunohistochemistry, other techniques showed significant tyrosinase overexpression in both melanotic and amelanotic melanomas [13].

The vaccination with plasmids carrying genes encoding tyrosinase family members demonstrated the ability to elicit both antibody and cytotoxic T cell responses leading to tumor rejection. This strategy offered a promising therapeutic approach and provided the basis for conducting the first trial of xenogeneic DNA vaccination in canine malignant melanoma using the human tyrosinase (hTyr) gene [14]. Following surgery, the vaccine was administered using a jet injection transdermal device (four biweekly doses followed by a booster every 6 months. Since this treatment was already approved by the US Department of Agriculture, the treatment is available, and additional studies were performed in diverse veterinary centers.

A retrospective study examined the outcomes of 69 canine oral melanoma patients with previous locoregional therapy treated with hTyr DNA vaccination [15]. A median survival time (MST) of 455 days, a figure comparable to earlier findings [16,17], was reported. Observing the responses in a group of 13 patients with macroscopic disease whose MST was 178 days, the authors proposed that the vaccine could serve as a palliative treatment for dogs with residual tumors or recurrence.

A new multi-institutional retrospective study offered new data about the clinical outcome of 131 dogs with oral melanoma (stages I to III) treated with hTyr DNA vaccination [18]. In this case, most dogs were also subjected to surgery and/or radiotherapy but not to chemotherapy. Again, a median survival time of 442 days was similar to earlier results [16,17], including the one described above [15]. As expected, a multivariable analysis indicated a negative correlation of clinical outcome with the stage of disease, while radiotherapy had a protective effect against tumor progression.

Even though a more recent retrospective small study focused on canine foot pad malignant melanoma included some patients (8/20) receiving hTyr DNA vaccination [19], there are very few data regarding its specific effects.

A small pilot study, comprising six oral melanoma canine patients, proposed DNA microseeding as an alternative delivery method for hTyr DNA vaccination [20]. This method was well tolerated with no significant toxicity detected and no signs of autoimmunity. Both local hTyr expression and humoral anti-human tyrosinase antibodies were detected. Further studies are necessary to assess the efficacy of this approach.

In general, the human tyrosinase DNA melanoma vaccine was well tolerated by canine patients [14–21]. Minimal adverse effects were described by owners, most being mild and self-limiting; the frequency of adverse reactions declined throughout the vaccination course. The lack of control groups and the diversity of concomitant treatments precluded the assessment of the clinical efficacy of the approach.

Similarly to dogs, malignant melanoma in cats is locally aggressive and highly metastatic, regardless of the primary site of origin [22,23]. The usual modalities, such as surgery and radiotherapy, often yield poor long-term results. Building on studies involving xenogeneic antigen vaccination for spontaneous canine malignant melanoma [21], research has reported the use of a transdermal, needle-free jet injection of a plasmid containing the hTyr gene. This plasmid's expression as a melanoma xenoantigen was tested against feline melanoma [22].

In a retrospective study, the safety of the canine melanoma DNA vaccine following surgery in 24 feline patients was assessed [22]. The vaccine appeared to be well tolerated, with a low number of reported adverse events. Since most cats ultimately died from this disease, prospective controlled studies are essential to assess the immunogenicity and efficacy of this vaccine, originally developed for canine melanoma, when used to treat feline melanoma. The use of the hTyr DNA melanoma vaccine remains controversial, as studies have yielded mixed results regarding its effectiveness in providing benefits or extending lifespan. While the vaccine appears to be safe, there does not appear to be evidence that it improves outcomes when used [24].

3.2. *Xenogeneic Chondroitin Sulfate Proteoglycan-4*

Chondroitin sulfate proteoglycan-4 (CSPG4) serves as an early cell surface marker of tumor progression and is linked to enhanced migration, invasion, and proliferation of tumor cells [25]. In a previous trial involving a xenogeneic immunogene therapeutic approach, a human CSPG4 DNA-based vaccine had been evaluated for canine oral malignant melanoma [26]. This treatment was proposed as an adjuvant to surgery and transferred by periodic electro-vaccination with a plasmid carrying the hCSPG4 gene, allowing the assessment of its immunogenicity, safety, and therapeutic efficacy.

In a later prospective study with hCSPG4-encoded plasmid for II/III-staged CSPG4-positive oral canine malignant melanoma, 42 dogs were subjected to en bloc resection [27]. Twenty-nine patients were electro-vaccinated and the remaining nineteen dogs were assigned to the control group. The median survival time (MST) was 684 days for the vaccinated group and 200 days for the unvaccinated controls. Smaller dogs (<20 kg) displayed longer MST. Local relapse and lung metastatic rate diminished in vaccinated dogs (34.8 and 39%) with respect to unvaccinated controls (42 and 79%).

A retrospective study evaluated 82 dogs treated with adjuvant CSPG4-DNA electro-vaccination with oral malignant melanomas (stages I to IV) [28]. Both median survival times in dogs surgically treated with a curative intent (51 dogs) and those marginally excised only (31 dogs) (MST: 1333 vs. 470 days) as well as the corresponding disease-free intervals were significantly different (DFI: 324 vs. 184 days). On the other hand, the corresponding 63 non-vaccinated controls that received alternative adjuvant therapies displayed no significant differences in median survival time between dogs surgically treated with a curative intent (37 dogs) and those marginally excised only (26 dogs) (MST: 594 vs. 458 days), disease-free intervals were significantly different (DFI: 232 vs. 183 days). The authors concluded that a curative intent surgical approach, when feasible, is advisable in an attempt to prolong both the DFI and survival.

In a parallel retrospective study with 39 dogs treated with adjuvant CSPG4-DNA electro-vaccination with oral malignant melanomas, the authors [29] did not find significant relationship between leukocytes ratios and histological parameters, CSPG4 expression, excision margin status, age, tumor size, and clinical stage. Pretreatment neutrophil to lymphocyte (NLR) and lymphocyte to monocyte (LMR) ratios did not show a prognostic impact on the survival time of the entire population.

A high percentage of oral canine melanomas express the CSPG4 antigen [30]. Being a self-antigen, autologous CSPG4 is poorly immunogenic. To overcome this issue a chimeric human–canine CSPG4 gene (HuDo-CSPG4) was designed and inserted in a plasmid for

DNA electro-vaccination. In a prospective study, 80 canine patients were included with surgically resected, CSPG4-positive, stage II–IV oral melanoma [31]. While 52 dogs received adjuvant HuDo-CSPG4 electro-vaccination, the remaining 28 were unvaccinated controls that only received conventional therapies. Dogs that received adjuvant vaccination demonstrated a significantly longer overall survival compared to the control group treated solely with conventional therapies. The median survival time (MST) was 653 days for the vaccinated group and 310 days for the control group. The implementation of this novel anti-CSPG4 mimicry strategy utilizing the hybrid HuDo-CSPG4 DNA vaccine proved to be safe and immunogenic, showing potential clinical benefits in extending survival.

Osteosarcoma represents approximately 85–90% of primary bone cancers in dogs. It is a common, highly metastatic cancer that predominantly affects large to giant breeds, primarily targeting the appendicular skeleton [32]. When adjuvant chemotherapy fails to control its disseminated form, there are no alternative treatment options available. Despite the surgical removal of the primary tumor before clinical detection of metastases, nearly all affected dogs eventually develop metastases in the lungs, bones, or other sites.

In a pilot veterinary study, 25 dogs with limb amputated, CSPG4-positive, stage I–III appendicular osteosarcoma were involved [33]. Patients also received standard chemotherapy to prevent the systemic spread of the disease. While 12 dogs received adjuvant HuDo-CSPG4 electro-vaccination, the remaining 13 were unvaccinated controls that only received conventional therapies. Dogs that received adjuvant vaccination demonstrated a significantly longer overall survival compared to the control group treated solely with conventional therapies. The median survival time (MST) was 484 days for the vaccinated group and 202 days for the control group. In addition, the disease-free interval (DFI) was 242 days for the vaccinated group and 160 days for the control group. In summary, the HuDo-CSPG4 vaccine has been shown to be safe and effective in inducing anti-CSPG4 immunity in dogs affected by osteosarcoma, leading to prolonged survival compared to control groups.

3.3. Other Antigens

Lymphoma is the most common hematopoietic cancer in dogs, accounting for approximately 15–20% of new canine cancer diagnoses [34]. The majority of these cases are malignant B-cell lymphomas, with diffuse large B-cell lymphoma (DLBCL) being the most prevalent subtype, often diagnosed at advanced stages (III–V) and typically exhibiting an aggressive clinical progression that necessitates prompt treatment. While multi-agent chemotherapy can achieve complete remission in some cases, the mortality rate for this neoplasm remains high.

A vaccine against dog telomerase reverse transcriptase (dTERT) was also proposed as a valid target for immunotherapy of diffuse large B-cell lymphoma [35]. The protocol, that includes a previous CHOP (cyclophosphamide, vincristine, doxorubicin, and prednisone) chemotherapy treatment, comprises an initial injection of an adenoviral vector containing the dTERT gene followed by periodic electro-vaccination boosts of a chimeric plasmid containing the dTERT fused at N-term with tissue plasminogen activator (TPA) leader sequence and at the C-term with B subunit of E. coli heat-labile enterotoxin (LTB). Seventeen canine patients with diffuse large B-cell lymphoma (DLBCL) were treated, and antibodies against the immunizing antigen were detected. The median survival time (MST) of 452 days observed in these patients was significantly longer than the 205 days recorded for control dogs treated only with COP chemotherapy (cyclophosphamide, vincristine, and prednisone) [36].

Horses and other equid species are commonly affected by skin tumors known as sarcoids, which are caused by bovine papillomavirus type 1 and/or 2 (BPV1, BPV2).

Although sarcoids do not metastasize, they pose a significant health challenge due to their resistance to BPV1/2-mediated treatments and their tendency to recur in a more severe and widespread form after accidental or medical trauma [37].

Recombinant influenza (A and B) virus carrying the sequences of BPV1oncoproteins E6 and E7 were proposed as an immunogene therapy [38]. Intratumoral injections were applied to 29 equine patients bearing sarcoids. The treatment was safe and effective: 10 patients achieved complete responses and 10 partial responses at the end of the trial. Among the 29 individuals treated, 9 with severe conditions and a history of unsuccessful therapeutic attempts either failed to respond (6 cases) or exhibited only temporary improvements (3 cases). The fact that, in some cases, untreated lesions also displayed antitumor effects indicated a systemic antitumor response.

In unspayed female dogs, mammary tumors are the most frequently observed type, comprising roughly 50% of all tumors, with about half of these tumors being malignant [39]. The p62 protein (SQSTM1) is crucial for selective macroautophagy and functions as a central hub for various signal transduction pathways. Notably, while p62 is dispensable for normal tissue function, it is essential for tumor development and survival [40]. A study tested the effects of an intramuscular p62 plasmid DNA vaccine on mammary tumors in six dogs [41]. Locally advanced tumors either reduced in size (5/6) or stabilized (1/6) without any observed toxic side effects. Moreover, p62 treatment enhanced the presence of intratumoral T cells. All patients treated with p62 remain tumor- and metastasis-free and continued with a good quality of life four years after undergoing mastectomy. This pilot study suggests that p62 DNA treatment has the potential to reprogram the tumor stroma and may serve as an adjuvant in cancer therapies, exerting its effects directly or indirectly through immune responses.

CD40 ligand (CD40L, also known as CD154) is a powerful activator of antigen-presenting cells, such as dendritic cells (DCs). Additionally, it promotes the M1 phenotype and diminishes tumor infiltration by myeloid-derived suppressor cells. As a result, adding CD40L to the tumor microenvironment can provoke tumor inflammation and stimulate antitumor immune responses. AdCD40L is a replication-deficient adenovirus that encodes the human CD40L gene [42].

In a study of local AdCD40L treatment, 32 cases of mucosal canine melanoma (stages I to IV) were intratumorally injected [43]. In 20 cases, this was combined with cytoreductive surgery while the remaining 12 cases only received immunotherapy. Tumor tissue was infiltrated with T and B lymphocytes after treatment, suggesting immune stimulation. The best overall responses based on results of immunotherapy were 7 complete, 5 partial responses, 5 stable, and 2 progressive diseases. This treatment offered a good safety profile and a median survival of 285 days to dogs with melanoma.

4. Cytokine Expression-Based Immunogene Therapy

Only four types of cytokine genes were used in veterinary clinical situations from 2017 to 2024: interleukin-12, interleukin-2, interferon-β, and granulocyte–macrophage colony-stimulating factor. Usually, these genes were employed combined with other standard therapies such as chemotherapy, radiotherapy, molecular targets [2,3]. In some cases, a cytokine gene was combined with another cytokine gene, a suicide gene, or eventually inserted in a conditionally replicative oncolytic virus.

4.1. Interleukin-12 (IL-12)

IL-12 is viewed as an auspicious cytokine for boosting antitumor immune responses; nevertheless, recombinant IL-12 has exhibited substantial toxicity and limited effectiveness in early clinical trials. Recently, many approaches for delivering IL-12 to tumor tissues have been developed, such as modifying IL-12 structure, employing viral vectors, non-viral vectors, and cellular vectors [44]. As one of the most potent proinflammatory cytokines, it was initially investigated for this purpose. However, initial murine and human studies in which IL-12 was administered systemically resulted in dangerous immunotoxicity associated with off-target immune activation [45].

Gene electrotransfer (GET) was the method of choice for delivering plasmids carrying the IL-12 gene into cells in vivo. In a pilot study, nine dogs with various spontaneous cancers underwent three consecutive intratumoral hIL-12 GET sessions to evaluate their clinical, immunological, and anti-angiogenic effects [46]. Both serum and tumor tissue showed temporary increases in canine interferon-γ (IFN-γ) and hIL-12 levels. Although intratumoral IL-12 GET demonstrated some anti-angiogenic and immunostimulatory activities along with transient objective responses, it did not yield clinically significant results. Sustained tumor regression could not be achieved during the trial. The therapy was confirmed to be safe, with no severe side effects observed.

The effects of combining three successive intratumoral hIL-12 gene electrotransfer treatments with metronomic cyclophosphamide were evaluated in six dogs with various spontaneous tumors [47]. This combination therapy resulted in tumor erythema and swelling, a transitory rise in IL-12 levels, and sustained increases in IFN-γ and thrombospondin-1, accompanied by an uninterrupted reduction in vascular endothelial growth factor (VEGF). The treatment improved body weight and quality of life while slowing tumor progress in the majority of patients (4 out of 6).

Mast cell tumors (MCTs) are among the most common cutaneous neoplasms in dogs, representing up to 21% of all canine skin tumors. While most cases can be effectively treated with local therapy, some exhibit biologically aggressive behavior, leading to local recurrence or metastasis [48].

Peritumoral hIL-12 gene electrotransfer combined with electro-chemotherapy (using cisplatin and/or bleomycin) was evaluated in 18 canine patients with MCTs [49]. This combination therapy achieved significant local tumor control within one month without causing side effects. The complete response rate rose to 72% over the observation period. Serum hIL-12 and/or IFN-γ levels were detectable in 78% of patients, demonstrating systemic immune activation. This study highlighted the high antitumor efficacy of this combined treatment in preventing recurrences and distant metastases while establishing its safety and feasibility.

As a follow-up to this study, histopathological analysis was performed on samples from 11 patients. The analysis confirmed that the combined electro-chemotherapy and IL-12 gene electrotransfer effectively triggered a cellular immune response against neoplastic cells. This was characterized by the recruitment of T-lymphocytes and macrophages, along with fibrotic proliferation and a reduction in microvessels [50].

In an additional trial, eight dogs (seven bearing mast cell tumors and one neurofibrosarcoma), were subjected to the combined i.v. bleomycin or i.t. cisplatin electro-chemotherapy and cIL-12 gene electro-transfer [51]. Dynamic contrast-enhanced ultrasound (DCE-US) serial examinations were conducted: before therapy, shortly after therapy, and long-term post-therapy. Tumors that achieved complete responses exhibited significantly lower capillary perfusion and perfusion heterogeneity compared to those that did not achieve complete responses. DCE-US proved to be a safe procedure, with no adverse effects observed following repeated administration. Moreover, it can provide a straightforward and repeatable method for assessing tumor perfusion during and after treatment. This diagnostic method could serve as a predictor of tumor responses, aiding in the evaluation of therapeutic options.

A prospective study recruited 77 dogs with spontaneous mast cell tumors (MCTs) and assigned them to three groups: one treated with a combination of ECT + GET p.t. (29 dogs), the second with the combination of ECT + GET i.t. (30 dogs), and the third with ECT alone (18 dogs) [52]. The results showed that local tumor control was significantly better in the ECT + GET i.t. group than in the ECT + GET p.t. or ECT groups. In addition, disease-free interval (DFI) and progression-free survival (PFS) were significantly longer in the ECT + GET i.t. group than in the other two groups. The data on local tumor response, DFI, and PFS were consistent with an increased percentage of circulating antitumor immune cells in the ECT + GET i.t. group, which also indicated the induction of a systemic immune response. No unwanted severe or long-lasting side effects were observed.

In a later trial involving 48 canine patients with 86 mast cell tumors (MCTs), blood samples were collected before ECT + GET i.t. treatment, and again at one- and six-months post-treatment, to identify trustworthy biomarkers predicting treatment response [53]. An increase in plasma nucleosome levels and serum lactate dehydrogenase (LDH) activity one month after treatment was related to a more robust local response, such as necrosis and swelling. These biomarkers displayed promise as early indicators of treatment success.

When assayed in canine oral malignant melanoma, a similar approach involved a combination of cytoreductive surgery, ECT with bleomycin, and p.t. cIL-12 GET [54]. While three dogs did not respond to treatment, six dogs displayed transient objective responses (4 CR, 2 PR) that later derived in PD; MST was 180 days. The observed reduction in circulating regulatory T cells (Treg) may be the result of a systemic antitumor response triggered by cIL-12 GET. This alternative combination therapy for advanced canine melanoma showed a good safety profile.

In general terms, candidate genes for electrotransfer, the delivery site, the dosage of plasmids encoding these genes, and the dose of chemotherapeutic agents require further investigation [55]. While peritumoral delivery of a plasmid encoding IL-12 via GET has shown promising results in treating various tumor types in dogs, the efficacy of intratumoral plasmid application remains to be determined.

Certain breeds of pet dogs naturally and sporadically develop high-grade gliomas. The most aggressive of these tumors resist the available limited treatment options, resulting in poor outcomes with median survival times of about 2 months in dogs [56]. A new approach proposed the combination of intracranial surgery, oncolytic virotherapy, and immunogene therapy [57]. Following surgery, a targeted conditionally replicating oncolytic virus carrying the hIL-12 gene was infused in the cavity. The trial included 21 dogs: 13 had high-grade gliomas, 5 had low-grade gliomas, and 3 were undetermined. The MST was 151 days and no significant adverse events were reported. The reported MST for symptomatic treatments was 65 days [58]. In a parallel study, it was demonstrated that this oncolytic herpes virus vector modulates the tumor-immune microenvironment in canine glioma patients [59]. All these early data support further trials for assessing the efficacy of the approach.

4.2. Interleukin-2 (IL-2)

IL-2 is a potent pleiotropic cytokine involved in exerting dual functions of driving effector T cell expansion and differentiation as well as mediating peripheral tolerance by inducing regulatory T cells. Due to its capacity to activate and expand cytotoxic effector cells, IL-2 immunotherapy has been examined in numerous clinical trials [60]. To overcome the systemic toxicity of this cytokine, several novel engineering strategies are being developed for IL-2 based cancer treatments, including approaches such as immunogene therapy.

The canarypox virus vector ALVAC bearing the feline IL-2 gene was tested for treating equine sarcoids [61]. In a prospective proof of concept study, 14 horses received periodical intratumor injections of the vector. The best responses were 7 CR, 5 PR. The median time to best response was 211 days. Adverse events were minimal. The results offer preliminary evidence that ALVAC-fIL2 is a safe and effective treatment for sarcoid tumors in horses.

Even though IL-2 gene was initially proposed as a single immunogene treatment, it was usually combined with other cytokine genes or other modalities like chemotherapy, ECT, or suicide gene therapy.

In a retrospective controlled trial, 30 dogs with inoperable stage III–IV oral canine melanoma were treated with bleomycin ECT (the drug i.v injected) [62]. One-third of these dogs also received GET with two different cytokines: proximal intra-tumor cIL-2 and distal intramuscular cIL-12. Both groups showed similar local response rates and overall survival times. However, progression-free survival (PFS) resulted significantly better in the ECT + GET group (180 d) than in the ECT group (120 d). Both groups demonstrated comparable local response rates and overall survival times. However, the ECT + GET group showed significantly improved progression-free survival. With AEs between low and mild, a larger trial could help to assess the efficacy of the immune systemic effects mediated by the expressed cytokines for metastatic disease control.

4.3. *Interferon-β (IFN-β)*

Type I interferons (IFNs) have been demonstrated to inhibit tumor growth directly and indirectly by acting upon tumors and immune cells, respectively. Furthermore, accumulating evidence indicates that endo- and exogenously enhancing type I interferons have a synergistic effect on antitumor immunity [63]. Interferon-β (IFN-β) exhibits anti-proliferative, anti-angiogenic, and immunomodulatory properties. It was one of the earliest cytokines to receive approval for clinical application. Despite its clinical efficacy, the continuation of IFN-β therapy is frequently hindered by its intrinsic systemic toxicity.

In a pilot assay, a recombinant vesicular stomatitis virus (VSV) carrying the canine or human IFN-β and the sodium–iodide symporter (NIS): VSV-cIFNβ-NIS and VSV-hIFN-NIS [64]. Ten dogs with different tumors were intravenously injected with the vector (5 and 5). The feasible safety of repeated intravenous VSV-IFNβ-NIS administration was demonstrated and some evidence of potential efficacy was supported by the 2 PR and 3 SD outcomes. In the context of this assay and the next one, the NIS gene was not considered for therapeutic purposes.

As a continuation of this assay, a new prospective trial with the recombinant vesicular stomatitis virus VSV-cIFNβ-NIS for canine appendicular osteosarcoma was performed [65]. In a neoadjuvancy setting, the recombinant oncolytic virus was intravenously injected 10 days before amputation. All dogs received standard carboplatin chemotherapy 14 days after amputation. Twenty-two dogs received systemic neoadjuvant intravenous VSV-cIFNβ-NIS and six received placebo. MST of treated patients was not significantly different from that of historical controls: most of them died due to disease progression. However, there was a group of seven long-term survivors (survival >479 days) that died of unrelated causes or were still alive at the end of this study. With a good safety profile, this study proved that pre-treatment immune infiltration is associated with longer overall survival.

4.4. *Granulocyte–Macrophage Colony-Stimulating Factor (GM–CSF)*

GM–CSF is a cytokine that drives the generation of myeloid cell subsets including neutrophils, monocytes, macrophages, and dendritic cells in response to stress, infections, and cancers [66]. By modulating the functions of innate immune cells that serve as a bridge to activate adaptive immune responses, GM–CSF globally impacts host immune surveillance under pathologic conditions. While little GM–CSF induces the appropriate production of innate immune cells and subsequent activation of adaptive anti-cancer immune responses, too much GM–CSF can exhaust immune cells and promote cancer growth.

Thirty-six canine mammary carcinoma patients underwent injections of lipoplexes carrying canine interferon-β (cIFN-β) and HSV-thymidine kinase (HSV-tk) genes ombined with ganciclovir (GCV) into the tumor bed directly following surgery [67]. Subsequently, patients received periodic subcutaneous injections of lipoplexes containing human granulocyte–macrophage colony-stimulating factor and interleukin-2 genes, mixed with allogeneic mammary carcinoma extracts. This combined therapeutic approach was safe and well-tolerated. Notably, among the 26 patients who underwent complete surgery, only 2 experienced local relapse, and none of the 29 stage II and III patients developed distant metastases, suggesting effective local and systemic antitumor activity. The most promising outcome was the extended survival times: 22 patients survived over 1 year, with 13 exceeding 2 years, and 4 surpassing 3 years, all while maintaining a good quality of life. In five additional patients with local disease, the inclusion of an extra HSV-tk/GCV plus cIFN-β gene treatment demonstrated local antitumor activity, resulting in four objective responses (one complete and three partial) and one case of stable disease. These results highlight the potential of this therapeutic strategy and support further studies with a control arm to validate its efficacy.

In a prospective controlled trial 364 canine malignant melanoma patients received after complete or partial surgery: (i) a combined treatment that involves intra/peri-tumoral injection of lipoplexes containing the HSV-tk/r suicide gene, ganciclovir, canine interferon-β gene and bleomycin intra and/or (ii) a subcutaneous vaccine composed by canine melanoma cell extract and lipoplexes containing hIL-2 and hGM-CSF genes [68]. The control group was 173 dogs treated only with surgery. The adjuvant treatment (AT) group was divided into three groups: (i) complete surgery plus vaccine (CS-V), (ii) complete surgery plus combined treatment (CS-CT), and (iii) partial surgery plus combined treatment (PS-CT). Compared to the surgery-only control group (So), both CS-CT and CS-V significantly improved outcomes. The proportion of patients free from local disease increased from 20% to 89% (CS-CT) and 74% (CS-V), while those without distant metastases (M0) increased from 45% to 87% (CS-CT) and 84% (CS-V). Although less effective than CS groups, the PS-CT arm still showed significantly better control of metastatic disease (M0: 80%) compared to So (M0: 44%). Adjuvant therapy also led to substantial increases in overall survival, with a 9.3-fold (CS-CT), 6.5-fold (CS-V), and 5.4-fold (PS-CT) improvement over their respective So controls. In general, AT transformed melanoma from a lethal disease into a chronic condition, with 70% of CS-CT, 51% of CS-V, and 14% of PS-CT patients succumbing to non-melanoma-related causes. These treatments successfully delayed or prevented post-surgical recurrence and metastasis, improved disease-free and overall survival rates, and preserved quality of life.

A significant increase in overall survival was observed in the PS-CT and CS-CT patients (415 and 880 days, respectively) compared to the survival rates in a similar trial without BLM (323 and 704 days, respectively) [69], supporting the use of concomitant local chemotherapy to achieve a better outcome.

In a variant of the treatment combining surgery with local suicide gene system plus IFN-β gene plus bleomycin, and subcutaneous vaccines stimulated by cytokine genes [68], the suicide gene system was replaced by locally applied 5-fluorouracil chemotherapy. Preliminary encouraging data from our team indicate safety and disease control in the treatment of various solid tumors as indicated by the following median survival time values: canine melanoma (844 d, n = 186), canine fibrosarcoma (500 d, n = 20), canine osteosarcoma (238 d, n = 11), feline melanoma (491 d, n = 10), feline squamous cell carcinoma (474 d,

$n = 47$) feline fibrosarcoma (545 d, $n = 43$), and equine melanoma (all alive, mean time under treatment > 1024 d, $n = 14$).

Beyond the encouraging results of the pilot studies and the studies without proper control arms, four prospective controlled trials [27,31,52,68] and one retrospective [28] demonstrated a statistically significant improvement in the antitumor effects (including in some cases extended survival) while maintaining a good quality of life. These findings provide strong evidence for further advancements in immunogene therapy.

Regarding the versatility of the methodology, non-viral systems involving IL-12 GET, CSPG4 GET, and the suicide gene system plus IFN-β (combined with cytokine-enhanced vaccines), have shown encouraging results in different types of canine tumors: mastocytoma and melanoma for the IL-12 gene [52,54], melanoma and osteosarcoma for the CSPG4 gene [31,33], and melanoma and mammary carcinoma for the suicide gene system [67,68].

5. Conclusions

Some studies were focused on validating gene transfer and expression methods by treating few patients with different kinds of tumors in the same study where the heterogeneity and the absence of control groups precluded conclusions about the efficacy of the treatment. Nevertheless, these studies were valuable as feasibility, proof of concept, and safety assessment.

On the other hand, larger studies involving higher numbers of patients with the same kind of tumor, often with suitable control groups, were able to generate statistically significant data to validate the efficacy of the proposed treatments.

For biosafety reasons (concerning both animals and humans), as well as the higher cost of viral vector production, most veterinary cancer gene therapy protocols have relied on non-viral vectors, with electrotransfer being the predominant delivery method from 2017 to 2024.

To date, more than 2054 clinical protocols for gene therapy for cancer have been registered [70], resulting in 20 treatments approved worldwide for marketing, according to the American Society of Gene and Cell Therapy (ASGCT). In veterinary oncology, the situation is much more modest, with around a hundred publications on the subject and only two treatments approved for commercialization: a plasmid vector with the human tyrosinase gene as a xenoantigen for canine melanoma (Oncept™ Canine Melanoma Vaccine, Duluth, GA, USA) and a canarypox virus vector with the feline interleukin-2 gene for feline fibrosarcoma (Oncept IL-2™).

Almost three decades have passed since the first report with clinical results of the application of genetic immunotherapy in companion animals [1]. Despite proposals for the implementation of veterinary clinical protocols for cancer therapies with a translational approach [6–8,71], the treatments approved so far for humans were not previously tested in companion animals.

On the other hand, some complex treatment approaches for autologous that involve genetic modifications of ex vivo cells, first went through clinical studies in humans than in companion animals. Although they may have high translational value, these technologies are currently very expensive to be applied in veterinary medicine.

Oncolytic virotherapy is emerging as a new tool against cancer in companion animals [72,73]. The update of those approaches that only used suicide genes and/or oncolytic viruses in companion animals would deserve a separate review.

In veterinary medicine, non-viral vectors probably offer a better opportunity to treat animal tumors in a more affordable way for owners. Greater investment in time and effort will be necessary to validate the results in prospective controlled studies and ultimately offer immunogene therapy to more patients who currently lack effective treatment options.

Table 1. Cancer immunogene therapy veterinary clinical trials 2017–2024.

#	Genes	Tumors	Vectors	Modes	Main Results	1st Authors/Year/ Main Country
1	htyr	MEL	Plasmid	t.d. jet-injection + SX ± RX ± CHT	Retrospective study stages I–III OMM (n = 56): MST = 455 d; DFI = 222 d.	Verganti et al., 2017, UK [15]
2	htyr	MEL (feline)	Plasmid	t.d. jet-injection + SX ± RX ± CHT	Retrospective study (n = 24). Manageable post-vaccination AEs: 11%. Safety verified. 42% died of unrelated causes or still alive.	Sarbu et al., 2017, USA [22]
3	htyr	MEL	Plasmid	i.d. micro-seeding njection + SX	Pilot study stages I–IV (n = 6): 3 NED (2 oral, 1 dermal MEL; survival > 1 year), 3 PD (2 oral, 1 dermal MEL). No significant AEs.	Zuleger et al., 2017, USA [20]
4	hIL-12	MCT	Plasmid	p.t. GET + i.t. cisplatin or i.t. bleomycin ECT	Prospective study. Stages I–III (n = 18): 13 CR, 2 PR, 1 SD, 2 PD. No major AEs.	Cemazar et al., 2017, Slovenia [49]
5	hIL-12	ADC, FSA, MEL, OSA, SCH	Plasmid	i.t. GET + metronomic p.o. cyclophosphamide CHT	Pilot study (n = 6): Combined therapy slowed down tumor progression and improved QoL. No significant unwanted side effects.	Cicchelero et al., 2017, Belgium [47]
6	hIL12	FSA, MAC, MCT, OSA, SCC, SCH	Plasmid	i,t, injection	Pilot study (n = 9): No significant clinical benefits. Safety studies and demonstration of immunogenic and anti-angiogenic effects.	Cicchelero et al., 2017, Belgium [46]
7	hCSPG4	MEL	Plasmid	SX and i.m. GET	Prospective controlled study. Surgically resected Stage II/III, LN (-) OMM, VAX (n = 23)/CTR (n = 19): MST: VAX 684 d/CTR 220 d DFI: VAX 477 d/CTR 180 d Safety and immunogenicity confirmed.	Piras et al., 2017, Italy [27]
8	cIFNβ + NIS or hIFNβ + NIS	ADC, MMA, LYP, MEL, OSA	Recombinant oncolytic vesicular stomatitis virus: VSV-IFNβ-NIS	i.v. injection	Pilot study. (n = 10): 2 PR, 5 SD, 3PD. Low or mild/transient AEs. Safety and preliminary evidence of efficacy.	Naik et al., 2018, USA [64]

Table 1. *Cont.*

#	Genes	Tumors	Vectors	Modes	Main Results	1st Authors/Year/ Main Country
9	HSV-tk + cIFN-β hIL-2 + hGM-CSF	MAC	Plasmid lipoplexes	SX and i.t. (SG+ GCV+IFN-β) and s.c. [(hIL-2+ hGM-CSF)+(TV)]	Prospective study Stages II to IV: Combined treatment (n = 36). Complete surgery arm (n = 26): Local disease-free patients: 92%. Metastasis-free patients: 89%. MST = 876 d; metastasis-free MST > 1498 d. Partial surgery arm (n = 10). MST = 241 d.	Finocchiaro et al., 2018, Argentina [67]
10	cTERT	LYP	Recombinant adenovirus and plasmid	i.m virus. injection and i.m. plasmid GET+CHOP-CHT	Prospective study (n = 17): No significant AEs. Improved MST = 452 d as compared with previous historical (n = 21) COP-CHT controls MST = 205 d.	Impellizieri et al., 2018, Italy [35]
11	cIL-12	MEL	Plasmid	PSX/p.t. GET + i.v. bleomycin ECT	Prospective study. Stages I–III OMM (n = 9): 4 CR, 2 PR, 3 PD. MST = 180 d. No major AEs.	Milevoj et al., 2019, Slovenia [54]
12	hIL-2 + hGM-CSF cIFN-β + HSV-tk	MEL	Plasmid lipoplexes	SX and i.t. injection (SG+ GCV+IFN-β+ bleomycin) and s.c. [(hIL-2+hGM-CSF)+(TV)]	Prospective controlled study. Stages I–IV: Combined treatment (n = 210)/Surgery controls (n = 173). Complete surgery arms: Local disease-free patients: 89%/20%. Metastasis-free patients: 89%/45%. MST > 1896 d/99 d. Metastasis-free MST > 1896 d/120 d. No significant AEs.	Finocchiaro et al., 2019, Argentina [68]
13	hp62	MAC	Plasmid	i.m. injection followed by mastectomy	Pilot study. (n = 6: 3 solid and 3 tubo-papillary mammary carcinomas): 5 PR, 1 SD. Good QoL > 4 years after surgery.	Venanzi et al., 2019, Italy [41]
14	htyr	MEL	Plasmid	t.d. jet-injection + SX ± RX	Retrospective study, stages I–III (n = 131): MEL-MST = 510 d/All causes MST 442 d.	Turek et al., 2020, USA [18]

Table 1. *Cont.*

#	Genes	Tumors	Vectors	Modes	Main Results	1st Authors/Year/ Main Country
15	hIL-12	ODG, ACT, GBM	Recombinant oncolytic HSV-1 virus: M032	i.c inoculation.	Prospective study (n = 21): MST = 151 d. No significant AEs and dose-limiting toxicities.	Omar et al., 2021, USA [57]
16	hCSPG4	MEL	Plasmid	en bloc SX + i.m. GET	Retrospective controlled study Stages I–IV: En bloc SX (n = 51): MST = 1333 d; DFI = 324 Marginal SX (n = 31): MST = 470 d; DFI = 184.	Giacobino et al., 2021, Italy [28]
17	cIL-12	MCT	Plasmid	i.t. GET + i.t. cisplatin or i.v. bleomycin ECT	Pilot study (n = 8) 12 tumors (11 MCT, 1 NFS): 7 CR,2 PR, 3 PD. Differences in DCE-US parameters between tumors with CR and non-CR tumors were found.	Brloznik et al., 2021, Slovenia [51]
18	CD40L	MEL	Recombinant adenovirus: AdCD40L	i.t. injection ± SX	Pilot study (n = 32): MTS = 285 d MTS (+SX) = 448 d MTS (-SX) = 80 d, 7 CR, 5 PR, 5 SD, 2 PD, 13 MR. Transient mild AEs. Low toxicity.	Saellstrom et al., 2021, Sweden [43]
19	BPV E6/E7	SAR (equine)	Recombinant influenza virus: iNSA/E6E7^{equ} and iNSB/E6E7^{equ}	i.t. injection	Prospective study (n = 29): 10 CR, 10 PR, 9 PD. Efficacy and safety were proven.	Jindra et al., 2021, Austria [37]
20	fIL-2	SAR (equine)	Recombinant canary pox virus. ALVAC-fIL2	i.t. injection	Pilot study (n = 14): 7 CR (50%), 5 PR (35%), 1 SD (7%), and 1 PD (7%) median time to best response = 211 d. Minimal AEs.	Saba et al., 2022, USA [61]
21	HuDo-CSPG4	MEL	Plasmid	en bloc SX + i.m. GET	Prospective controlled study. Stages II–IV: VAX Treated (n = 52): MST = 653 d Control (n = 28) MST = 310 d. Safe and effective.	Riccardo et al., 2022 Italy [31]

Table 1. *Cont.*

#	Genes	Tumors	Vectors	Modes	Main Results	1st Authors/Year/ Main Country
22	cIL-2 + cIL-12	MEL	Plasmid	i.t cIL-2 GET i.m.cIL-12 GET + bleomycin ECT	Retrospective controlled study. Stages III–IV GET + ECT (n = 10): MST = 165 d/PFS = 165 d. ECT (n = 20): MST = 180 d PFS = 120 d.	Tellado et al., 2023, Argentina [62]
23	cIL-12	MCT	Plasmid	i.t or p.t. GET+ i.t. cisplatin or i.v. bleomycin-ECT	Prospective controlled study: ECT (n = 18): CR = 69%; ECT + GETp.t. (n = 29): CR = 88%; ECT + GETi.t. (n = 30): CR = 94%. DFI and PFS: ECT + GETi.t. > ECT + GETp.t. ≈ ECT.	Lampreht et al., 2023, Slovenia [52]
24	HuDo-CSPG4	OSA	Plasmid	i.m.HuDo-CSPG4 + limb amputation + i.v. carboplatin-CHT	Pilot study: VAX + CHT treated (n = 12): MST = 484 d DFI = 242 d; historical CHT controls (n = 13): MST = 202 d; DFI = 160 d.	Tarone et al., 2023, Italy [33]
25	cIFNβ + NIS	OSA	Recombinant oncolytic vesicular stomatitis virus: VSV-cIFNβ-NIS	i.v. rec virus + i.v. carboplatin CHT	Prospective study (n = 28): 35% survival > 479 d. Safety documented.	Makielski et al., 2023, USA [65]
26	cIL-12	MCT	Plasmid	i.t GET + i.t./i.v. bleomycin ± i.t. cisplatin ECT	Prospective study (n = 48): after 1 month: CR = 43%; PR = 41%; PD = 0. After 6 months: CR = 77%; PR = 12%; PD = 6%. Two possible efficacy markers were defined.	Vilfan et al., 2024, Slovenia [53]

Clinical trials appear ordered according to their publication date. Unless otherwise indicated all tumor types are canine. All gene transfers were performed in vivo. TUMORS: ACT, astrocytoma; ADC, adenocarcinoma; FSA, fibrosarcoma; GBM, glioblastoma; LYP, lymphoma, MAC, mammary adenocarcinoma; MCT, mast cell tumor; MEL: melanoma; MMA, multiple mieloma; NFS, neurofibrosarcoma; ODG, oligodendroglioma; OMM, oral malignant melanoma; OSA, osteosarcoma; SAR, sarcoid; SCC, squamous cell carcinoma; SCH, schwannoma//GENES: BPV E6/E7, bovine papilloma virus E6/E7; CD40L, CD40 ligand; CSPG4, chondroitin sulfate proteoglycan-4; GM-CSF, granulocyte macrophage colony-stimulating factor; HSV-tk, herpes simplex thymidine kinase; IFN-β, interferon-β; IL-12, interleukin-12; IL-2, interleukin-2; NIS, sodium-iodide symporter; p62, protein 62; TERT, telomerase reverse transcriptase; tyr, tyrosinase//GENE PREFIXES: c, canine; f, feline; h, human; HuDo, human-dog chimeric//RESPONSES: AEs, adverse events; CR, complete response; CTR, control; DFI: disease-free interval, MR, mixed responses; MST, median survival time; NED, no evidence of disease; PD, progressive disease; PR, partial response; QoL, quality of life; SD, stable disease//TREATMENTS: CHT, chemotherapy; COP, cyclophosphamide, vincristine, and prednisone chemotherapy; CHOP, cyclophosphamide, vincristine doxorubicin and prednisone chemotherapy; ECT, electro-chemotherapy; GCV, ganciclovir; GET, gene electrotransfer; RX, radiotherapy; SG, suicide gene; SX, surgical excision; TV, tumor vaccine; VAX, vaccinated//ROUTES: i.c., intracranial; i.d, intradermal; i.m., intramuscular; i.t., intratumoral; p.o., oral; p.t., peritumoral; s.c., subcutaneous; t.d, transdermal//DCE-US, dynamic contrast-enhanced ultrasound//d, days.

Author Contributions: Conceptualization, G.C.G. and L.M.E.F.; writing—original draft preparation, G.C.G. and L.M.E.F.; writing—review and editing, G.C.G. and L.M.E.F. All authors have read and agreed to the published version of the manuscript.

Funding: This research received no external funding.

Institutional Review Board Statement: Not applicable.

Informed Consent Statement: Not applicable.

Data Availability Statement: Not applicable.

Acknowledgments: G.C.G. and L.M.E.F. are, respectively, investigators of the Consejo Nacional de Investigaciones Científicas y Técnicas (CONICET, Argentina) and the Instituto de Oncología "Ángel H. Roffo", Universidad de Buenos Aires (UBA, Argentina).

Conflicts of Interest: The authors declare no conflicts of interest.

References

1. Quintin-Colonna, F.; Devauchelle, P.; Fradelizi, D.; Mourot, B.; Faure, T.; Kourilsky, P.; Roth, C.; Mehtali, M. Gene therapy of spontaneous canine melanoma and feline fibrosarcoma by intratumoral administration of histoincompatible cells expressing human interleukin-2. *Gene Ther.* **1996**, *3*, 1104–1112. [PubMed]
2. Glikin, G.C.; Finocchiaro, L.M. Clinical trials of immunogene therapy for spontaneous tumors in companion animals. *Sci. World J.* **2014**, *2014*, 718520. [CrossRef]
3. Finocchiaro, L.M.E.; Glikin, G.C. Recent clinical trials of cancer immunogene therapy in companion animals. *World J. Exp. Med.* **2017**, *7*, 42–48. [CrossRef]
4. Ruzzi, F.; Riccardo, F.; Conti, L.; Tarone, L.; Semprini, M.S.; Bolli, E.; Barutello, G.; Quaglino, E.; Lollini, P.L.; Cavallo, F. Cancer vaccines: Target antigens, vaccine platforms and preclinical models. *Mol. Aspects Med.* **2025**, *101*, 101324. [CrossRef]
5. Zhou, Y.; Wei, Y.; Tian, X.; Wei, X. Cancer vaccines: Current status and future directions. *J. Hematol. Oncol.* **2025**, *18*, 18. [CrossRef]
6. Jacob, J.A. Researchers Turn to Canine Clinical Trials to Advance Cancer Therapies. *JAMA* **2016**, *315*, 1550–1552. [CrossRef]
7. Garden, O.A.; Volk, S.W.; Mason, N.J.; Perry, J.A. Companion animals in comparative oncology: One Medicine in action. *Vet. J.* **2018**, *240*, 6–13. [CrossRef]
8. Oh, J.H.; Cho, J.Y. Comparative oncology: Overcoming human cancer through companion animal studies. *Exp. Mol. Med.* **2023**, *55*, 725–734. [CrossRef]
9. Konduri, V.; Halpert, M.M.; Baig, Y.C.; Coronado, R.; Rodgers, J.R.; Levitt, J.M.; Cerroni, B.; Piscoya, S.; Wilson, N.; DiBernardi, L.; et al. Dendritic cell vaccination plus low-dose doxorubicin for the treatment of spontaneous canine hemangiosarcoma. *Cancer Gene Ther.* **2019**, *26*, 282–291. [CrossRef] [PubMed]
10. Panjwani, M.K.; Atherton, M.J.; MaloneyHuss, M.A.; Haran, K.P.; Xiong, A.; Gupta, M.; Kulikovsaya, I.; Lacey, S.F.; Mason, N.J. Establishing a model system for evaluating CAR T cell therapy using dogs with spontaneous diffuse large B cell lymphoma. *Oncoimmunology* **2019**, *9*, 1676615. [CrossRef] [PubMed]
11. Forsberg, E.M.V.; Riise, R.; Saellström, S.; Karlsson, J.; Alsén, S.; Bucher, V.; Hemminki, A.E.; Olofsson Bagge, R.; Ny, L.; Nilsson, L.M.; et al. Treatment with Anti-HER2 Chimeric Antigen Receptor Tumor-Infiltrating Lymphocytes (CAR-TILs) Is Safe and Associated with Antitumor Efficacy in Mice and Companion Dogs. *Cancers* **2023**, *15*, 648. [CrossRef]
12. Pazzi, P.; Steenkamp, G.; Rixon, A.J. Treatment of Canine Oral Melanomas: A Critical Review of the Literature. *Vet. Sci.* **2022**, *9*, 196. [CrossRef]
13. Bergman, P.J. Cancer Immunotherapies. *Vet. Clin. N. Am. Small Anim. Pract.* **2019**, *49*, 881–902. [CrossRef]
14. Bergman, P.J.; McKnight, J.; Novosad, A.; Charney, S.; Farrelly, J.; Craft, D.; Wulderk, M.; Jeffers, Y.; Sadelain, M.; Hohenhaus, A.E.; et al. Long-term survival of dogs with advanced malignant melanoma after DNA vaccination with xenogeneic human tyrosinase: A phase I trial. *Clin. Cancer Res.* **2003**, *9*, 1284–1290. [PubMed]
15. Verganti, S.; Berlato, D.; Blackwood, L.; Amores-Fuster, I.; Polton, G.A.; Elders, R.; Doyle, R.; Taylor, A.; Murphy, S. Use of Oncept melanoma vaccine in 69 canine oral malignant melanomas in the UK. *J. Small Anim. Pract.* **2017**, *58*, 10–16. [CrossRef]
16. McLean, J.L.; Lobetti, R.G. Use of the melanoma vaccine in 38 dogs: The South African experience. *J. S. Afr. Vet. Assoc.* **2015**, *86*, 1246. [CrossRef]
17. Treggiari, E.; Grant, J.P.; North, S.M. A retrospective review of outcome and survival following surgery and adjuvant xenogeneic DNA vaccination in 32 dogs with oral malignant melanoma. *J. Vet. Med. Sci.* **2016**, *78*, 845–850. [CrossRef]
18. Turek, M.; LaDue, T.; Looper, J.; Nagata, K.; Shiomitsu, K.; Keyerleber, M.; Buchholz, J.; Gieger, T.; Hetzel, S. Multimodality treatment including ONCEPT for canine oral melanoma: A retrospective analysis of 131 dogs. *Vet. Radiol. Ultrasound.* **2020**, *61*, 471–480. [CrossRef]

19. Jeon, M.D.; Leeper, H.J.; Cook, M.R.; McMillan, S.K.; Bennett, T.; Murray, C.A.; Tripp, C.D.; Curran, K.M. Multi-institutional retrospective study of canine foot pad malignant melanomas: 20 cases. *Vet. Comp. Oncol.* **2022**, *20*, 854–861. [CrossRef]
20. Zuleger, C.L.; Kang, C.; Ranheim, E.A.; Kurzman, I.D.; Macklin, M.D.; Newton, M.A.; Wolchok, J.D.; Vail, D.M.; Eriksson, E.; Albertini, M.R. Pilot study of safety and feasibility of DNA microseeding for treatment of spontaneous canine melanoma. *Vet. Med. Sci.* **2017**, *3*, 134–145. [CrossRef]
21. Grosenbaugh, D.A.; Leard, A.T.; Bergman, P.J.; Klein, M.K.; Meleo, K.; Susaneck, S.; Hess, P.R.; Jankowski, M.K.; Jones, P.D.; Leibman, N.F.; et al. Safety and efficacy of a xenogeneic DNA vaccine encoding for human tyrosinase as adjunctive treatment for oral malignant melanoma in dogs following surgical excision of the primary tumor. *Am. J. Vet. Res.* **2011**, *72*, 1631–1638. [CrossRef]
22. Sarbu, L.; Kitchell, B.E.; Bergman, P.J. Safety of administering the canine melanoma DNA vaccine (Oncept) to cats with malignant melanoma—A retrospective study. *J. Feline Med. Surg.* **2017**, *19*, 224–230. [CrossRef]
23. Polton, G.; Borrego, J.F.; Clemente-Vicario, F.; Clifford, C.A.; Jagielski, D.; Kessler, M.; Kobayashi, T.; Lanore, D.; Queiroga, F.L.; Rowe, A.T.; et al. Melanoma of the dog and cat: Consensus and guidelines. *Front. Vet. Sci.* **2024**, *11*, 1359426. [CrossRef]
24. Pellin, M.A. The Use of Oncept Melanoma Vaccine in Veterinary Patients: A Review of the Literature. *Vet. Sci.* **2022**, *9*, 597. [CrossRef]
25. Chen, X.; Habib, S.; Alexandru, M.; Chauhan, J.; Evan, T.; Troka, J.M.; Rahimi, A.; Esapa, B.; Tull, T.J.; Ng, W.Z.; et al. Chondroitin Sulfate Proteoglycan 4 (CSPG4) as an Emerging Target for Immunotherapy to Treat Melanoma. *Cancers* **2024**, *16*, 3260. [CrossRef]
26. Riccardo, F.; Iussich, S.; Maniscalco, L.; Lorda Mayayo, S.; La Rosa, G.; Arigoni, M.; De Maria, R.; Gattino, F.; Lanzardo, S.; Lardone, E.; et al. CSPG4-specific immunity and survival prolongation in dogs with oral malignant melanoma immunized with human CSPG4 DNA. *Clin. Cancer Res.* **2014**, *20*, 3753–3762. [CrossRef]
27. Piras, L.A.; Riccardo, F.; Iussich, S.; Maniscalco, L.; Gattino, F.; Martano, M.; Morello, E.; Lorda Mayayo, S.; Rolih, V.; Garavaglia, F.; et al. Prolongation of survival of dogs with oral malignant melanoma treated by en bloc surgical resection and adjuvant CSPG4-antigen electrovaccination. *Vet. Comp. Oncol.* **2017**, *15*, 996–1013. [CrossRef]
28. Giacobino, D.; Camerino, M.; Riccardo, F.; Cavallo, F.; Tarone, L.; Martano, M.; Dentini, A.; Iussich, S.; Lardone, E.; Franci, P.; et al. Difference in outcome between curative intent vs marginal excision as a first treatment in dogs with oral malignant melanoma and the impact of adjuvant CSPG4-DNA electrovaccination: A retrospective study on 155 cases. *Vet. Comp. Oncol.* **2021**, *19*, 651–660. [CrossRef]
29. Camerino, M.; Giacobino, D.; Iussich, S.; Ala, U.; Riccardo, F.; Cavallo, F.; Martano, M.; Morello, E.; Buracco, P. Evaluation of prognostic impact of pre-treatment neutrophil to lymphocyte and lymphocyte to monocyte ratios in dogs with oral malignant melanoma treated with surgery and adjuvant CSPG4-antigen electrovaccination: An explorative study. *Vet. Comp. Oncol.* **2021**, *19*, 353–361. [CrossRef]
30. Mayayo, S.L.; Prestigio, S.; Maniscalco, L.; Rosa, G.; Aricò, A.; Maria, R.; Cavallo, F.; Ferrone, S.; Buracco, P.; Iussich, S. Chondroitin sulfate proteoglycan-4: A biomarker and a potential immunotherapeutic target for canine malignant melanoma. *Vet. J.* **2011**, *190*, e26–e30. [CrossRef]
31. Riccardo, F.; Tarone, L.; Camerino, M.; Giacobino, D.; Iussich, S.; Barutello, G.; Arigoni, M.; Conti, L.; Bolli, E.; Quaglino, E.; et al. Antigen mimicry as an effective strategy to induce CSPG4-targeted immunity in dogs with oral melanoma: A veterinary trial. *J. Immunother. Cancer.* **2022**, *10*, e004007. [CrossRef]
32. Poon, A.C.; Matsuyama, A.; Mutsaers, A.J. Recent and current clinical trials in canine appendicular osteosarcoma. *Can. Vet. J.* **2020**, *61*, 301–308. [PubMed]
33. Tarone, L.; Giacobino, D.; Camerino, M.; Maniscalco, L.; Iussich, S.; Parisi, L.; Giovannini, G.; Dentini, A.; Bolli, E.; Quaglino, E.; et al. A chimeric human/dog-DNA vaccine against CSPG4 induces immunity with therapeutic potential in comparative preclinical models of osteosarcoma. *Mol. Ther.* **2023**, *31*, 2342–2359. [CrossRef]
34. Bennett, P.; Williamson, P.; Taylor, R. Review of Canine Lymphoma Treated with Chemotherapy-Outcomes and Prognostic Factors. *Vet. Sci.* **2023**, *10*, 342. [CrossRef]
35. Impellizeri, J.A.; Gavazza, A.; Greissworth, E.; Crispo, A.; Montella, M.; Ciliberto, G.; Lubas, G.; Aurisicchio, L. Tel-eVax: A genetic vaccine targeting telomerase for treatment of canine lymphoma. *J. Transl. Med.* **2018**, *16*, 349. [CrossRef]
36. Gavazza, A.; Lubas, G.; Fridman, A.; Peruzzi, D.; Impellizeri, J.A.; Luberto, L.; Marra, E.; Roscilli, G.; Ciliberto, G.; Aurisicchio, L. Safety and efficacy of a genetic vaccine targeting telomerase plus chemotherapy for the therapy of canine B-cell lymphoma. *Hum. Gene Ther.* **2013**, *24*, 728–738. [CrossRef]
37. Jindra, C.; Hainisch, E.K.; Brandt, S. Immunotherapy of Equine Sarcoids-From Early Approaches to Innovative Vaccines. *Vaccines* **2023**, *11*, 769. [CrossRef]
38. Jindra, C.; Hainisch, E.K.; Rümmele, A.; Wolschek, M.; Muster, T.; Brandt, S. Influenza virus vector iNS1 expressing bovine papillomavirus 1 (BPV1) antigens efficiently induces tumour regression in equine sarcoid patients. *PLoS ONE* **2021**, *16*, e0260155. [CrossRef]

39. Nosalova, N.; Huniadi, M.; Horňáková, Ľ.; Valenčáková, A.; Horňák, S.; Nagoos, K.; Vozar, J.; Cizkova, D. Canine Mammary Tumors: Classification, Biomarkers, Traditional and Personalized Therapies. *Int. J. Mol. Sci.* **2024**, *25*, 2891. [CrossRef]
40. Sabbieti, M.G.; Marchegiani, A.; Sufianov, A.A.; Gabai, V.L.; Shneider, A.; Agas, D. P62/SQSTM1 beyond Autophagy: Physiological Role and Therapeutic Applications in Laboratory and Domestic Animals. *Life* **2022**, *12*, 539. [CrossRef]
41. Venanzi, F.M.; Gabai, V.; Mariotti, F.; Magi, G.E.; Vullo, C.; Sufianov, A.A.; Kolesnikov, S.I.; Shneider, A. p62-DNA-encoding plasmid reverts tumor grade, changes tumor stroma, and enhances anticancer immunity. *Aging* **2019**, *11*, 10711–10722. [CrossRef]
42. Irenaeus, S.; Hellström, V.; Wenthe, J.; Krause, J.; Sundin, A.; Ahlström, H.; Tufveson, G.; Tötterman, T.H.; Loskog, A.; Ullenhag, G.J. Intratumoral immunostimulatory AdCD40L gene therapy in patients with advanced solid tumors. *Cancer Gene Ther.* **2021**, *28*, 1188–1197. [CrossRef]
43. Saellstrom, S.; Sadeghi, A.; Eriksson, E.; Segall, T.; Dimopoulou, M.; Korsgren, O.; Loskog, A.S.; Tötterman, T.H.; Hemminki, A.; Ronnberg, H. Adenoviral CD40 Ligand Immunotherapy in 32 Canine Malignant Melanomas-Long-Term Follow Up. *Front. Vet. Sci.* **2021**, *8*, 695222. [CrossRef]
44. Dong, C.; Tan, D.; Sun, H.; Li, Z.; Zhang, L.; Zheng, Y.; Liu, S.; Zhang, Y.; He, Q. Interleukin-12 Delivery Strategies and Advances in Tumor Immunotherapy. *Curr. Issues Mol. Biol.* **2024**, *46*, 11548–11579. [CrossRef]
45. Geils, C.; Kathrein, K.L. Augmentation of Solid Tumor Immunotherapy with IL-12. *J. Gene Med.* **2024**, *26*, e70000. [CrossRef]
46. Cicchelero, L.; Denies, S.; Haers, H.; Vanderperren, K.; Stock, E.; Van Brantegem, L.; de Rooster, H.; Sanders, N.N. Intratumoural interleukin 12 gene therapy stimulates the immune system and decreases angiogenesis in dogs with spontaneous cancer. *Vet. Comp. Oncol.* **2017**, *15*, 1187–1205. [CrossRef]
47. Cicchelero, L.; Denies, S.; Vanderperren, K.; Stock, E.; Van Brantegem, L.; de Rooster, H.; Sanders, N.N. Immunological, anti-angiogenic and clinical effects of intratumoral interleukin 12 electrogene therapy combined with metronomic cyclophosphamide in dogs with spontaneous cancer: A pilot study. *Cancer Lett.* **2017**, *400*, 205–218. [CrossRef]
48. Bellamy, E.; Berlato, D. Canine cutaneous and subcutaneous mast cell tumours: A narrative review. *J. Small Anim. Pract.* **2022**, *63*, 497–511. [CrossRef]
49. Cemazar, M.; Ambrozic Avgustin, J.; Pavlin, D.; Sersa, G.; Poli, A.; Krhac Levacic, A.; Tesic, N.; Lampreht Tratar, U.; Rak, M.; Tozon, N. Efficacy and safety of electrochemotherapy combined with peritumoral IL-12 gene electrotransfer of canine mast cell tumours. *Vet. Comp. Oncol.* **2017**, *15*, 641–654. [CrossRef]
50. Salvadori, C.; Svara, T.; Rocchigiani, G.; Millanta, F.; Pavlin, D.; Cemazar, M.; Lampreht Tratar, U.; Sersa, G.; Tozon, N.; Poli, A. Effects of Electrochemotherapy with Cisplatin and Peritumoral IL-12 Gene Electrotransfer on Canine Mast Cell Tumors: A Histopathologic and Immunohistochemical Study. *Radiol. Oncol.* **2017**, *51*, 286–294. [CrossRef]
51. Brloznik, M.; Kranjc Brezar, S.; Boc, N.; Knific, T.; Cemazar, M.; Milevoj, N.; Sersa, G.; Tozon, N.; Pavlin, D. Results of Dynamic Contrast-Enhanced Ultrasound Correlate With Treatment Outcome in Canine Neoplasia Treated With Electrochemotherapy and Interleukin-12 Plasmid Electrotransfer. *Front. Vet. Sci.* **2021**, *8*, 679073. [CrossRef]
52. Lampreht Tratar, U.; Milevoj, N.; Cemazar, M.; Znidar, K.; Ursic Valentinuzzi, K.; Brozic, A.; Tomsic, K.; Sersa, G.; Tozon, N. Treatment of spontaneous canine mast cell tumors by electrochemotherapy combined with IL-12 gene electrotransfer: Comparison of intratumoral and peritumoral application of IL-12. *Int. Immunopharmacol.* **2023**, *120*, 110274. [CrossRef]
53. Vilfan, M.; Lampreht Tratar, U.; Milevoj, N.; Nemec Svete, A.; Čemažar, M.; Serša, G.; Tozon, N. Comparison of Nucleosome, Ferritin and LDH Levels in Blood with Clinical Response before and after Electrochemotherapy Combined with IL-12 Gene Electrotransfer for the Treatment of Mast Cell Tumours in Dogs. *Animals* **2024**, *14*, 438. [CrossRef]
54. Milevoj, N.; Tratar, U.L.; Nemec, A.; Brožič, A.; Žnidar, K.; Serša, G.; Čemažar, M.; Tozon, N. A combination of electrochemotherapy, gene electrotransfer of plasmid encoding canine IL-12 and cytoreductive surgery in the treatment of canine oral malignant melanoma. *Res. Vet. Sci.* **2019**, *122*, 40–49. [CrossRef]
55. Nemec, A.; Milevoj, N.; Lampreht Tratar, U.; Serša, G.; Čemažar, M.; Tozon, N. Electroporation-Based Treatments in Small Animal Veterinary Oral and Maxillofacial Oncology. *Front. Vet. Sci.* **2020**, *7*, 575911. [CrossRef]
56. Miller, A.D.; Miller, C.R.; Rossmeisl, J.H. Canine Primary Intracranial Cancer: A Clinicopathologic and Comparative Review of Glioma, Meningioma, and Choroid Plexus Tumors. *Front. Oncol.* **2019**, *9*, 1151. [CrossRef]
57. Omar, N.B.; Bentley, R.T.; Crossman, D.K.; Foote, J.B.; Koehler, J.W.; Markert, J.M.; Platt, S.R.; Rissi, D.R.; Shores, A.; Sorjonen, D.; et al. Safety and interim survival data after intracranial administration of M032, a genetically engineered oncolytic HSV-1 expressing IL-12, in pet dogs with sporadic gliomas. *Neurosurg. Focus.* **2021**, *50*, E5. [CrossRef]
58. Hu, H.; Barker, A.; Harcourt-Brown, T.; Jeffery, N. Systematic Review of Brain Tumor Treatment in Dogs. *J. Vet. Intern. Med.* **2015**, *29*, 1456–1463. [CrossRef]
59. Chambers, M.R.; Foote, J.B.; Bentley, R.T.; Botta, D.; Crossman, D.K.; Della Manna, D.L.; Estevez-Ordonez, D.; Koehler, J.W.; Langford, C.P.; Miller, M.A.; et al. Evaluation of immunologic parameters in canine glioma patients treated with an oncolytic herpes virus. *J. Transl. Genet. Genom.* **2021**, *5*, 423–442. [CrossRef]
60. Rokade, S.; Damani, A.M.; Oft, M.; Emmerich, J. IL-2 based cancer immunotherapies: An evolving paradigm. *Front. Immunol.* **2024**, *15*, 1433989. [CrossRef]

61. Saba, C.; Eggleston, R.; Parks, A.; Peroni, J.; Sjoberg, E.; Rice, S.; Tyma, J.; Williams, J.; Grosenbaugh, D.; Leard, A.T. ALVAC-fIL2, a feline interleukin-2 immunomodulator, as a treatment for sarcoids in horses: A pilot study. *J. Vet. Intern. Med.* **2022**, *36*, 1179–1184. [CrossRef]
62. Tellado, M.; De Robertis, M.; Montagna, D.; Giovannini, D.; Salgado, S.; Michinski, S.; Signori, E.; Maglietti, F. Electrochemotherapy Plus IL-2+IL-12 Gene Electrotransfer in Spontaneous Inoperable Stage III–IV Canine Oral Malignant Melanoma. *Vaccines* **2023**, *11*, 1033. [CrossRef]
63. Yu, R.; Zhu, B.; Chen, D. Type I interferon-mediated tumor immunity and its role in immunotherapy. *Cell Mol. Life Sci.* **2022**, *79*, 191. [CrossRef]
64. Naik, S.; Galyon, G.D.; Jenks, N.J.; Steele, M.B.; Miller, A.C.; Allstadt, S.D.; Suksanpaisan, L.; Peng, K.W.; Federspiel, M.J.; Russell, S.J.; et al. Comparative Oncology Evaluation of Intravenous Recombinant Oncolytic Vesicular Stomatitis Virus Therapy in Spontaneous Canine Cancer. *Mol. Cancer Ther.* **2018**, *17*, 316–326. [CrossRef]
65. Makielski, K.M.; Sarver, A.L.; Henson, M.S.; Stuebner, K.M.; Borgatti, A.; Suksanpaisan, L.; Preusser, C.; Tabaran, A.F.; Cornax, I.; O'Sullivan, M.G.; et al. Neoadjuvant systemic oncolytic vesicular stomatitis virus is safe and may enhance long-term survivorship in dogs with naturally occurring osteosarcoma. *Mol. Ther. Oncolytics.* **2023**, *31*, 100736. [CrossRef]
66. Kumar, A.; Taghi Khani, A.; Sanchez Ortiz, A.; Swaminathan, S. GM-CSF: A Double-Edged Sword in Cancer Immunotherapy. *Front. Immunol.* **2022**, *13*, 901277. [CrossRef]
67. Finocchiaro, L.M.E.; Spector, A.I.M.; Agnetti, L.; Arbe, M.F.; Glikin, G.C. Combination of Suicide and Cytokine Gene Therapies as Surgery Adjuvant for Canine Mammary Carcinoma. *Vet. Sci.* **2018**, *5*, 70. [CrossRef]
68. Finocchiaro, L.M.E.; Agnetti, L.; Fondello, C.; Glikin, G.C. Combination of cytokine-enhanced vaccine and chemo-gene therapy as surgery adjuvant treatments for spontaneous canine melanoma. *Gene Ther.* **2019**, *26*, 418–431. [CrossRef]
69. Finocchiaro, L.M.; Fondello, C.; Gil-Cardeza, M.L.; Rossi, Ú.A.; Villaverde, M.S.; Riveros, M.D.; Glikin, G.C. Cytokine-Enhanced Vaccine and Interferon-β plus Suicide Gene Therapy as Surgery Adjuvant Treatments for Spontaneous Canine Melanoma. *Hum. Gene Ther.* **2015**, *26*, 367–376. [CrossRef]
70. Ginn, S.L.; Mandwie, M.; Alexander, I.E.; Edelstein, M.; Abedi, M.R. Gene therapy clinical trials worldwide to 2023-an update. *J Gene Med.* **2024**, *26*, e3721. [CrossRef]
71. Khanna, C.; London, C.; Vail, D.; Mazcko, C.; Hirschfeld, S. Guiding the optimal translation of new cancer treatments from canine to human cancer patients. *Clin. Cancer Res.* **2009**, *15*, 5671–5677. [CrossRef] [PubMed]
72. Sánchez, D.; Cesarman-Maus, G.; Amador-Molina, A.; Lizano, M. Oncolytic Viruses for Canine Cancer Treatment. *Cancers* **2018**, *10*, 404. [CrossRef] [PubMed]
73. Gentschev, I.; Patil, S.S.; Petrov, I.; Cappello, J.; Adelfinger, M.; Szalay, A.A. Oncolytic virotherapy of canine and feline cancer. *Viruses* **2014**, *6*, 2122–2137. [CrossRef]

veterinary sciences

MDPI

Article

Comparative Echocardiographic Evaluation of Right Pulmonary Artery Dimensions and Right Pulmonary Artery Distensibility Index in Dogs with Heartworm Disease

Jorge Isidoro Matos [1], Alicia Caro-Vadillo [2], Eva Mohr-Peraza [1], Sara Nieves García-Rodríguez [1,*], José Alberto Montoya-Alonso [1] and Elena Carretón [1]

1 Internal Medicine, Veterinary Medicine and Therapeutic Research Group, Faculty of Veterinary Science, Research Institute of Biomedical and Health Sciences (IUIBS), University of Las Palmas de Gran Canaria, 35017 Las Palmas de Gran Canaria, Spain; jorge.matos@ulpgc.es (J.I.M.); eva.mohr@ulpgc.es (E.M.-P.); alberto.montoya@ulpgc.es (J.A.M.-A.); elena.carreton@ulpgc.es (E.C.)

2 Department of Animal Medicine and Surgery, Faculty of Veterinary Medicine, Complutense University, 28040 Madrid, Spain; aliciac@vet.ucm.es

* Correspondence: saranieves.garcia@ulpgc.es

Simple Summary: The existence of two methods for obtaining the right pulmonary artery distensibility index (RPADi) has complicated the standardization and use of echocardiographic measurements in determining pulmonary hypertension (PH) in heartworm disease and other cardiorespiratory conditions. This study aimed to determine the accuracy and agreement between RPADVisser and RPADVenco in diagnosing PH and to assess the differences in pulmonary arterial diameters in dogs with heartworm. RPADVisser and RPADVenco were measured in 248 dogs, along with the values for RPADmax, RPADmin, RPADmax:Ao, and RPADmin:Ao for each method. The dogs were grouped into three categories: healthy, infected without PH, and infected with PH. Statistical analyses revealed significant differences between these groups for RPADmax, RPADmin, RPADmax:Ao, and RPADmin:Ao. Additionally, this study found high similarity and agreement between RPADVisser and RPADVenco. However, 5.6% of the cases showed differences in the diagnosis for PH, indicating that these methods cannot be used interchangeably.

Abstract: The existence of several methods to obtain the right pulmonary artery distensibility index (RPADi) has complicated the standardization and use of echocardiographic measurement for the determination of pulmonary hypertension (PH) in heartworm disease and other cardiorespiratory pathologies. Therefore, the objective was to determine the precision and agreement between RPADVisser and RPADVenco in the diagnosis of PH and to evaluate the differences in pulmonary arterial diameters in dogs with heartworm. The RPADVisser and RPADVenco were determined in 248 dogs and the determinations of RPADmax, RPADmin, RPADmax:Ao, and RPADmim:Ao were also obtained from each of the methods. Through the presence/absence of heartworm disease and the presence of PH (RPADi < 30% in both methods), three groups were created: healthy animals, infected dogs without PH, and animals infected with PH. Statistical analyzes determined significant differences between the analyzed groups for RPADmax, RPADmin, RPADmax:Ao, and RPADmim:Ao. Likewise, a high similarity and agreement was reported between RPADVisser and RPADVenco; however, 5.6% of the cases showed a different diagnosis for the presence of PH in the animals analyzed, showing that these methods cannot be used interchangeably.

Vet. Sci. **2025**, *12*, 246 https://doi.org/10.3390/vetsci12030246

Keywords: veterinary; cardiology; pulmonary hypertension; echocardiography; pulmonary artery; RPADi; heartworm; *Dirofilaria immitis*

1. Introduction

Canine cardiopulmonary dirofilariosis or heartworm disease (*Dirofilaria immitis*), is a globally spread vector-borne disease. In dogs, the presence of adult parasites causes endothelial damage in the pulmonary arteries, leading to the appearance of proliferative pulmonary endarteritis. The chronicity of endothelial lesions and the perpetuation of adult parasites trigger phenomena of pulmonary hypertension (PH) of precapillary origin [1–3].

The presence of PH is a common and frequent sign in canine heartworm, affecting around 30–60% of the parasitized animals [4,5], causing cardiorespiratory symptoms and right congestive heart failure. The clinical importance in endemic areas and in those where the disease is expanding make the diagnosis of PH produced by *D. immitis* of mandatory interest [3–5].

The right pulmonary artery distensibility index (RPADi) is an echocardiographic measure used to assess the probability of suffering from PH in dogs. Traditionally, the echocardiographic measurement of the RPADi, has proven to be useful for the diagnosis and staging of PH in heartworm disease, especially when there is no possibility of measuring tricuspid or pulmonary regurgitation flows [6–9]. Likewise, it has been shown to be an adequate index in the study of other cardiorespiratory pathologies that generate PH, in which, as in heartworm disease, dilation of the right pulmonary artery occurs, in addition to a loss of compliance during cardiac systole and diastole [9–12]. The cut-off point of <30% has been widely used in previous studies involving the diagnosis of moderate or severe PH (>50 mmHg), both in heartworm disease and in other pathologies [5,8,11–15]. However, the existence of two performance methods to obtain the RPADi complicates the standardization and use of the echocardiographic measurement at present. On the one hand, the use of study level through a long axis right parasternal view using monodimensional mode (M-Mode) developed in veterinary medicine following the indications of Venco et al. [8], and on the other hand the use of a short axis right parasternal view at of the bifurcation of the branches of the pulmonary trunk using a bidimensional mode (2D mode) developed by Visser et al. [11].

Recently, a study assessed the differences between these methods for determining the RPADi, reporting differences between them and concluding that cannot be used interchangeably in the detection of PH [10]. However, a larger sample size and other echocardiographic variables involved in the determination of RPADi in dogs with heartworm must be evaluated to confirm the previous findings. Therefore, the aim of the present research was to ascertain the precision and concordance between both methods of RPADi determination in the diagnosis of PH in canine heartworm disease.

2. Materials and Methods

Animals

The study subjects consisted of 248 privately owned dogs attending at the Veterinary Teaching Hospital of the University of Las Palmas de Gran Canaria (Canary Islands, Spain). All dogs lived in a hyperendemic area and without prophylactic measures against heartworm infection [16] and attended a heartworm screening campaign. A complete record of each animal was kept, including age, sex, breed, weight, presence of cardiorespiratory symptoms associated with heartworm disease (dyspnea, cough, exercise intolerance, weakness, wight loss, and syncope), and also signs of right-sided congestive heart failure

(R-CHF), mainly based on evidence of ascites, pleural effusion, jugular pulse and cava vein distention. Animals receiving any medication for previous cardiorespiratory diseases were excluded from this study.

Of the included dogs, 71.37% (177/248) were diagnosed with heartworm disease using a commercial immunochromatographic test kit (Urano test Dirofilaria, Urano Vet SL, Barcelona, Spain), while 28.63% (71/248) were considered healthy based on the absence of clinical signs, history, physical examination, thoracic radiographs, cardiovascular and echocardiographic evaluation, and negative *D. immitis* antigen detection test results. Dogs with evidence of cardiac or respiratory conditions coexisting with heartworm disease were excluded from this study.

Echocardiography

All dogs included in this study underwent echocardiographic examination, which were performed between September 2021 and October 2023 by the same veterinary cardiologist using a Vivid Iq ultrasonographer (General Electrics, Boston, MA, USA) equipped with 2.5–12.5 MHz phased array transducers. No dogs were sedated or anesthetized, and an electrocardiographic monitoring was maintained throughout the examination. The average measurement of three consecutive cardiac cycles in sinus rhythm was determined to avoid any influence of the respiratory and cardiac cycles.

The determination of the presence of PH was based on the two different methods of obtaining the RPADi through the use of echocardiographic M-mode and 2D mode, being a value < 30% correlated in both cases with the presence of moderate or severe PH (>50 mmHg) [8,11]. To determine the RPADi using the Visser et al., method [11], the right parasternal short-axis view was applied at the level of the pulmonary trunk bifurcation, clearly observing the progression of the right and left branches. An adequate two-dimensional image with direct visualization of the longitudinal axis of the right pulmonary artery was obtained in all cases. The minimum diastolic and maximum systolic internal diameters were measured with the "trailing edge to leading edge" (te-le) method, using the same location of the right pulmonary artery (RPADVisser). Special care was taken to take measurements of the farthest portion, avoiding including the aorta (Ao) in the image (Figure 1). On the other hand, the determination of the RPADi using the Venco et al. method [8] was based on the right parasternal long-axis view of four cardiac chambers. M-mode was used in this case on the transverse axis of the right pulmonary artery when the left atrium, and pulmonary artery and vein were accurately observed. The right pulmonary artery diameters were measured using the "leading edge to leading edge" (le-le) measurement convention (RPADVenco) (Figure 2). In both cases, systolic diameter was measured at the maximum (T wave) and diastolic diameter at the smallest (Q wave) dimensions. The RPADi was calculated as the percentage difference in right pulmonary artery diameter in systole and diastole, using the following formula: $RPADi = [(RPAmax - RPAmin)/RPAmax] \times 100$, as previously described [10].

Additionally, in each of the methods evaluated (RPADVisser and RPADVenco) the maximum (RPAmax) and minimum (RPAmin) diameters of the right pulmonary artery were described and compared. Likewise, the RPAmax and RPAmin of each of the methods were indexed to the diameter of the Ao measured from a right parasternal long-axis view and optimized for the left ventricular outflow tract. Measurement of the Ao was performed during early systole at the base of the aortic leaflets, visible at maximum open aortic valve. In this way, the RPAmax:Ao and RPAmin:Ao ratios were obtained for each of the RPADi methods [17].

Finally, the estimated burden of adult worms (I–IV) was classified according to Venco et al. [18]. Low parasite burden (I–II) was determined when no parasites were echocardiographically observed or when they were observed only in the right pulmonary branch, and a high parasite burden (III–IV) was determined when adult worms were observed in the pulmonary trunk or when adult parasites were observed in the right heart chambers.

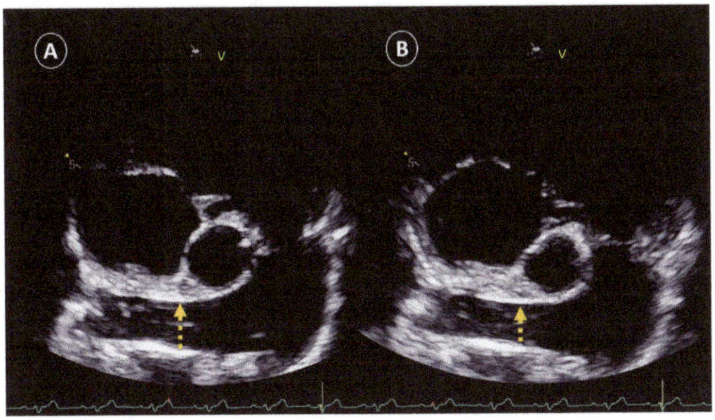

Figure 1. Representative image of the right pulmonary artery distensibility index (RPADi) measurement using the method of Visser et al. [10] in a dog with heartworms and pulmonary hypertension. The right parasternal short-axis view was used at the level of the pulmonary trunk bifurcation. An adequate two-dimensional image with direct visualization of the longitudinal axis of the right pulmonary artery was obtained. Systolic diameter (**A**) was measured at the maximum dimension (T wave), and diastolic diameter (**B**) at the smallest dimension (Q wave). Internal diameters were measured with the "trailing edge to leading edge" method, using the same location of the right pulmonary artery.

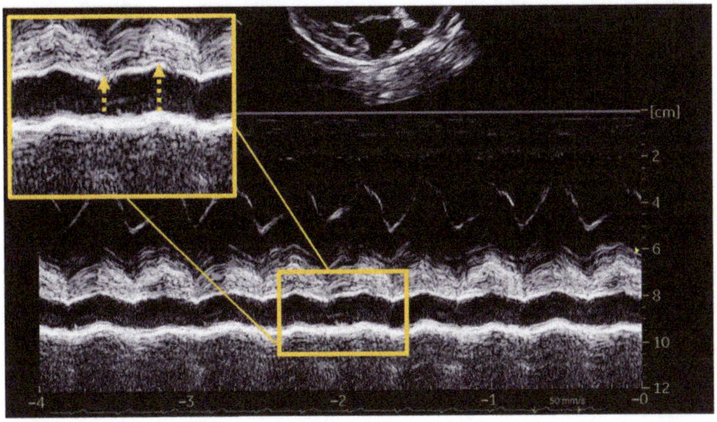

Figure 2. Representative image of the right pulmonary artery distensibility index (RPADi) measurement using the method of Venco et al. [6] in a dog infected by *Dirofilaria immitis* and with pulmonary hypertension. The right parasternal long-axis view of four cardiac chambers was optimized and M-mode was used on the transverse axis of the right pulmonary artery when the left atrium and pulmonary artery and vein were accurately observed. Systolic diameter was measured at the maximum dimension (T wave) and diastolic diameter at the smallest dimension (Q wave). The right pulmonary artery diameters were measured using the "leading edge to leading edge" measurement convention.

Statistical analysis

For categorical variables, frequencies, and percentages were shown. For continuous variables, the descriptive of the mean, standard deviation, median, and interquartile range were shown. The differences in parameters between groups were evaluated, if continuous, using Mann–Whitney (non-parametric) or T-student (parametric) tests based on the normality of the variables to be evaluated using the Shapiro–Wilks test and, if the variables were categorical, using Pearson's Chi^2 test. All multiple comparisons were adjusted by Bonferroni correction. All contrasts were accompanied by the effect size estimator to complete the interpretation of the results: Cramer's V for categorical variables and Cohen's d for continuous variables. The criterion for Cohen's d for classifying the magnitude of the effect was as follows: small (d = 0.2–0.4), medium (d = 0.5–0.8), and large (d = greater than 0.8); For Cramer's V it was the following: 0.00–0.09 as negligible, 0.10–0.29 as low, 0.30–0.49 as medium and from 0.50 as high. For agreement between methods, the Pearson correlation and the Intraclass Correlation Coefficient (ICC) were applied, which allowed measuring the agreement between two or more quantitative assessments obtained with different measuring instruments or evaluators. In addition, the McNemar test compared the percentages of change established by both methods (PH with one method and non-PH with the other and vice versa). To complement the agreement between methods, the ROC curve was calculated for each method with respect to the Yes/No PH classification obtained by the other method. Choosing the value of the parameter that maximizes the Youden index (sensitivity + specificity − 1) as the cut-off point, the best possible approximation to the real classification was obtained. Finally, the Kappa concordance index was used to evaluate the degree of concordance between the results provided by both observers. The Kappa index is a parameter whose maximum possible value is 1 (total concordance). The level of significance used in the analysis was 5% ($\alpha = 0.05$).

Ethical statement

All owners were informed and gave their consent to participate in this study. Ethical approval was not required for this study, as it was a purely observational study with voluntary enrollment and did not involve additional invasive clinical diagnostic procedures. The evaluation of this study included ethical considerations and legal aspects regarding animal protection and welfare and was carried out in accordance with the current Spanish and European legislation on animal protection.

3. Results

All controls were healthy without presence of PH and, among the patients with heartworm, 43.5% had PH based on RPADVisser and 48% based on RPADVenco. According to the results, dogs were divided into three groups: Group A ($n = 71$) included healthy animals, Group B ($n = 100$ for RPADVisser and $n = 92$ for RPADVenco) were heartworm-infected animals without PH, and Group C ($n = 77$ for RPADVisser and $n = 85$ for RPADVenco) were dogs with heartworm and PH.

Table 1 shows the descriptive results for the three groups of dogs for each method. The groups were fairly homogeneous, with no significant differences in the results with respect to sex, race, weight or age between the groups studied. Respiratory symptoms and the presence of R-CHF were significantly related to the disease and the presence of PH in both RPADi methods. The parasite burden also increased with the PH in both methods: all dogs in Group A showed absence of parasites; 27% (RPADVisser) – 29% (RPADVenco) of dogs in Group B had high parasite burden; and 72% (RPADVisser) – 74% (RPADVenco) of dogs in Group C had high parasite burden. Moreover, the measurements of RPADmax, RPADmin, RPADmax:Ao, and RPADmim:Ao reported significant differences between RPADVisser and RPADVenco.

The large sample size of this study allowed assuming the normality of the two methods to be compared by applying the Central Limit Theorem. The Pearson correlation between the two ratios was very high: 0.963 (p-value 0.000) (Figure 3).

Table 1. Epidemiological, clinical, and echocardiographic parameters of the dogs studied (n = 248). Data represent median and standard deviation unless otherwise indicated. Group A: healthy dogs; Group B: dogs with heartworms and absence of pulmonary hypertension (PH); Group C: dogs with heartworms and with PH. Legend: RPADVisser: right pulmonary artery distensibility index measured following the Visser et al. method [10]; RPADVenco: right pulmonary artery distensibility index measured following the Venco et al. method [6]; R-CHF: right congestive heart failure; HR: heart rate; RPADmim: minimum diameter of the right pulmonary artery; RPADmax: maximum diameter of the right pulmonary artery; Ao: Aorta. Results for female, mongrel, and R-CHF, high parasite burden and respiratory symptoms are expressed as n (%). Significant differences were considered when the p value < 0.05.

	RPADVisser (n = 248)				RPADVenco (n = 248)			
Parameter	**Group A (n = 71)**	**Group B (n = 100)**	**Group C (n = 77)**	**p-Value (Cohen's d or Cramer's V Value)**	**Group A (n = 71)**	**Group B (n = 92)**	**Group C (n = 85)**	**p-Value (Cohen's d or Cramer's V Value)**
Age (years)	8.32 ± 3.78	8.24 ± 3.78	8.07 ± 3.78	0.176	8.32 ± 3.78	8.08 ± 2.94	8.06 ± 3.58	0.234
Body weight (kg)	16.60 ± 10.17	16.60 ± 10.17	16.22 ± 10.29	0.106	16.60 ± 10.17	16.56 ± 9.72	16.31 ± 10.20	0.232
Female: number (%)	39 (54.9%)	56 (56%)	39 (50.6%)	0.222	39 (54.9%)	51 (55.4%)	42 (49.4%)	0.199
Mongrel: number (%)	21 (29.6%)	51 (51%)	40 (51.9%)	0.268	21 (29.6%)	48 (52.2%)	43 (50.6%)	0.119
Respiratory symptoms: number (%)	0 (0%)	35 (35%)	64(83.1%)	0.000 (0.66)	0 (0%)	28 (30.4%)	71 (83.5%)	0.000 (0.69)
R-CHF: number (%)	0 (0%)	0 (0%)	30 (39%)	0.000 (0.54)	0 (0%)	0 (0%)	30 (35.3%)	0.000 (0.50)
High parasite burden: number (%)	0 (0%)	29 (29%)	57 (74%)	0.000 (1.23)	0 (0%)	25 (27.2%)	61 (71.8%)	0.000 (1.16)
RPADmin (mm)	5.07 ± 1.82	5.14 ± 1.94	11.64 ± 1.84	0.000 (1.27)	3.77 ± 1.37	3.86 ± 1.43	10.14 ± 1.46	0.000 (1.56)
RPADmax (mm)	8.64 ± 3.15	8.86 ± 3.22	15.07 ± 3.19	0.000 (1.31)	7.11 ± 2.36	7.02 ± 2.31	13.71 ± 2.41	0.000 (1.22)
RPADmin:Ao	0.41 ± 0.22	0.38 ± 0.27	0.69 ± 0.24	0.000 (1.12)	0.35 ± 0.31	0.30 ± 0.38	0.61 ± 0.36	0.000 (1.13)
RPADmax:Ao	0.70 ± 0.16	0.67 ± 0.21	0.91 ± 0.18	0.000 (1.05)	0.51 ± 0.22	0.54 ± 0.27	0.82 ± 0.25	0.000 (1.10)

p < 0.01 = ANOVA and Cohen's d value in brackets; p < 0.01 = Mann–Whitney/Kruskall–Wallis and Cohen's d value in brackets; p < 0.01 = Chi^2 test and Cramer's V in brackets.

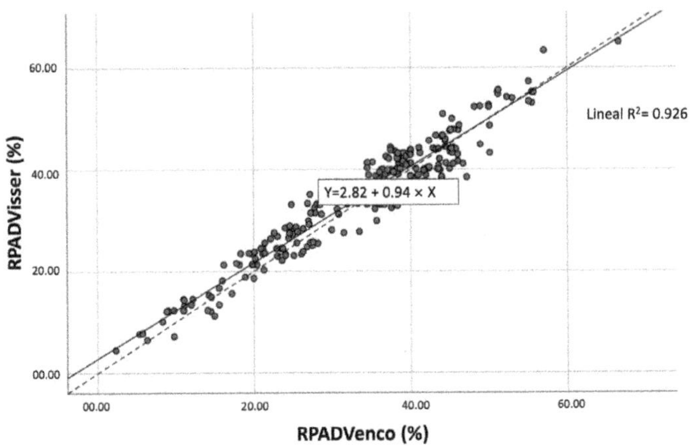

Figure 3. Scatter plots showing significant (p-value < 0.001) correlations (R^2 = 0.926) between the RPADVisser and RPADVenco measurements. The solid line within each scatterplot represents the line of best fit. The Pearson correlation between the two ratios was very high: 0.963 (p-value < 0.001).

To assess the agreement between the methods, the intraclass correlation coefficient (ICC) was calculated. Both the ICC of absolute agreement and that of consistency were excellent with values > 0.95 [ICC absolute agreement: 0.979 (p-value 0.000); ICC consistency: 0.981 (p-value 0.000)]. However, the p-value of the t-student test for paired samples was 0.000 which indicated significant differences in the values of the two indices.

The Bland–Altman representation investigated how the differences between the methods behave (Figure 4). The bias (mean of the differences) between methods was −0.9069 (solid line) with a t-student p-value of 0.000, which meant that on average the RPADVenco and RPADVisser indices did not measure statistically equal values. Approximately 95%

of the differences between methods were between −7.1673 (−0.9069 − 1.96 × SD) and 5.3535 (−0.9069 + 1.96 × SD)—dashed lines. The regression line calculated for the differences was practically horizontal and almost overlapped with the bias line, so there was no proportional bias but a constant bias, that is, the cloud of points was distributed quite homogeneously around zero, except for some values below 30 where RPADVisser scored much higher, and some above 30 where RPADVenco scored much higher (points outside the solid lines). In other words, there was no pattern of discordance, but sometimes one index scored higher and sometimes the other, without following any order.

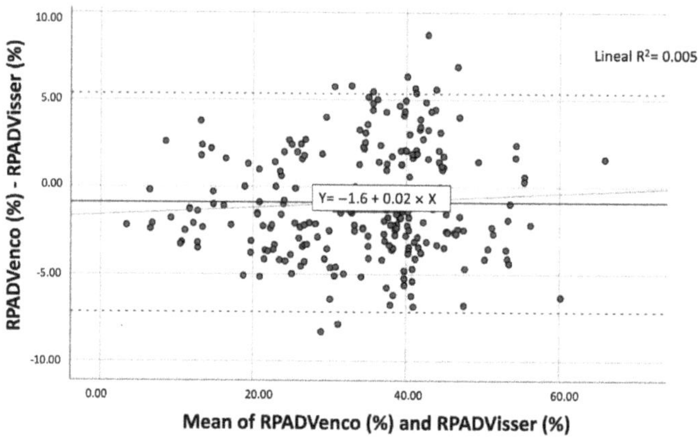

Figure 4. Scatter plots showing the Bland–Altman representation of the RPADVisser and RPADVenco measurements. The point cloud was distributed quite homogeneously around zero except for some values below 30 where RPADVisser scored much higher, and some above 30 where the RPADVenco scored much higher.

This comparative study, by establishing a cut-off point for RPADi < 30% [PH No (≥30%)/Yes (<30%)], allowed a comparison the classification of dogs by using both methods: 1.8% (n = 3) of the dogs detected as normotensive by RPADVenco were classified as dogs with PH by RPADVisser and, on the contrary, 13% (n = 11) of the dogs with PH by RPADVenco were classified as normotensive dogs by RPADVisser. On the other hand, 6.4% (n = 11) of the normo-tensive dogs detected by RPADVisser were classified as dogs with PH by RPADVenco and, conversely, 2.9% (n = 3) of the dogs with PH by RPADVisser were normotensive dogs by RPADVenco (Table 2).

The statistical values of both measurements were represented depending on the presence or absence of PH through the mean, standard deviation, median, 25th quartile, and 75th quartile. The values were around the cut-off point of 30% with interquartile ranges with limits between 27% and 33% (Table 3).

To verify the diagnostic capacity of both indices, two ROC CURVES were applied—one for each index—in which the events to be studied were as follows: suffering from PH (RPADi < 30%) determined with RPADVisser for the RPADVenco and suffering from PH (RPADi < 30%) determined with RPADVenco for the RPADVisser (Table 4). The results showed that the areas under the curve (AUC) of the two parameters were excellent (>0.95): Any value ≤ 27.7 of RPADVisser led to PH (RPADVenco < 30%) in 94.8% of cases and a value > 27.7 led to normotension (RPADVenco ≥ 30%) in 97.1% of cases. Any value ≤ 31.83 of RPADVenco led to PH (RPADVisser < 30%) in 92.8% of cases and a value > 31.83 led to normotension (RPADVisser ≥ 30) in 96.9% of cases.

Table 2. Discordant pulmonary hypertension (PH) diagnosis groups between the RPADVisser and RPADVenco measurements in the studied dogs (n = 248).

			RPADVenco (%)		Total
			No PH	Yes PH	
RPADVisser (%)	No PH	Count	160	11	171
		% of total	64.5%	4.4%	69.0%
	Yes PH	Count	3	74	77
		% of total	1.2%	29.8%	31.0%
Total		Count	163	85	248
		% of total	65.7%	34.3%	100.0%

Table 3. Descriptive statistical values between the RPADVisser and RPADVenco methods in the dogs studied (n = 248) depending on the presence or absence of pulmonary hypertension (PH).

		Presence PH RPADVenco (<30%)			
		No		Yes	
		Presence PH RPADVisser		Presence PH RPADVisser	
		No	Yes	No	Yes
RPADVenco (%)	Valid N	160	3	11	74
	Mean	41.15	31.68	27.60	19.02
	Standard deviation	6.04	1.75	1.20	6.45
	Median	40.00	31.55	27.80	20.30
	25th percentile	37.01	30.00	27.00	14.51
	75th percentile	44.47	33.50	28.50	24.44
RPADVisser (%)	Valid N	160	3	11	74
	Mean	41.66	27.80	32.60	20.37
	Standard deviation	6.39	0.28	1.51	6.56
	Median	40.28	27.75	33.10	23.05
	25th percentile	37.60	27.55	31.44	14.50
	75th percentile	44.08	28.11	33.33	25.50

Table 4. Results of area under the curve (AUC), confidence interval 95% (CI95%), sensitivity (Se), specificity (Sp), and Youden index of cut-off points of RPADVisser y RPADVenco measurements for predicting pulmonary hypertension (PH) for detecting RPADiVenco and RPADVisser <30%, respectively.

Method	Event	AUC	CI 95%	Cut-Off	Sen	Sp	YOUDEN Index (Sen + Sp − 1)
RPADVisser (%)	RPADVenco < 30%	0.989	(0.981, 1.000)	≤27.70	0.948	0.971	0.919
RPADVenco (%)	RPADVisser < 30%	0.994	(0.988, 1.000)	≤31.83	0.918	0.969	0.887

Finally, the Kappa index for the concordance of both methods in the classification of hypertension was 0.872 ($p < 0.001$), indicating very good concordance.

4. Discussion

The RPADi measurement is an echocardiographic determination widely used in both human and veterinary medicine in the diagnosis of PH, both of precapillary [5–9,12,15,19–26] and postcapillary [11,13,14,20,22–24] origin. In particular, it is considered a measure of choice when Doppler tracing of pulmonary or tricuspid regurgitation is not available [1,21]. The origin of PH seems to be of particular relevance to the usefulness of RPADi; in general, previous studies have found that RPADi values are lower in animals with PH affected by precapillary causes, with both RPADVenco and RPADVisser echocardiographic measurement, and a direct correlation with the measurement of tricuspid regurgitation if present [12,13,24–26]. This is probably due to direct damage to the pulmonary arterial vasculature in precapillary PH, versus retrograde damage from increased pressure in the pulmonary venous vasculature due to left heart pathology in postcapillary PH.

More specifically, in the case of heartworm disease, a good correlation of RPADi with measurements made by invasive direct right heart catheterization (gold standard) has been described for the estimation of PH [8] and the use of RPADi has been the most widely used measure to determine the presence of PH in dogs parasitized by *D. immitis* [5–9,15], generally RPADVenco. In fact, several studies have investigated the RPADi in dogs with heartworm and other diseases using the RPADVenco, determining that the lowest values were found in animals infected with *D. immitis* [13,23,25]. Since pulmonary vascular lesions generated by the presence of adult worms cause endothelial damage, dilatation, loss of structure, tortuosity, dilatation of damaged pulmonary arteries, and PH [22], their determination by RPADi is considered to be of special interest [5–9].

However, in veterinary medicine there is no standardized protocol for RPADi determination [10]. There are studies performed by longitudinal right parasternal view [3,6–8,11,13,23,24] and others by transverse view [6,7,11,12,20,24], as well as 2D mode [5,7,11,12,20,24] or M-mode has also been used for RPADi measurement [5,8,9,13–15,25,26]. Variations have also been observed regarding the determination of the diameters in systole or diastole of the right pulmonary artery (trailing edge-to-leading Edge [6,7,11,12,20,24] or leading edge-to-leading Edge [5,8,9,13,15,25,26]). Therefore, it was necessary to carry out studies comparing the different methodologies described.

In the present study, RPADi measurement was investigated in 248 dogs, 71.4% of which had heartworm disease. Statistical analysis between the RPADVisser and RPADVenco methods showed that the two methods were not interchangeable with each other, although a good correlation between the methods was recorded. These results were similar to those reported by Basile et al. [10], who also evaluated in 46 healthy and heartworm-infected dogs the comparison between methods of estimating PH using the RPADVisser and RPADVenco methods, in addition to some modifications for both methods. The authors concluded that, although the Bland–Altman test showed statistical agreement between the different methods used to calculate the RPADi, the results of both methods varied, so the method of measurement used when evaluating an animal should always be specified.

On the other hand, the study developed by Vezzosi et al. [23] determined that both methods could be used interchangeably to assess RPADi, by comparing the two echocardiographic views (short axis and a long axis) to obtain the RPADi, but using M-mode in both measurements; the Pearson test showed a strong positive linear correlation between the RPADi values obtained with both methods (r^2 = 0.9346) and the Bland–Altman test showed good agreement between the two methods in the estimation of RPADi, the mean difference between the two methods being 0.51%. This difference in results could be due to the sample size, the pathologies involved in the study—presumably causing PH of post-capillary origin—and the difference in the methodology used for RPADi measurement, due to the fact that M-mode was used in both cases.

The RPADVenco and RPADVisser methods, although measuring the same vascular structures, do so in completely different ways. The results obtained and previously published references reinforce the idea that they are useful measures but should be used independently and not interchangeably to determine the presence of PH in veterinary medicine, especially in the determination of PH in heartworm disease [10].

As in previous studies performed on heartworm disease, no relationship between the presence of PH and age, breed, weight, or sex was observed [5,8,9,15]. Also, respiratory symptoms and R-CHF were directly related to the presence of PH, as previously described [7,8,13,27]. The results showed that high parasitic burden does appear to be a predisposing factor for PH which may be explained by increased endothelial damage, as well as increased ability to impede blood flow and increased risk of thromboembolism, caused by a higher relative presence of worms [9,15]. However, in contrast to the present study, other authors observed that there were no significant differences in the presence or absence of PH between animals with low or high parasite burden, as estimated by the RPADi [5,6] or other echocardiographic parameters [27], and that the chronicity of the disease, the intensity of exercise and the immune response of the host to the parasite may exert a greater influence on the development of PH [8,28]. Nevertheless, in the conducted research, other echocardiographic determinations routinely used in the diagnosis of PH were not included, which can be considered a limitation in interpreting the obtained results. Likewise, the results obtained must be analyzed while considering the specific conditions of the patients included in this study. Factors such as the wide variation in body weight, age, and breed were not analyzed, and their impact on the results remains unknown.

To the authors' knowledge, RPADmax:Ao and RPADmin:Ao measurements have not been previously evaluated in heartworm disease; therefore, this study described those values based on the two RPADi methods for the first time. RPADmax, RPADmin, RPADmax:Ao, and RPADmin:Ao measurements were slightly lower with the PRADVenco method compared to RPAD-Visser. These differences may be explained by the different location of the determination of the systolic and diastolic diameters of the right pulmonary artery, as well as the difference in the use of M-mode or 2D-mode and whether or not the arterial vascular wall was added to the measurement. Recently, Grosso et al. [17] showed reference values for RPADmax, RPADmin, RPADmax:Ao, and RPADmin:Ao for healthy dogs using the RPADVisser method. The results for the healthy dogs in the present study showed values within these ranges for both RPADVisser and RPADVenco.

5. Conclusions

The present study showed that the main methods of the RPADi determination, RPAD-Visser and RPADVenco, are highly similar to each other in determining the presence of PH in dogs with heartworm disease. However, slight but statistically significant differences have been found between them, which indicate that they are not interchangeable and must be used independently of each other. Similarly, different values were reported for RPADmax, RPADmin, RPADmax:Ao, and RPADmim:Ao between the methods analyzed, highlighting the differences between the technique used to determine the RPADi.

Author Contributions: J.I.M., J.A.M.-A. and E.C. designed the study; J.I.M., A.C.-V. and E.C. wrote the manuscript; J.I.M., E.M.-P. and S.N.G.-R. performed the fieldwork, collected the data, and performed the experiments. All authors participated in the discussion of the results and corrected, read, and approved the final manuscript. All authors have read and agreed to the published version of the manuscript.

Funding: The presented study was supported by funds from the Internal Medicine Service FULP/ULPGC (SD-240/030/0026). S.N.R.-G. weas supported by the "Grants for the financing of predoctoral contracts" program of the Universidad de Las Palmas de Gran Canaria (PIFULPGC-2020-2-CCSALUD-2).

Institutional Review Board Statement: Ethical review and approval were not required for the animal in this study. All echocardiographic measures were routinely collected for prescribed diagnostic purposes or official monitoring studies and subsequently made available to this study. All of the dog owners were informed about the present study and consented to participate. This study was carried out in accordance with the current Spanish and European legislation on animal protection (Spanish Royal Decree 53/2013 and 2010/63/UE Directive).

Informed Consent Statement: Informed consent was obtained from all subjects involved in the study.

Data Availability Statement: All data generated or analyzed during this study are included in this article. The datasets used and/or analyzed during the present study are available from the corresponding author upon reasonable request.

Acknowledgments: The authors would like to thank Uranovet for kindly providing the Urano test Dirofilaria®. Also, they would like to thank to the Hospital Clínico Veterinario of the Universidad de Las Palmas de Gran Canaria (ULPGC) for their support.

Conflicts of Interest: The authors declare no conflicts of interest.

References

1. Reinero, C.; Visser, L.C.; Kellihan, H.B.; Masseau, I.; Rozanski, E.; Clercx, C.; Williams, K.; Abbott, J.; Borgarelli, M.; Scansen, B.A. ACVIM consensus statement guidelines for the diagnosis, classification, treatment, and monitoring of pulmonary hypertension in dogs. *J. Vet. Intern. Med.* **2020**, *34*, 549–573. [CrossRef]
2. Pajas, A.M.G.; Acorda, J.A. Echocardiographic, electrocardiographic and thoracic radiographic findings in dogs with dirofilariasis. *Philipp. J. Vet. Med.* **2018**, *55*, 71–84.
3. Simón, F.; Siles-Lucas, M.; Morchón, R.; González-Miguel, J.; Mellado, I.; Carretón, E.; Montoya-Alonso, J.A. Human and Animal Dirofilariasis: The Emergence of a Zoonotic Mosaic. *J. Clin. Microbiol.* **2012**, *25*, 507–544. [CrossRef]
4. Falcon-Cordon, Y.; Montoya-Alonso, J.A.; Caro-Vadillo, A.; Matos-Rivero, J.I.; Carreton, E. Persistence of pulmonary endarteritis in canine heartworm infection 10 months after the eradication of adult parasites of *Dirofilaria immitis*. *Vet. Parasitol.* **2019**, *273*, 1–4. [CrossRef] [PubMed]
5. Maerz, I. Clinical and diagnostic imaging findings in 37 rescued dogs with heartworm disease in Germany. *Vet. Parasitol.* **2020**, *283*, 109156. [CrossRef] [PubMed]
6. Serrano-Parreño, B.; Carretón, E.; Caro-Vadillo, A.; Falcon-Cordon, Y.; Falcon-Cordon, S.; Montoya-Alonso, J.A. Evaluation of pulmonary hypertension and clinical status in dogs with heartworm by right pulmonary artery distensibility index and other echocardiographic parameters. *Parasit. Vectors* **2017**, *10*, 106. [CrossRef]
7. Romano, A.E.; Saunders, A.B.; Gordon, S.G.; Wesselowsky, S. Intracardiac heartworms in dogs: Clinical and echocardiographic characteristics in 72 cases. *J. Vet. Intern. Med.* **2021**, *35*, 88–97. [CrossRef] [PubMed]
8. Venco, L.; Mihaylovab, L.; Boon, J.A. Right Pulmonary Artery Distensibility Index (RPAD Index). A field study of an echocardiographic method to detect early development of pulmonary hypertension and its severity even in the absence of regurgitant jets for Doppler evaluation in heartworm-infected dogs. *Vet. Parasitol.* **2014**, *206*, 60–66.
9. Serrano-Parreño, B.; Carretón, E.; Caro-Vadillo, A.; Falcón-Cordón, S.; Falcón-Cordón, Y.; Montoya-Alonso, J.A. Pulmonary hypertension in dogs with heartworm before and after the adulticide protocol recommended by the American Heartworm Society. *Vet. Parasitol.* **2017**, *236*, 34–37. [CrossRef]
10. Basile, A.; Napoli, E.; Brianti, E.; Venco, L. Right Pulmonary Artery Distensibility Index in Heartworm Infected Dogs: Are the Different Methods Leading to Same Results? *Animals* **2023**, *13*, 418. [CrossRef]
11. Visser, L.C.; Im, M.K.; Johnson, L.R.; Stern, J.A. Diagnostic value of right pulmonary artery distensibility index in dogs with pulmonary hypertension: Comparison with Doppler echocardiographic estimates of pulmonary arterial pressure. *J. Vet. Intern. Med.* **2016**, *30*, 543–552. [CrossRef]
12. Yuchi, Y.; Suzuki, R.; Saito, T.; Yasumura, Y.; Teshima, T.; Matsumoto, H.; Koyama, H. Echocardiographic characteristics of dogs with pulmonary hypertension secondary to respiratory diseases. *J. Vet. Intern. Med.* **2023**, *37*, 1656–1666. [CrossRef] [PubMed]
13. Nijhawan, D.; Chawla, R.; Saxena, M.; Dixi, S.; Sharma, V. To Assess the Clinical Value of the RPAD Index in Dogs with Varying Degrees of Pulmonary Hypertension. *Asian J. Med. Res.* **2019**, *8*, AN17–AN19. [CrossRef]
14. Basava, R.K.; Ansar-Kamran, C.; Ramesh, P.T.; Upendra, H.A.; Rao, S.; Mahesh, V. Importance of Right Pulmonary Artery Distensibility Index in the Diagnosis and Management of Pulmonary Arterial Hypertension Secondary to Canine Dilated Cardiomyopathy. *Indian J. Anim. Res.* **2022**, *1*, 1–5.

15. Matos, J.I.; Caro-Vadillo, A.; Falcón-Cordón, Y.; García-Rodríguez, S.N.; Costa-Rodríguez, N.; Carretón, E.; Montoya-Alonso, J.A. Echocardiographic Assessment of the Pulmonary Vein to Pulmonary Artery Ratio in Canine Heartworm Disease. *Animals* **2023**, *13*, 703. [CrossRef] [PubMed]
16. Montoya-Alonso, J.A.; Morchón, R.; García-Rodríguez, S.N.; Falcón-Cordón, Y.; Costa-Rodríguez, N.; Matos, J.I.; Escolar, I.R.; Carretón, E. Expansion of canine heartworm in Spain. *Animals* **2022**, *12*, 1268. [CrossRef]
17. Grosso, G.; Tognetti, R.; Domenech, O.; Marchesotti, F.; Patata, V.; Vezzosi, T. Echocardiographic reference intervals of the dimensions of the main pulmonary artery and the right pulmonary artery: A prospective study in 269 healthy dogs. *J. Vet. Cardiol.* **2023**, *50*, 29–38. [CrossRef]
18. Venco, L.; Genchi, C.; Vigevani-Colson, P.; Kramer, L. Relative utility of echocardiography, radiography, serologic testing and microfilariae counts to predict adult worm burden in dogs naturally infected with heartworms. In *Recent Advances in Heartworm Disease. Symposium'01*; American Heartworm Society: Batavia, IL, USA, 2004; pp. 24–111.
19. Malhotra, R.; Dhakal, B.P.; Eisman, A.S.; Pappagianopoulos, P.P.; Dress, A.; Weiner, R.B.; Baggish, A.L.; Semigran, M.J.; Lewis, G.D. Pulmonary Vascular Distensibility Predicts Pulmonary Hypertension Severity, Exercise Capacity, and Survival in Heart Failure. *Circ. Heart Fail.* **2016**, *9*, 10. [CrossRef]
20. Ferraro, A.; Hartnack, S.; Schwarzwald, C.C. Diagnostic value of two-dimensional echocardiographic measurements of the pulmonary artery diameter and the pulmonary artery distensibility index to detect pulmonary hypertension in horses. *J. Vet. Cardiol.* **2023**, *49*, 52–66. [CrossRef]
21. Lyssens, A.; Lekane, M.; Gommeren, K.; Merveille, A.C. Focused Cardiac Ultrasound to Detect Pre-capillary Pulmonary Hypertension. *Front. Vet. Sci.* **2022**, *9*, 83–91. [CrossRef]
22. Birettoni, F.; Caivano, D.; Patata, V.; Moïse, N.S.; Guglielmini, C.; Rishniw, M.; Porciello, F. Canine pulmonary vein-to-pulmonary artery ratio: Echocardiographic technique and reference intervals. *J. Vet. Cardiol.* **2016**, *18*, 326–335. [CrossRef] [PubMed]
23. Vezzosi, T.; Marchesotti, F.; Tognetti, R.; Venco, L.; Domenech, O. Comparison of two echocardiographic views for evaluating the right pulmonary artery distensibility index in dogs. *J. Vet. Intern. Med.* **2015**, *29*, 426.
24. Oleynikov, D.; Yi, M. Echocardiographic Pulmonary to Left Atrial Ratio in Dogs (ePLAR): A Differential Marker of Pre- and Postcapillary Pulmonary Hypertension. *Am. J. Anim. Vet. Sci.* **2022**, *17*, 42–52. [CrossRef]
25. Roels, E.; Merveille, A.C.; Moyse, E.; Gomart, S.; Clercx, C.; Mc Entee, K. Diagnostic value of the pulmonary veinto-right pulmonary artery ratio in dogs with pulmonary hypertension of precapillary origin. *J. Vet. Cardiol.* **2019**, *24*, 85–94. [CrossRef]
26. Chan, I.P.; Weng, M.C.; Hsueh, T.; Lin, Y.C.; Lin, S.L. Prognostic value of right pulmonary artery distensibility in dogs with pulmonary hypertension. *J. Vet. Sci.* **2019**, *20*, e34. [CrossRef] [PubMed]
27. Venco, L.; Bertazzolo, W.; Giordano, G.; Paltrinieri, S. Evaluation of C-reactive protein as a clinical biomarker in naturally heartworm-infected dogs: A field study. *J. Vet. Parasitol.* **2014**, *206*, 48–54. [CrossRef]
28. Dillon, R.A.; Brawner, W.R.; Hanrahan, L. Influence of number of parasites and exercise on the severity of heartworm disease in dogs. In Proceedings of the Heartworm Symposium '95; American Heartworm Society: Vatavia, IL, USA, 1995; pp. 113–116.

veterinary sciences

MDPI

Article

Feline Erythrocytic Osmotic Fragility in Normal and Anemic Cats—A Preliminary Study

Purin Lophaisankit [1], Kunanon Boonyok [2], Jaruwan Khonmee [1], Chatchanok Udomtanakunchai [3], Chollada Sodarat [1], Kannika Phongroop [1] and Worapat Prachasilchai [1,4,*]

1 Faculty of Veterinary Medicine, Chiang Mai University, Chiang Mai 50100, Thailand; purin_lo@cmu.ac.th (P.L.); jaruwan.khonmee@cmu.ac.th (J.K.); chollada.s@cmu.ac.th (C.S.); kannika.p@cmu.ac.th (K.P.)

2 Faculty of Veterinary Medicine, Kasetsart University, Bangkok 10900, Thailand; kunanonboonyok@gmail.com

3 Department of Radiologic Technology, Faculty of Associated Medical Sciences, Chiang Mai University, Chiang Mai 50200, Thailand; chatchanok.u@cmu.ac.th

4 Research Center in Bioresources for Agriculture, Industry and Medicine, Chiang Mai University, Chiang Mai 50100, Thailand

* Correspondence: worapat.p@cmu.ac.th or liverpoolpat@yahoo.com

Simple Summary: This study assessed red blood cell membrane strength and characteristics in non-anemic and anemic cats using the osmotic fragility test and flow cytometry. Blood samples from 18 non-anemic and 18 anemic cats, divided into adults and seniors, were analyzed for complete blood counts, blood chemistry, the osmotic fragility test, and flow cytometry. The findings showed no significant differences in osmotic fragility between non-anemic and anemic cats in either age group. In flow cytometry analysis, normal senior cats had higher forward scatter than senior anemic cats, while normal adult cats had higher side scatter than anemic adult cats. These results suggest that, while osmotic fragility remains consistent, red blood cell size and density differences vary with age and health status, highlighting the need for further research on red blood cell characteristics in different diseases.

Abstract: Erythrocyte osmotic fragility is an excellent parameter for evaluating the red blood cell (RBC) membrane, which may be abnormal in several pathological conditions. The flow cytometer is a powerful tool that analyzes a single cell in a solution and can detect alterations in RBCs, providing key differential diagnostic information. Both the osmotic fragility test (OFT) and flow cytometry are valuable diagnostic tools in veterinary medicine, but their diagnostic usefulness in anemic cats has not yet been determined. This study aimed to evaluate RBC membrane strength using an OFT in non-anemic and anemic cats and to compare RBC size and density using a flow cytometer in non-anemic and anemic cats. A total of 18 cats in the non-anemic group and 18 cats in the anemic group, divided into adults and seniors, were included in this study. Blood samples were collected for a complete blood count (CBC) and blood chemistry. The remaining blood was used for OFT to evaluate 50% hemolysis from the hemolysis curve and for the flow cytometer to measure forward scatter characteristics (FSC) and side scatter characteristics (SSC). The result of OFT showed no significant difference in OF between normal and anemic cats in the adult and senior groups. In terms of flow cytometry analysis, normal and anemic cats in the adult group showed no significant difference in the FSC ($p = 0.769$). On the other hand, the FSCs of normal senior cats were significantly higher than those of anemic cats ($p = 0.0486$). The SSCs of normal cats were significantly higher than those of anemic cats in the adult group ($p = 0.048$). However, the SSCs of the senior group showed no significant difference ($p = 0.074$). Based on these results, we concluded that, in the senior group, normal cats had higher FSCs than anemic cats, and in the adult group, normal cats had higher SSCs than anemic cats. However, both normal and anemic cats exhibited similar osmotic fragility. Further studies on various diseases are suggested.

Keywords: erythrocyte osmotic fragility; anemia; red blood cell; forward scatter characteristics; side scatter characteristics

1. Introduction

Red blood cells (RBCs) are unique structures in vertebrates that, upon maturation, lose the ability to perform ribosomal protein synthesis or mitochondrial phosphorylation. Their fatty acid and phospholipid compositions vary across species [1]. Evaluating the RBC membrane is critical for understanding its functional integrity, particularly in pathological conditions. Osmotic fragility (OF) is the resistance of RBC hemolysis to osmotic changes used to assess RBC friability, the susceptibility of red blood cells to break or rupture under mechanical stress or osmotic pressure, which may be abnormal in various diseases [2–4]. Several methods exist for measuring OF, including the classic OF test, the eosin-5′-maleimide dye-binding test, the flow cytometric OF test, and the cryohemolysis test [5]. The classic OF test, a widely used method, involves incubating whole blood in a hypotonic sodium chloride (NaCl) solution at room temperature [6]. Exposure to the hypotonic solution causes water to diffuse into RBCs, leading to swelling, leakage, and eventual lysis. Hemoglobin (Hgb) released into the supernatant is measured spectrophotometrically, with the results expressed as the percentage of completely lysed blood in distilled water. The normal OF curve is sigmoid, and the mean osmotic fragility (MOF)—the NaCl concentration at which 50% of RBCs are lysed—typically occurs around 0.5% NaCl in healthy animals. OF is generally inversely correlated with the normal size of RBCs across species [7] and can be used to evaluate the rate of hemolysis [8].

Poikilocytosis, the presence of abnormally shaped RBCs in the blood, provides significant diagnostic information [9]. Alterations in RBC morphology can aid in identifying underlying diseases in both humans and animals [10–12]. In addition to pathological causes, erythrocyte morphology can change due to physiological factors, such as aging [13]. Flow cytometry offers another valuable tool for analyzing RBC populations by utilizing lasers to produce scattered light signals. These signals are detected and analyzed, with forward scatter (FSC) indicating relative cell size and side scatter (SSC) reflecting internal complexity or granularity [14].

Feline anemia is a common condition that can develop among many other conditions, such as hemorrhage, hemolysis, bone marrow suppression, chronic inflammation, tumors, endocrine diseases, and nephropathies, among others. Previous studies have found that increased erythrocytic OF, leading to hemolysis, has been reported in various conditions, including immune-mediated hemolytic anemia in cats [15], anemia in domestic shorthair and other cat breeds [3], and hereditary RBC membrane defects in Abyssinian and Somali cats [16]. Fragile RBCs have also been associated with conditions such as chronic intermittent hemolysis, macrocytic-hypochromic and variably regenerative anemia, splenomegaly, hyperglobulinemia, lymphocytosis, hyperbilirubinemia, stress, hyperlipidemia, hyperglycemia, microcytosis due to intestinal parasitism, intoxication with beta-acetylphenylhydrazine, and elevated serum hepatic enzyme activities [3,8,17–22]. This study aims to evaluate RBC membrane strength using the classic osmotic fragility test in various groups of cats, including healthy and anemic individuals. Additionally, the study seeks to compare RBC size and complexity across these groups, contributing valuable insights into feline hematology and the diagnostic potential of RBC characteristics in identifying underlying conditions.

2. Materials and Methods

2.1. Sample Population

A total of 36 client-owned domestic shorthair cats of all genders more than 1 year old from the Small Animal Hospital, Chiang Mai University were included in our study between May and December 2023. Animals were divided into 2 groups: normal (non-anemic) cats (n = 18) and anemic cats (n = 18). Each group was divided into 2 subgroups: adults (1–5 years old, n = 9) and seniors (>6 years old, n = 9). Therefore, the cats were categorized into four groups: normal adult (NA), normal senior (NS), anemic adult (AA), and anemic senior (AS).

All cats underwent a physical examination, including temperature measurement, observation of general appearance, hydration status, body condition score, mentation, posture, thoracic auscultation, abdominal palpation, and blood collection. If the physical examination results showed no clinical signs and hematologic parameters were within normal limits [23], cats were categorized into the non-anemic group. There was no infectious disease screening (e.g., FeLV, FIV, or other region-specific pathogens) of individuals as part of the initial study. Some further investigations, including diagnostic imaging or more specific tests, were not performed for every individual with abnormalities.

Cats were included in the anemic group if at least one of the following clinical signs were present: pale mucous membranes, loss of appetite, weight loss, depression, tachycardia, hematochezia, and melena, along with laboratory findings showing a hematocrit less than 30%.

This study received ethical approval from the Faculty of Veterinary Medicine, Chiang Mai University Animal Care and Use Committee (FVM—ACUC). The reference number is S11/2566.

2.2. Sample Collection

Blood samples (1.5 mL) were collected from non-fasting, well-hydrated cats and then divided into 0.5 mL in an ethylenediaminetetraacetic acid (EDTA) tube and 1 mL in a heparin tube. Blood samples were chilled to 4 °C after collection and shipped to the laboratory for analysis within 24 h after collection.

2.3. Laboratory Testing

Routine clinical laboratory techniques were used to perform complete blood counts (Dymind DF56VET, Dymind Biotechnology, Shenzhen, China) and microscopic blood smear (Dip Quick Stain, M&P impex, Bangkok, Thailand) evaluations, including RBC morphology, platelet counts, and blood parasites, performed by the scientists of the Small Animal Hospital laboratory using the EDTA tube. The blood in the heparin tube was analyzed for blood chemistry panels (Sysmex DX-3010, Sysmex corporation, Kobe, Japan), including alkaline phosphatase (ALP), alanine transaminase (ALT), blood urea nitrogen (BUN), creatinine, triglycerides (TG), total cholesterol (Chol), and blood glucose. A blood smear was performed to evaluate blood parasites, such as *Mycoplasma haemofelis*, *Babesia felis*, etc., to evaluate any abnormalities of erythrocyte morphology and to evaluate platelet counts. The platelet count was assessed using a blood smear examined under a 100× oil immersion lens. Platelet numbers were estimated using the following formula: platelet count (per μL) = number of cells per 100× oil field × 20,000. A platelet count within the normal range was considered adequate.

2.4. Osmotic Fragility Test

The leftover blood from the EDTA tube was centrifuged at 7000 rpm for 1 min. The classic OF test of the erythrocyte was performed by preparing a series of 1 mL NaCl tubes containing concentrations of 0.9, 0.7, 0.6, 0.5, 0.4, 0.2, and 0.1% NaCl. Packed red cells (10 μL) were diluted with 300 μL of 0.9% NaCl. A 10 μL volume of the diluted solution was added to each tube, followed by incubation at room temperature for 3 min and centrifugation at 7000 rpm for 1 min. The supernatant's optical density was measured using a spectrophotometer (Shimadzu UV-2700, UV-Vis spectrophotometer, Kyoto, Japan) at 415 nm. For each tube, the hemolysis percentage was calculated. The mean OF was determined from a hemolysis curve [3,16].

2.5. Flow Cytometry Test

The flow cytometry test was performed by diluting 5 μL of packed red cells with 1 mL of 0.9%, and then forward scatter characteristics (FSC) and side scatter characteristics (SSC), which refer to size and density, respectively, were measured by flow cytometry (CytoFLEX Flow Cytometer, Beckman Coulter, Brea, CA, USA).

2.6. Statistical Analysis

The following data were collected in Excel (Microsoft Corporation, Redmond, Washington). Descriptive statistics were used for analysis. Continuous variables were expressed as mean ± standard deviation (SD). The statistical analysis was performed using OriginPro 2015 software (OriginLab Corporation, Northampton, MA, USA). An independent t-test was used to compare continuous variables between groups and within groups. The correlation between blood profiles and laboratory results was determined by linear regression. A p-value < 0.05 indicated statistical significance. A correlation coefficient (r) > 0.7 was considered a strong relationship, between 0.4 to 0.7 was considered a moderate relationship, and <0.4 was considered a weak relationship [24].

3. Results

3.1. Animals

A total of 36 client-owned domestic shorthair cats from the Small Animal Hospital, Chiang Mai University were included in this study between May and December 2023. All cats underwent clinical assessment for dehydration prior to blood collection, and none were found to exhibit signs of dehydration. The NA group (n = 9) comprised five males and four females, with an average age of 2.2 ± 1.2 years. The SA group (n = 9) included three males and six females, with an average age of 8.1 ± 2.3 years. The AA group (n = 9) consisted of four males and five females, with an average age of 2.0 ± 1.4 years. Lastly, the SA group (n = 9) comprised three males and six females, with an average age of 9.8 ± 2.8 years. The average ages were expressed as mean ± SD.

Several etiologies contributing to anemia were considered within this study. In the anemic adult group, anemia was predominantly caused by infectious disease (55.55%, 5/9). Conversely, in the anemic senior group, anemia was prominently associated with kidney disease (44.44%, 4/9), as shown in Table 1.

Table 1. The cause of anemia in the anemic adult (n = 9) and anemic senior (n = 9) groups.

Group	Disease	n
Anemic adult	Polycystic kidney disease (PKD)	1
	Chronic kidney disease (CKD)	1
	Blood loss (hit by car)	1
	Feline leukemia virus (FeLV)	3
	Blood parasite (*Mycoplasma* spp.)	1
	Feline immunodeficiency virus (FIV)	1
	Lymphoma (CHOP treatment)	1
Anemic senior	Chronic kidney disease (CKD)	4
	Blood loss (hit by car)	1
	Blood loss (post-operative anemia)	1
	Blood parasite (*Mycoplasma* spp.)	2
	Unknown cause	1

3.2. Laboratory Tests

All normal cats exhibited hematologic parameters that were mostly within normal ranges [23], except for the platelet count, which appeared to decrease on the analyzer but was adequate upon blood smear evaluation. On the other hand, anemic cats exhibited a diverse range of hematologic parameters, with six cats (33.33%, 6/18) displaying a decreased platelet count upon platelet examination.

Mean hematologic parameters, including hematocrit (Hct), hemoglobin (Hgb), red blood cell counts (RBCs), mean corpuscular volume (MCV), mean corpuscular hemoglobin (MCH), mean corpuscular hemoglobin concentration (MCHC), white blood cell counts (WBCs), and platelet counts (PLTs) of cats from each group are displayed in Table 2. All groups of cats showed a decreased platelet count when assessed by the analyzer. However, blood smear analysis revealed an adequate platelet quantity in all cats from the normal group (100%, 18/18) and in 12 cats from the anemia group (66.66%, 12/18).

Table 2. Mean hematologic parameters of normal adult, normal senior, anemic adult, and anemic senior cats.

Parameter (Unit)	Normal Adult	Normal Senior	Anemic Adult	Anemic Senior	References Interval
Hct (%)	43.33 ± 3.42^{a}	42.88 ± 1.81^{a}	19.60 ± 8.76^{b}	22.00 ± 6.93^{b}	30–45
Hgb (g/dL)	14.40 ± 1.24^{a}	14.30 ± 0.90^{a}	6.08 ± 2.85^{b}	8.50 ± 5.27^{b}	9.8–15.4
RBCs ($\times 10^6$/uL)	8.67 ± 0.60^{a}	8.47 ± 0.89^{a}	3.89 ± 1.84^{b}	4.84 ± 2.14^{b}	5–10
MCV (fL)	50.23 ± 4.38	51.04 ± 3.89	51.07 ± 6.28	51.89 ± 13.38	39–55
MCH (pg)	16.62 ± 0.83	17.01 ± 0.89	15.62 ± 2.11	18.89 ± 9.27	13–17
MCHC (g/dL)	33.21 ± 2.33	33.39 ± 1.23	30.63 ± 2.81	29.33 ± 6.87	30–36
WBCs ($\times 10^3$/uL)	11.43 ± 6.44^{a}	9.73 ± 2.81^{a}	24.76 ± 9.55^{b}	32.69 ± 29.04^{b}	5.5–19.5
PLTs ($\times 10^3$/uL)	255.77 ± 54.33^{a}	235.88 ± 51.53^{a}	143.00 ± 116.83^{b}	171.44 ± 133.45^{b}	300–800

Data are expressed as mean ± standard deviation. Different letters mean statistically significant difference (p-value < 0.05). Abbreviations: Hct = hematocrit, Hgb = hemoglobin, RBCs = red blood cells, MCV = mean corpuscular volume, MCH = mean corpuscular hemoglobin, MCHC = mean corpuscular hemoglobin concentration, WBCs = white blood cells, and PLTs = platelets.

The majority of normal cats (94.44%, 17/18) exhibited blood chemistry parameters within the normal range; however, an exception was observed in the normal senior group, where one cat displayed mildly elevated ALT levels.

Six anemic cats (33.33%, 6/18) displayed elevated levels of BUN and creatinine. Notably, within the anemic senior group, one cat exhibited a severe elevation of ALT.

Mean blood chemistry parameters, including BUN, creatinine, ALT, ALP, total protein, albumin, cholesterol, triglycerides, and glucose, of the cats in each group are displayed in Table 3.

Table 3. Mean blood chemistry parameters of normal adult, normal senior, anemic adult, and anemic senior cats.

Parameter (Unit)	Normal Adult	Normal Senior	Anemic Adult	Anemic Senior	Reference Interval
BUN (mg/dL)	31.17 ± 5.23	33.10 ± 9.88	53.77 ± 60.25	59.50 ± 43.36	19–34
Creatinine (mg/dL)	1.48 ± 0.10	1.60 ± 0.32	2.60 ± 2.66	3.90 ± 5.13	0.9–2.2
ALT (U/L)	72 ± 29.09	69.57 ± 42.47	88.50 ± 88.80	89.00 ± 80.58	25–97
ALP (U/L)	48.22 ± 16.38	45.88 ± 19.07	19.30 ± 10.67	42.56 ± 38.77	0–45
Total protein (g/dL)	8.96 ± 0.81	8.39 ± 0.59	8.94 ± 2.75	8.88 ± 1.55	6–7.9
Albumin (g/dL)	3.94 ± 0.16	3.78 ± 0.29	3.24 ± 0.44	2.99 ± 0.39	2.8–3.9
Cholesterol (mg/dL)	110.44 ± 20.82	143.13 ± 47.97	144.20 ± 45.35	156.11 ± 57.61	71–156
Triglyceride (mg/dL)	116.56 ± 49.59	148.25 ± 115.69	120.40 ± 93.92	100.89 ± 50.23	27–94

Data are expressed as mean ± standard deviation. Abbreviations: BUN = blood urea nitrogen, ALT = alanine transaminase, and ALP = alkaline phosphatase.

3.3. Osmotic Fragility Test

The 50% osmotic fragility value revealed no statistically significant difference between groups, and the mean 50% osmotic fragility is shown in Figure 1.

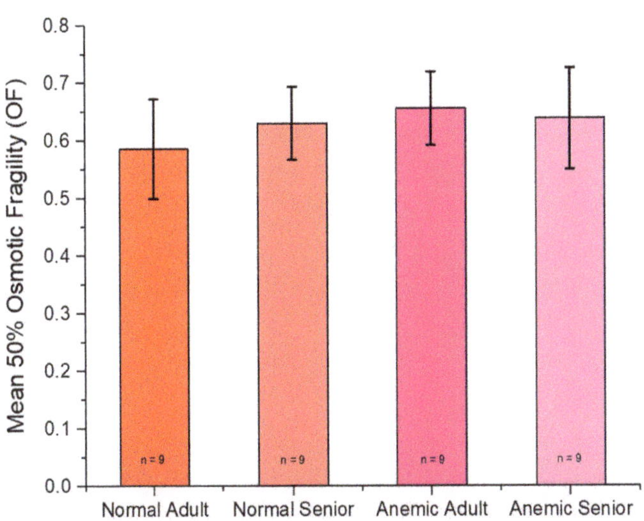

Figure 1. Mean 50% osmotic fragility (OF) of the normal adult, normal senior, anemic adult, and anemic senior groups. Error bar represents SD.

3.4. Flow Cytometry Test

The forward scatter characteristics (FSCs) of each group showed significant differences in the mean FSC between the normal senior group and the normal adult group ($p = 0.02864$), with the normal senior group exhibiting higher FSC values, and the normal senior group also demonstrated higher FSCs than the anemic senior group ($p = 0.0486$), as shown in Figure 2.

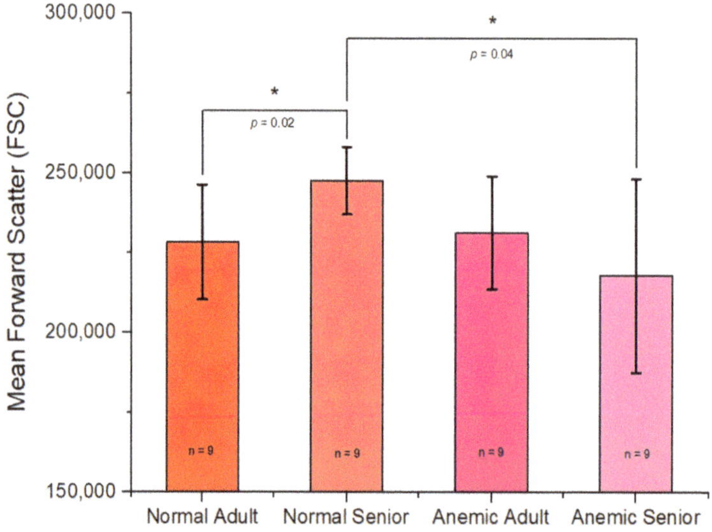

Figure 2. Mean forward scatter characteristics (FSC) of the normal adult, normal senior, anemic adult, and anemic senior groups. Error bar represents SD. * Statistically significant (p-value < 0.05).

The side scatter characteristic (SSC) of each group showed a significant difference in mean SSC between the normal adult group and the anemic adult group ($p = 0.048$), with the normal adult group exhibiting higher SSC values than the anemic adult group, as shown in Figure 3.

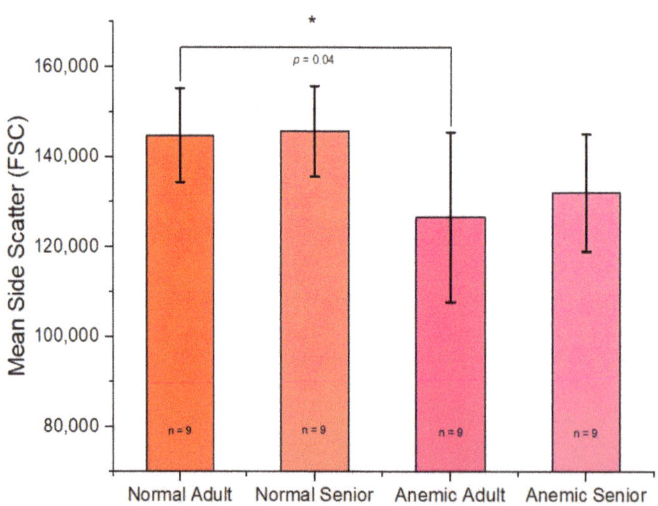

Figure 3. Mean side scatter characteristics (SSC) of the normal adult, normal senior, anemic adult, and anemic senior groups. Error bar represents SD. * Statistically significant (p-value < 0.05).

Linear regression analyses demonstrated the relationship between osmotic fragility and the forward scatter characteristic of each group, revealing a statistically significant positive correlation in the anemic senior group (AS), with $p = 0.0103$, r = 0.79, and $R^2 = 0.58$, as shown in Figure 4.

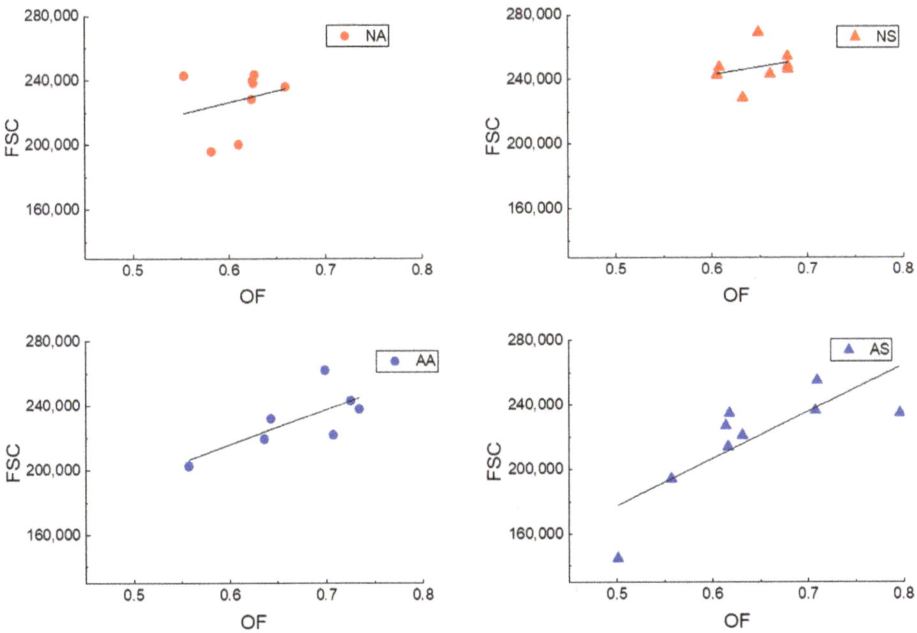

Figure 4. Linear regression analyses demonstrated the relationship of osmotic fragility and forward scatter characteristic of the normal adult group (NA), normal senior group (NS), anemic adult group (AA), and anemic senior group (AS).

4. Discussion

A total of 36 client-owned domestic shorthair cats were enrolled in the study and categorized into two groups: non-anemic cats and anemic cats. All the cats in the non-anemic group exhibited no clinical signs and demonstrated nearly all typical hematologic and blood chemistry parameters within normal ranges, except one cat in the senior group that displayed mildly elevated ALT. These findings suggest the possibility of underlying conditions that were not fully explored. Every group of cats showed thrombocytopenia and increased triglycerides. However, upon examination of platelet smears, an adequate quantity of platelets was observed, indicating that the deviation in the platelet count may not be indicative of an actual platelet deficiency, and these results should be interpreted with caution because the fasting statuses of patients at the time of sample collection were not standardized in this study, which may explain the elevated triglyceride levels observed across groups. Within the anemic group, several abnormalities were observed in hematologic and blood chemistry parameters. Notably, three cats that exhibited a decrease in platelet quantity upon examination were associated with *Mycoplasma* spp. infection. These findings are consistent with those reported in previous studies [25–27]. However, a major limitation of this study is the lack of confirmation of mycoplasmosis through polymerase chain reaction (PCR). Diagnoses in three cats were based on blood smear evaluation, which has low sensitivity and specificity. As a result, some cats classified as negative for mycoplasmosis may have had low parasitemia. We recommend that future studies incorporate PCR testing to accurately confirm mycoplasmosis status.

In terms of blood chemistry, six cats in the anemic group (33.33%, 6/18) demonstrated azotemia, indicating that a notable number of them were associated with kidney disease. A significant limitation of this study is the absence of infectious disease testing (e.g., FeLV, FIV, or other region-specific pathogens) in the initial study design, which could have affected the inclusion of cats with latent or subclinical infections.

Our research findings revealed a mean osmotic fragility (OF) of 0.586 ± 0.086 in the normal adult group, while the normal senior group exhibited a higher mean OF of 0.630 ± 0.063. Remarkably, our results indicate a higher mean OF than reported in a previous study [3,16]. Conversely, within the anemic group, the mean OF for anemic adults was 0.656 ± 0.064, and anemic seniors had a mean OF of 0.638 ± 0.088, in close agreement with a prior study [3,16]. Nevertheless, our results on osmotic fragility revealed no statistically significant differences between groups, as shown in Figure 1. This variation may be attributed to the diversity observed in osmotic fragility within the normal group, particularly in the normal senior group, which exhibited higher osmotic fragility than the normal adult cats. Contrarily, in the anemic group, some cats demonstrated very high osmotic fragility, while others exhibited low osmotic fragility. This variation suggests that osmotic fragility differs between diseases. Therefore, we suggest that further studies are necessary, and it is advisable to increase the population size for each anemia-inducing disease.

A thorough review of the literature shows that kidney disease leads to imbalances in calcium phosphate, acid-base chemistry, and electrolyte levels due to uremia, which, in turn, affects red blood cell shape and survival [28,29]. Secondary hyperparathyroidism, associated with CKD, arises from underlying hyperphosphatemia, low calcium, and reduced calcitriol levels. Parathyroid hormone (PTH) affects the osmotic fragility of RBCs by enhancing calcium entry into cells [30]. This increased calcium influx stimulates Ca^{2+}-activated ATPase, resulting in ATP depletion and erythrocyte fragmentation. Elevated PTH levels also increase RBC calcium concentrations, leading to the cross-linking of membrane proteins, which disrupts red blood cell structure and stability, ultimately causing cell lysis [31]. Secondary immune-mediated hemolytic anemia has been documented in cats with FeLV infection, hemobartonellosis, and lymphoma [32–34], all of which are associated with increased OF. Moreover, FeLV infection, inflammation, and neoplastic disease [35–37] may reduce the lifespan of RBCs in cats.

Using flow cytometry analysis, we found that the normal senior group exhibited a higher mean FSC than the normal adult group. This result contrasts with a previously conducted study in humans [38], which reported that the normal adult group had a higher FSC than the normal senior group. In terms of SSC, our findings revealed that the normal adult group exhibited a higher SSC than the anemic adult group. This finding is related to a prior study that reported a correlation between side scatter characteristics and the hemoglobin concentration of red blood cells [39]. In the senior group, we observed insignificant differences, which could be due to the predominance of kidney disease in the anemic senior group, whereas the anemic adult group was primarily associated with infectious disease, which might have a more pronounced effect on the hemoglobin concentration.

Linear regression analyses revealed a positive correlation between FSC and OF, indicating that a larger RBC size is associated with higher osmotic fragility. This finding is similar to those of a previous human study, which reported that larger RBCs are associated with increased fragility, rendering them more susceptible to damage and subsequent lysis [38,40]. In contrast with previous studies [17–19], our investigation did not identify a significant relationship between OF and triglycerides, cholesterol, or glucose levels.

The results could contribute to building a database of OF in various diseases for future studies. Additionally, these findings could aid in the development of a rapid OF test, a screening tool for each cause of anemia in patients, and the correlation between OF and RBC size and density may be used to create a predictive model for OF.

5. Conclusions

Our study revealed a higher mean OF in normal cats than reported in previous studies. However, both normal and anemic cats exhibited similar osmotic fragilities. We found that a larger RBC size is associated with higher osmotic fragility. Further studies of various anemia-inducing diseases are suggested.

Author Contributions: Conceptualization, P.L. and K.B.; methodology, C.U.; software, P.L. and K.B.; validation, P.L., K.B. and C.U.; formal analysis, W.P.; investigation, P.L. and K.B.; resources, C.U.; data curation, P.L. and K.B.; writing—original draft preparation, P.L. and K.B.; writing—review and editing, C.S., K.P. and W.P.; visualization, P.L., K.B. and C.U.; supervision, C.U., J.K. and W.P.; project administration, W.P.; funding acquisition, W.P. All authors have read and agreed to the published version of the manuscript.

Funding: This study was financially supported by the Faculty of Veterinary Medicine, Chiang Mai University.

Institutional Review Board Statement: The animal study protocol was approved by the Faculty of Veterinary Medicine Chiang Mai University Animal Care and Use Committee (FVM—ACUC) (S11/2566, 12 June 2023).

Informed Consent Statement: Informed consent was obtained from the owners of all animals involved in the study.

Data Availability Statement: The data presented in this study are available in the article. Further inquiries can be directed to the corresponding author.

Acknowledgments: The authors are deeply indebted to the owners of the feline participants for their collaboration and support. The authors sincerely thank the veterinarians and assistants at the Small Animal Hospital, Chiang Mai University, for their invaluable clinical assistance. The authors also extend their gratitude to the laboratory of the Department of Radiologic Technology, Faculty of Associated Medical Sciences, Chiang Mai University, for providing essential equipment, techniques, and expertise that were crucial to the study's success.

Conflicts of Interest: The authors declare no conflicts of interest.

References

1. Bernhardt, I.; Ellory, J. *Red Cell Membrane Transport in Health and Disease*; Springer Science & Business Media: Berlin/Heidelberg, Germany, 2003.
2. Slappendel, R.J. Abnormal Osmotic Fragility of Erythrocytes in Dogs and Cats. *Vet. Q.* **1998**, *20*, S38–S39. [CrossRef] [PubMed]
3. Tritschler, C.; Mizukami, K.; Raj, K.; Giger, U. Increased erythrocytic osmotic fragility in anemic domestic shorthair and purebred cats. *J. Feline Med. Surg.* **2016**, *18*, 462–470. [CrossRef] [PubMed]
4. Beutler, E. *Hematology*; McGraw Hill Publishing Co.: New York, NY, USA, 1990.
5. Park, S.H.; Park, C.J.; Lee, B.R.; Cho, Y.U.; Jang, S.; Kim, N.; Koh, K.N.; Im, H.J.; Seo, J.J.; Park, E.S.; et al. Comparison Study of the Eosin-5′-Maleimide Binding Test, Flow Cytometric Osmotic Fragility Test, and Cryohemolysis Test in the Diagnosis of Hereditary Spherocytosis. *Am. J. Clin. Pathol.* **2014**, *142*, 474–484. [CrossRef]
6. Udomtanakunchai, C.; Mernsri, S.; Jeejai, S.; Intachai, N.; Ruengdit, C.; Pornprasert, S. Effect of low dose X-ray on membrane fluidity of thalassemic red blood cells. *Int. J. Radiat. Res.* **2021**, *19*, 75–80. [CrossRef]
7. Perk, K.; Hort, I.; Perri, H. The degrees of swelling and osmotic resistance in hypotonic solutions of erythrocyte from various domestic animals. *Refuah Vet.* **1964**, *20*, 122.
8. Sultana, F.; Khatoon, F.; Hazari, A.H.; Qudsya, S.M.; Farheen, A. Effect of stress on erythryocytic osmotic fragility in healthy individuals. *Indian. J. Clin. Anat. Physiol.* **2017**, *4*, 30–33.

9. Wahed, A.; Quesada, A.; Dasgupta, A. Chapter 3—Red blood cell disorders. In *Hematology and Coagulation*, 7th ed.; Academic Press: Cambridge, MA, USA, 2020; pp. 31–50.
10. Ford, J. Red blood cell morphology. *Int. J. Lab. Hematol.* **2013**, *35*, 351–357. [CrossRef]
11. Newman, A.W.; Rishniw, M.; Behling-Kelly, E. Reporting and interpreting red blood cell morphology: Is there discordance between clinical pathologists and clinicians? *Vet. Clin. Pathol.* **2014**, *43*, 487–495. [CrossRef]
12. Maksimov, G.V.; Mamaeva, S.N.; Antonov, S.R.; Munkhalova Ya, A.; Kononova, I.V.; Sheikin, I.Y. Measuring Erythrocyte Morphology by Electron Microscopy to Diagnose Hematuria. *Meas. Tech.* **2016**, *59*, 327–330. [CrossRef]
13. Strijkova-Kenderova, V.; Todinova, S.; Andreeva, T.; Bogdanova, D.; Langari, A.; Danailova, A.; Krumova, S.; Zlatareva, E.; Kalaydzhiev, N.; Milanov, I.; et al. Morphometry and Stiffness of Red Blood Cells-Signatures of Neurodegenerative Diseases and Aging. *Int. J. Mol. Sci.* **2021**, *23*, 227. [CrossRef]
14. McKinnon, K.M. Flow Cytometry: An Overview. *Curr. Protoc. Immunol.* **2018**, *120*, 5.1.1–5.1.11. [CrossRef] [PubMed]
15. Kohn, B.; Weingart, C.; Eckmann, V.; Ottenjann, M.; Leibold, W. Primary immune-mediated hemolytic anemia in 19 cats: Diagnosis, therapy, and outcome (1998–2004). *J. Vet. Intern. Med.* **2006**, *20*, 159–166. [PubMed]
16. Kohn, B.; Goldschmidt, M.H.; Hohenhaus, A.E.; Giger, U. Anemia, splenomegaly, and increased osmotic fragility of erythrocytes in Abyssinian and Somali cats. *J. Am. Vet. Med. Assoc.* **2000**, *217*, 1483–1491. [CrossRef] [PubMed]
17. Jain, S.K. Hyperglycemia Can Cause Membrane Lipid Peroxidation and Osmotic Fragility in Human Red Blood Cells*. *J. Biol. Chem.* **1989**, *264*, 21340–21345. [CrossRef]
18. Cooper, R.A.; Leslie, M.H.; Knight, D.; Detweiler, D.K. Red cell cholesterol enrichment and spur cell anemia in dogs fed a cholesterol-enriched atherogenic diet. *J. Lipid Res.* **1980**, *21*, 1082–1089. [CrossRef]
19. Olorunshola, K.V.; Eze, K.O.; Achie, L.N. Effect of road transport stress on erythrocyte osmotic fragility (EOF) of healthy young adult Nigerians during the harmattan season. *Niger. J. Physiol. Sci.* **2012**, *27*, 157–164.
20. Akuzawa, M.; Matumoto, M.; Okamoto, K.; Nakashima, F.; Shinozaki, M.; Morizono, M. Hematological, osmotic, and scanning electron microscopic study of erythrocytes of dogs given beta-acetylphenylhydrazine. *Vet. Pathol.* **1989**, *26*, 70–74. [CrossRef]
21. Jain, N.C. Osmotic fragility of erythrocytes of dogs and cats in health and in certain hematologic disorders. *Cornell Vet.* **1973**, *63*, 411–423.
22. Makinde, M.O.; Bobade, P.A. Osmotic fragility of erythrocytes in clinically normal dogs and dogs infected with parasites. *Res. Vet. Sci.* **1994**, *57*, 343–348. [CrossRef]
23. Stickle, J.E. Duncan & Prasse's Veterinary Laboratory Medicine: Clinical Pathology, 4th edition. *Vet. Clin. Pathol.* **2003**, *32*, 218.
24. Schober, P.; Boer, C.; Schwarte, L.A. Correlation Coefficients: Appropriate Use and Interpretation. *Anesth. Analg.* **2018**, *126*, 1763–1768. [CrossRef] [PubMed]
25. Lobetti, R.G.; Tasker, S. Diagnosis of feline haemoplasma infection using a real-time PCR assay. *J. S. Afr. Vet. Assoc.* **2004**, *75*, 94–99. [CrossRef] [PubMed]
26. Lobetti, R.; Lappin, M.R. Prevalence of Toxoplasma gondii, Bartonella species and haemoplasma infection in cats in South Africa. *J. Feline Med. Surg.* **2012**, *14*, 857–862. [CrossRef] [PubMed]
27. Raimundo, J.; Guimaraes, A.; Rodrigues, R.; Botelho, C.; Peckle, M.; Pires, M.; Machado, C.; Santos, H.; Massard, C.; André, M.; et al. Hematological changes associated with hemoplasma infection in cats in Rio de Janeiro, Brazil. *Rev. Bras. De Parasitol. Veterinária* **2016**, *25*, 441–449. [CrossRef]
28. Christopher, M.M. Of human loss and erythrocyte survival: Uremia and anemia in chronic kidney disease. *Isr. J. Vet. Med.* **2008**, *63*, 4.
29. Wu, S.-G.; Jeng, F.-R.; Wei, S.-Y.; Su, C.-Z.; Chung, T.-C.; Chang, W.-J.; Chang, H.-W. Red Blood Cell Osmotic Fragility in Chronically Hemodialyzed Patients. *Nephron* **1997**, *78*, 28–32. [CrossRef]
30. Soldati, L.; Adamo, D.; Zerbi, S.; Caumo, A.; Spaventa, R.; Bianchi, G.; Vezzoli, G. Erythrocyte voltage-dependent calcium influx is reduced in hemodialyzed patients. *Kidney Int.* **1999**, *56*, 190–197. [CrossRef]
31. Panda, S.; Mishra, A.; Jena, M.; Rout, S.B.; Mohapatra, S. Study of Red Cell Fragility in Different Stages of Chronic Kidney Disease in Relation to Parathyroid Hormone. *J. Clin. Diagn. Res.* **2017**, *11*, BC29–BC32. [CrossRef]
32. Werner, L.L.; Gorman, N.T. Immune-mediated disorders of cats. *Vet. Clin. N. Am. Small Anim. Pract.* **1984**, *14*, 1039–1064. [CrossRef]
33. Maede, Y.; Hata, R. Studies on feline haemobartonellosis. II. The mechanism of anemia produced by infection with Haemobartonella felis. *Nihon Juigaku Zasshi* **1975**, *37*, 49–54. [CrossRef]
34. Madewell, B.R.; Feldman, B.F. Characterization of anemias associated with neoplasia in small animals. *J. Am. Vet. Med. Assoc.* **1980**, *176*, 419–425. [CrossRef] [PubMed]
35. Mackey, L.; Jarrett, W.; Jarrett, O.; Laird, H. Anemia associated with feline leukemia virus infection in cats. *J. Natl. Cancer Inst.* **1975**, *54*, 209–217. [CrossRef] [PubMed]
36. Weiss, D.J.; Krehbiel, J.D.; Lund, J.E. Studies of the pathogenesis of anemia of inflammation: Mechanism of impaired erythropoiesis. *Am. J. Vet. Res.* **1983**, *44*, 1832–1835. [CrossRef]

37. Madewell, B.R.; Jain, N.C.; Weller, R.E. Hematologic abnormalities preceding myeloid leukemia in three cats. *Vet. Pathol.* **1979**, *16*, 510–519. [CrossRef] [PubMed]
38. Huang, Y.X.; Wu, Z.J.; Mehrishi, J.; Huang, B.T.; Chen, X.Y.; Zheng, X.J.; Liu, W.J.; Luo, M. Human red blood cell aging: Correlative changes in surface charge and cell properties. *J. Cell Mol. Med.* **2011**, *15*, 2634–2642. [CrossRef]
39. Tycko, D.H.; Metz, M.H.; Epstein, E.A.; Grinbaum, A. Flow-cytometric light scattering measurement of red blood cell volume and hemoglobin concentration. *Appl. Opt.* **1985**, *24*, 1355. [CrossRef]
40. Canham, P.B. Difference in geometry of young and old human erythrocytes explained by a filtering mechanism. *Circ. Res.* **1969**, *25*, 39–45. [CrossRef]

veterinary sciences

MDPI

Article

Quantitative Analysis of 137 MRI Images in Hydrocephalic Dogs

Hao Zhuang [1], Qiqing Yang [2], Lin Zhang [3], Xiaosong Xiang [4], Dandan Geng [4], Qiyun Xie [1] and Changmin Hu [1,5,*]

[1] Department of Clinical Veterinary Medicine, College of Veterinary Medicine, Huazhong Agricultural University, Wuhan 430070, China; zhuanghao2019@163.com (H.Z.); xieqy0326@126.com (Q.X.)
[2] Shanghai AiLv Animal Hospital, Shanghai 200235, China; 13901782616@163.com
[3] Shanghai ChongQin Animal Hospital, Shanghai 200235, China; shcqpet@163.com
[4] Shanghai Miero Pet Hospital, Shanghai 200235, China; xiaosongxiang@sina.com (X.X.); dandangeng1124@sina.com (D.G.)
[5] The Veterinary Teaching Hospital, College of Veterinary Medicine, Huazhong Agricultural University, Wuhan 430070, China
* Correspondence: hcm@mail.hzau.edu.cn

Simple Summary: The purpose of this study was to evaluate the lateral ventricles of hydrocephalic patients using quantitative analysis methods. We measured the height, area, and volume of the lateral ventricles and brain in 154 dogs (17 normal and 137 hydrocephalus cases) using magnetic resonance imaging (MRI), and calculated the percentage of the ventricle measurements relative to the brain. Since there were no significant differences in area and volume between normal animals and those with hydrocephalus, while a significant difference in height was observed, we chose the maximum percentage of the ventricle height to brain height (H-max%) as the core reference indicator. In hydrocephalus cases, H-max% was correlated with the area (A-max%) and volume (V-max%).

Abstract: With the increasing popularity of dogs as pets, cases of hydrocephalus have risen significantly. Due to the ongoing challenges in the diagnosis and treatment of hydrocephalus, advancements in magnetic resonance imaging (MRI) technology have greatly enhanced the diagnostic capabilities in small animal clinical practice. Assessing ventricular size is crucial for the clinical management of hydrocephalus and other neurological disorders. However, methods for quantifying ventricular size and evaluating the severity of hydrocephalus requires further optimization. This study involved 137 hydrocephalus and 17 normal dogs. In hydrocephalus cases, the maximum percentage of the ventricle height to brain height (H-max%) was correlated with the area (A-max%) and volume (V-max%). Equations were calculated based on these findings, showing that the percentage of height can effectively represent the percentage of area and volume, which can indicate the diagnosis and monitoring of hydrocephalus prognosis.

Keywords: quantitative analysis; magnetic resonance imaging; hydrocephalus; lateral ventricles

1. Introduction

The animal brain consists of several major structures, including the cerebrum, brainstem, cerebellum, and ventricular system [1]. The ventricular system includes the left and right lateral ventricles, the third ventricle, and the fourth ventricle. In dogs, cerebrospinal fluid is produced by these four ventricles at a rate of 0.047 mL/min [2]. Hydrocephalus results from a dynamic imbalance between the production and absorption of cerebrospinal fluid, enlarging the ventricles and the subarachnoid space [3]. In 1919, Walter E. Dandy developed a preliminary classification of hydrocephalus using the dye method. Today, hydrocephalus can be categorized into congenital or acquired, communicating or obstructive, and intra-ventricular or extra-ventricular types, as referenced in humans [4].

The primary clinical symptoms of hydrocephalus include abnormal head enlargement, thinning of the skull, expansion of the fontanelles, and separation of the sutures, resulting from increased intracranial pressure. These symptoms may manifest as ataxia, vision deficits, obtundation, pain, strabismus, skull conformation, behavioral changes, disorientation, head tilt, nystagmus, and others [5]. Diagnosing hydrocephalus typically requires imaging examinations for confirmation, such as magnetic resonance imaging (MRI) [6–8]. Additionally, MRI allows for detailed studies of blood flow and cerebrospinal fluid dynamics [9,10].

Generally, a height ratio of 15% is used as a diagnostic criterion for hydrocephalus [11]. In 2001, Esteve-Ratsch studied the percentage of the ventricular area in the cerebral hemisphere of Yorkshire Terriers, classifying the asymmetry of the left versus right ventricles into three categories [12]. In 2010, Woo et al. proposed the ventricular-to-brain volume (VV/BV) percentage. Normal dogs exhibited a ventricular brain ratio (VBHR) of less than 25%, a ventricular brain area ratio (VBAR) of less than 7%, and a ventricular brain volume ratio (VBVR) of less than 5%. In contrast, dogs with hydrocephalus had a VBHR greater than 20%, a VBAR greater than 7%, and a VBVR greater than 5% [13]. In 2015, Steffi Laubner showed that a VBA index (the ratio of the ventricular width to the brain width in the coronal image) of more than 0.6, accompanied by elevation of the corpus callosum, deformation of the thalamic adhesion, periventricular edema, expansion of the olfactory recess, thinning of the sulci and subarachnoid space, or rupture of the internal capsule near the tail nucleus, is highly suggestive of clinically significant lateral ventricular expansion caused by increased intracranial pressure [14]. In humans, the Evans index is primarily used, which refers to the ratio of the widest diameter of the frontal horn of the lateral ventricle to the widest diameter of the skull in the same plane on a coronal image [15].

Previous studies have primarily focused on a limited number of samples (10–20 cases) and concentrated on specific breeds, such as German Shepherds or Yorkshire Terriers [12–14]. In this study, we conducted a retrospective analysis of 137 cases of hydrocephalus. We performed a quantitative analysis of the height, area, and volume percentage at the thalamic adhesion level using MRI images, without differentiating by the breed, sex, or age of the animals. Our findings demonstrate a strong correlation and provide new insights into the diagnosis of this condition.

2. Materials and Methods

2.1. Patient Selection and Data Acquisition

MRI images of dogs diagnosed with hydrocephalus were retrospectively analyzed from four referral central animal hospitals in Shanghai and Wuhan, China. The diagnosis was confirmed by a neurologist and imaging specialist certified by the Chinese Veterinary Medical Association. The inclusion criteria required a definitive diagnosis of hydrocephalus based on clinical symptoms (e.g., typical neurological signs such as circling or forebrain dysfunction), neurological examination, and MRI images (TR 615 ms; TE 9.5 ms; slice thickness, 3 mm/5 mm). Ventricular enlargement was assessed according to the criteria described by Laubner et al. (2015), which noted a greater proportion of intracranial volume occupied by the lateral ventricles in cases of ventriculomegaly. In total, 17 normal and 137 hydrocephalus brain MRI images, along with clinical symptoms and basic information, were collected. The study was approved by the Huazhong Agricultural University Animal Ethical and Welfare Committee, approval number HZAUDO-2024-0028. All animal owners signed the informed consent forms [14].

2.2. MRI Image Height Processing of Normal and Hydrocephalus Cases

T1 images in the DICOM (Digital Imaging and Communications in Medicine) format were imported into RadiAnt Viewer 2020.2 (64-bit) to identify the transverse images at the thalamic adhesion level (Figure 1B). The images were used to measure the height of the brain, denoted as BH (brain height); the height of the left ventricle, recorded as VHl (left ventricular height); and the height of the right ventricle, recorded as VHr (right ventricular height).

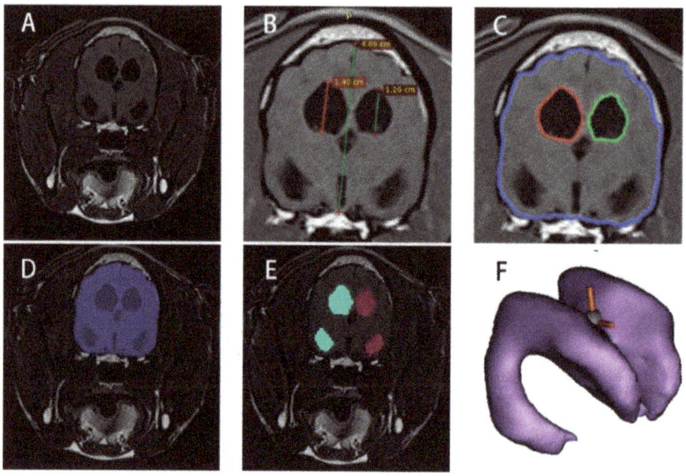

Figure 1. MRI image analysis procedure. (**A**): original MRI image; (**B**): measured heights at the thalamic adhesion level; (**C**): drawing of the ventricle and brain areas; (**D**): ROI extraction of the brain; (**E**): ROI extraction of left and right ventricles; (**F**): 3D model of the lateral ventricle.

2.3. MRI Image Area Processing of Normal and Hydrocephalus Cases

The DICOM-format T1 images were imported into ImageJ 1.52a to identify transverse images at the thalamic adhesion level (Figure 1C). The "Wand Tool" function was utilized, with the tolerance set to approximately 10 (pixel threshold algorithm). The region of interest (ROI) for the ventricles and the brain was outlined. The "Brush Selections" function was employed for detailed processing of the ROI, achieving a relatively accurate delineation of the brain area (BA) and ventricle areas (VAl, VAr). The corresponding data were calculated using the "Analyze" and "Measure" functions.

2.4. MRI Image Volume Processing of Normal and Hydrocephalus Cases

The DICOM files were imported into Mimics (Materialise Co., Ltd., Leuven, Belgium). A new mask was created using the "Threshold" to select the ROI. The ROI range was refined by editing the mask, and modification tools were employed to accurately outline the ROI, as illustrated in Figure 1D,E. The "Boolean Operations", "Region Growing", "Smart Fill", and "Intelligent Expansion" functions were used next for image modification. A 3D model was then generated and saved as an "xxx.sty" file (Figure 1F). This model was utilized to create segments and perform volume measurements for the entire brain (BV), left lateral ventricle (VVl), and right lateral ventricle (VVr).

2.5. Quantitative Analysis of Normal and Hydrocephalus Cases

The one-dimensional height, two-dimensional area, and three-dimensional volume of the left and right ventricles were quantitatively analyzed using MRI image segmentation. The percentages of the left and right ventricles to the brain were calculated and recorded as

VHl%, VHr%, VAl%, VAr%, VVl%, and VVr% for both normal and hydrocephalus cases. Additionally, the maximum percentages of height (H-max%), area (A-max%), and volume (V-max%) were determined for the hydrocephalus group.

2.6. Regression Analysis of Hydrocephalus Cases

For statistical analysis, the hydrocephalus group and the normal group were first compared using the Mann–Whitney U test for non-parametric data analysis, as the data did not meet the assumptions of normal distribution. Correlation analysis was performed using Spearman's rank correlation coefficient, for example, to assess the relationships between the maximum percentages of ventricular height, area, and volume. Regression equations were then formulated based on the strongest correlations identified. All analyses were conducted using SPSS version 26.0 and GraphPad Prism version 8.0, with $p < 0.05$ representing a significant difference.

3. Results

3.1. Retrospective Analysis of Normal and Hydrocephalus Cases

There were no significant differences in age and body weight between the normal group and the hydrocephalus group ($p > 0.05$). Similarly, there were no significant differences in the proportions of gender and the breed group. The top three clinical symptoms in the hydrocephalus group were as follows: disorientation (27.74%), ataxia (14.60%), and head tilt (9.49%) (Table 1).

Table 1. Retrospective analysis.

	Normal (n = 17)	Hydrocephalus (n = 137)	*p*-Value *
Year	6.24 ± 4.41	5.85 ± 4.77	*0.751340*
Body Weight	7.71 ± 2.82	6.13 ± 6.49	*0.325334*
Sex	Female: 8 Male: 9	Female: 73 Male: 64	
Clinical Symptoms		Disorientation: 27.74% (Dogs: 38) Ataxia: 14.60% (Dogs: 20) Head tilt: 9.49% (Dogs: 13)	

*: $p < 0.05$.

3.2. Left and Right Ventricle and Brain Heights in Normal and Hydrocephalus Cases

In a series of hydrocephalic cases, the distribution characteristics of the heights of the left and right lateral ventricles at the thalamic adhesion level were as follows: the height of the left lateral ventricle (VHl) was 8.99 ± 5.42 mm, while the height of the right lateral ventricle (VHr) was 8.59 ± 5.08 mm. The height of the brain (BH) was 31.81 ± 12.48 mm. The data for the normal group are shown in Table 2.

Table 2. Heights of left and right ventricles and brain.

	Normal (n = 17)	Hydrocephalus (n = 137)	*p*-Value *
VHl (mm)	3.17 ± 0.96	8.99 ± 5.42	*0.000020*
VHr (mm)	3.21 ± 0.92	8.59 ± 5.08	*0.000026*
BH (mm)	36.35 ± 5.20	31.81 ± 12.48	*0.140816*

Note: VHl: left ventricular height; VHr: right ventricular height; BH: brain height. *: $p < 0.05$.

3.3. Left and Right Ventricles and Brain Areas in Normal and Hydrocephalus Cases

In the hydrocephalus group, the distribution characteristics of the left and right lateral ventricle areas and the brain area at the thalamic adhesion level were as follows: the area of the left lateral ventricle (VAl) was 123.10 ± 215.00 mm^2; the area of the right lateral ventricle (VAr) was 119.80 ± 211.70 mm^2; and the area of the brain (VA) was 1747.00 ± 985.30 mm^2. The data for the normal group are presented in Table 3.

Table 3. Areas of left and right ventricles and brain.

	Normal (n = 17)	Hydrocephalus (n = 137)	*p*-Value *
VAl (mm^2)	18.13 ± 8.31	123.10 ± 215.00	0.046458
VAr (mm^2)	17.52 ± 8.59	119.80 ± 211.70	0.048792
BA (mm^2)	1279.00 ± 482.00	1747.00 ± 985.30	0.055857

Note: VAl: left ventricular area; VAr: right ventricular area; BA: brain area. *: $p < 0.05$.

3.4. Left and Right Ventricle and Brain Volumes in Normal and Hydrocephalus Cases

In the hydrocephalus group, the distribution characteristics of the volumes of the left and right lateral ventricles and the brain were as follows: the volume of the left lateral ventricle (VVl) was 2755.00 ± 6247.00 mm^3; the volume of the right lateral ventricle (VVr) was 2646.00 ± 6194.00 mm^3; and the volume of the brain (VV) was 50,906.00 ± 31,346.00 mm^3. The data for the normal group are presented in Table 4.

Table 4. Volumes of left and right ventricles and brain.

	Normal (n = 17)	Hydrocephalus (n = 137)	*p*-Value *
VVl (mm^3)	506.50 ± 386.20	2755.00 ± 6247.00	*0.141055*
VVr (mm^3)	493.10 ± 370.30	2646.00 ± 6194.00	*0.155121*
BV (mm^3)	61,134.00 ± 21,383.00	50,906.00 ± 31,346.00	*0.193463*

Note: VVl: left ventricular volume; VVr: right ventricular volume; BV: brain volume. *: $p < 0.05$.

3.5. Quantitative Analysis Results for Normal and Hydrocephalus Cases

In animals with hydrocephalus, the percentage of left ventricle height to brain height (VHl%) was 28.56 ± 10.58%, while the percentage of right ventricle height to brain height (VHr%) was 27.46 ± 9.67%. The percentage of left ventricle area to brain area (VAl%) was 6.93 ± 6.94%, and the percentage of right ventricle area to brain area (VAr%) was 6.91 ± 7.05%. For volume, the percentage of left ventricle volume to brain volume (VVl%) was 5.36 ± 5.96%, and the percentage of right ventricle volume to brain volume (VVr%) was 5.38 ± 6.60%. There were significant differences between normal and hydrocephalus cases ($p < 0.001$; Figure 2).

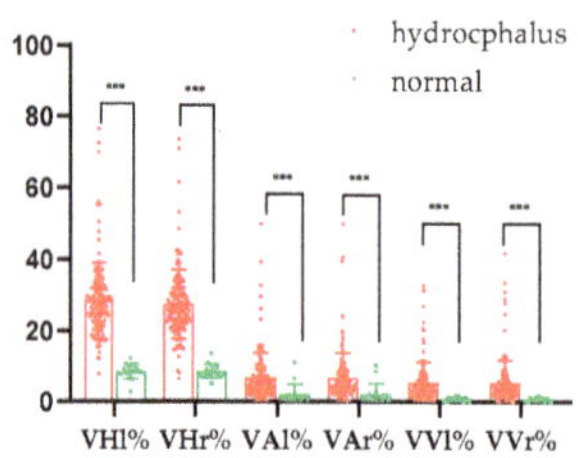

Figure 2. Percentages of height, area, and volume. VHl%: the percentage of left ventricle height to brain height; VHr%: the percentage of right ventricle height to brain height; VAl%: the percentage of left ventricle area to brain area; VAr%: the percentage of right ventricle area to brain area. VVl%: the percentage of left ventricle volume to brain volume; VVr%: the percentage of right ventricle volume to brain volume. ***: $p < 0.001$.

In animals with hydrocephalus, the maximum percentage of the ventricle height to brain height (H-max%) was 28.56 ± 10.58% (range: 15.14~76.44%), the maximum percentage of the ventricle area to brain area (A-max%) was 7.94 ± 7.54% (range: 0.76~50.00%), and the maximum percentage of the ventricle volume to brain volume (V-max%) was 6.10 ± 6.80% (range: 0.30~41.75%). Conversely, the normal animals' H-max% was 9.44 ± 1.70% (range: 6.51–13.78%), their A-max% was 2.36 ± 2.96% (range: 0.60–11.30%), and their V-max% was 0.86 ± 0.58% (range: 0.12–1.62%). There were significant differences between the normal and hydrocephalus cases ($p < 0.001$; Figure 3).

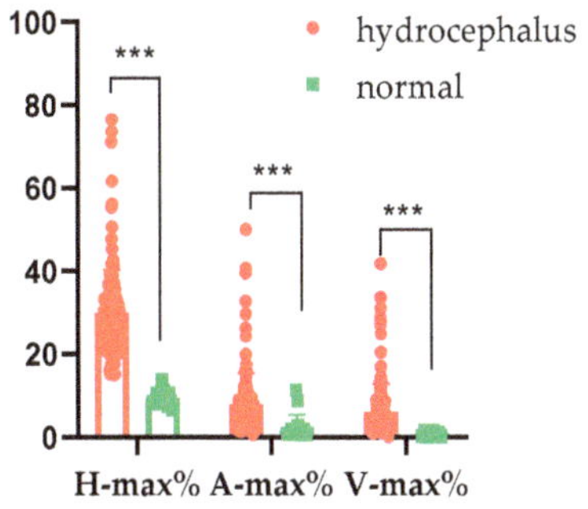

Figure 3. Maximum percentages of hydrocephalus and normal brains. H-max%: the maximum percentage of the ventricle height to brain height; A-max%: the maximum percentage of ventricle area to brain area; V-max%: the maximum percentage of the ventricle volume to brain volume. ***: $p < 0.001$.

3.6. Regression Analysis Result for the Hydrocephalus Cases

The data analysis of hydrocephalic animals showed that the correlation coefficient between the maximum percentage of height and area was R = 0.894, while the correlation coefficient between the maximum percentage of height and volume was R = 0.792. Compared with other correlation coefficients, these values represent an optimal combination (Table 5).

Table 5. Spearman correlation.

Correlation			H-Max%	A-Max%	V-Max%
Spearman	H-max%	Correlation coefficient	*1*	*0.894 ** *	*0.792 ** *
		Sig. (two-tailed)	*0.000*	*0.000*	*0.000*
		N	137	137	137
	A-max%	Correlation coefficient	*0.883 ** *	*1*	*0.777 ** *
		Sig. (two-tailed)	*0.000*	*0.000*	*0.000*
		N	137	137	137
	V-max%	Correlation coefficient	*0.792 ** *	*0.796 ** *	*1*
		Sig. (two-tailed)	*0.000*	*0.000*	*0.000*
		N	137	137	137

** Correlation is significant at the 0.01 level (two-tailed).

A linear regression model was established (Figure 4) describing the relationship between the maximum percentage of the ventricle height to brain height (X = H-max%) and the maximum percentage of the ventricle area to brain area (Y = A-max%). The established equation was as follows: $Y = -2.723 + 0.1451X + 0.006262X^2$. Additionally, the regression equation for the relationship between the maximum percentage of the ventricle height to brain height (X = H-max%) and the maximum percentage of the ventricle volume to brain volume (Y = V-max%) was $Y = 7.027 - 0.7862X + 0.03074X^2 - 0.0002134X^3$.

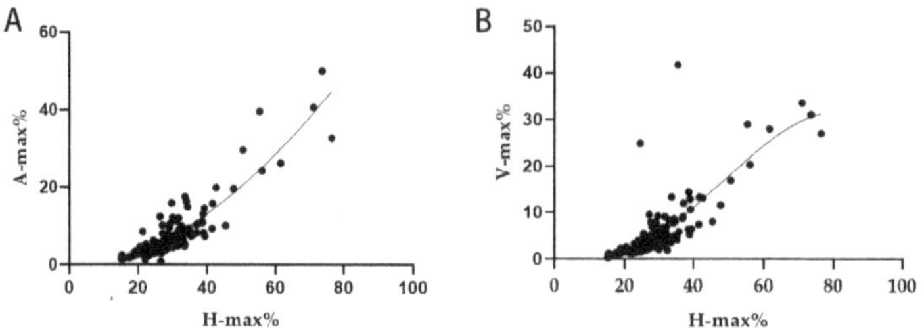

Figure 4. Regression plots of height versus area and height versus volume. H-max%: the maximum percentage of height; A-max%: the maximum percentage of area; V-max%: the maximum percentage of volume. (**A**): Height and area regression plot; (**B**): Height and volume regression plots.

4. Discussion

Hydrocephalus presents with a range of clinical symptoms, primarily characterized by neurological signs and behavioral abnormalities [16]. In the hydrocephalus group in this study, the three most common clinical symptoms were disorientation (27.74%), ataxia (14.60%), and head tilt (9.49%). The balance between cerebrospinal fluid (CSF) production and absorption was disrupted, leading to ventricular dilation, the compression of neural structures, and subsequent functional impairments. The stages of hydrocephalus development and its etiology vary, resulting in different clinical symptoms [5,14,17]. The clinical symptoms observed in this study align with findings from previous research [2,5]; however, the incidence of strabismus in our data was relatively rare. Reketa (2011) noted that hydrocephalus caused by choroid plexus tumors is defined as communicating, while others are obstructive [18]. In this retrospective study, no cases of communicating hydrocephalus were found, which may be due to its relative rarity. In obstructive hydrocephalus, the ventricles are the first to dilate, particularly in the lateral ventricular region, and this dilation may subsequently extend to the anterior and posterior horns, potentially affecting the olfactory bulbs and optic nerves [14]. In this study, disorientation (27.74%) was attributed to the dilation of the lateral ventricles compressing neural structures. As CSF flows, it accumulates in the fourth ventricle, potentially elevating the cerebellar peduncles. This accumulation ultimately affects the third ventricle, leading to reduced and flattened thalamic adhesion. Vestibular signs are associated with the dilation of the fourth ventricle [19], which accounts for the ataxia (14.60%) and head tilt (9.49%) cases in this study, indicating significant clinical relevance. Abnormalities at the cranial–cervical junction are also associated with the fourth ventricle. Additionally, severe dilation of the lateral ventricles can compress the optic nerve pathways [3]. However, no significant damage to the optic nerves was observed in this study. Increased intracranial pressure may result in neural injury, manifesting as more pronounced clinical symptoms, but intracranial pressure was not investigated in this study. Therefore, the clinical manifestations of hydrocephalus are closely related to its etiology and the location of the lesion.

Hydrocephalus can be diagnosed using various imaging techniques, including ultrasound, X-ray, CT, and MRI [16]. Ultrasound is an effective method for examining open fontanelles in animals [19]. X-ray is another useful examination for assessing skull structure, but it can sometimes produce misleading results. CT is appropriate for initial screening [19,20], while MRI, with its superior soft tissue resolution, is essential for accurate diagnosis and treatment strategy formulation [19]. Therefore, our study focused on analyzing MRIs of hydrocephalus cases to minimize errors in hydrocephalus research and improve diagnostic accuracy. In human medicine, the Evans index is a key indicator in diagnosing hydrocephalus, particularly when its value exceeds 0.3 [15,21], as it accurately reflects the enlargement of the ventricular system. Additionally, enhanced indicators such as the Z-EI, VBA, and the angle of the corpus callosum offer a more comprehensive perspective for monitoring changes in the lateral ventricular system [21,22]. However, these indicators primarily target frontal horn expansion and require specific positioning for accurate measurement. Furthermore, cases involving the isolated expansion of the fourth ventricle are often at risk of being misdiagnosed [23]. In veterinary clinical practice, these indicators are challenging to implement, and isolated expansion of the fourth ventricle is common in animal patients. Therefore, our study underscored the necessity of developing and applying diagnostic tools and indicators tailored to the unique characteristics of animals in the veterinary field to enhance the accuracy and comprehensiveness of hydrocephalus diagnosis.

This study quantitatively analyzed the lateral ventricles and measured the middle region of the left and right lateral ventricles (thalamic adhesion). Although this method may reduce sensitivity, it enhances specificity [22,24]. Using the Mimics software 21.0, we accurately segmented the images of the lateral ventricles and proposed a novel perspective: irrespective of factors such as breed, weight, and age, the height and area-to-volume ratio at the thalamic adhesion level exhibit a significant relationship, which can indirectly indicate the degree of lateral ventricular expansion [25,26]. This finding offers a new approach to the quantitative assessment of hydrocephalus. Due to the absence of standardized protocols for animal MRI scans, this study aligned the coronal plane parallel to the skull base and positioned the sagittal plane perpendicularly to measure the height and volume relationship of the ventricles at the mid-position. Out of 137 cases of hydrocephalus, no significant differences were observed in area or volume compared with normal animals, making it difficult to distinguish hydrocephalus based on these parameters (Tables 3 and 4). However, a significant difference in height was noted (Table 2). The simple measurement of height alone did not sufficiently capture the differences; therefore, the maximum percentage of ventricle height to brain height was chosen as the primary indicator [13]. In line with previous studies, the ratio of ventricular height to brain height was classified as normal size (0–14% Vh/Bh), moderately enlarged (15% to 25% Vh/Bh), or severely enlarged (>25% Vh/Bh) [11,27]. In this study, the range of maximum height percentages for hydrocephalus (15.14% to 76.44%) was significantly different from that of the normal group (6.5% to 13.18%), which is consistent with prior research. To establish a more accurate diagnostic score, a large-scale investigation of the height percentages of hydrocephalus in animals that closely resemble normal specimens is necessary. This will facilitate the construction of a receiver operating characteristic (ROC) curve to develop diagnostic standards for animals.

Based on the study by Esteve-Ratsch et al. (2001), the asymmetry of the left versus right ventricles was classified into three categories: slight (1:1.1 to 1.5), moderate (1.6 to 2:1), and extensive (>2:1) asymmetry [12]. In this study, extensive asymmetry (>2:1) was noted in 16 of the 137 cases. This finding indicates asymmetry in hydrocephalic cases, likely related to the site of blockage, the location of local lesions, and the developmental characteristics of the animals. In this study, the hydrocephalus group exhibited a significant difference in height compared with the normal group; however, the differences in area and volume were not pronounced, which is consistent with the findings of Woo et al. (2010). Compared with Woo's equation [13], the data from this study exhibit a greater skew toward the *X*-axis, and the sample size is larger than in previous studies, resulting in enhanced specificity. A significant difference can be observed at a height ratio of 15% between normal and hydrocephalic animals.

The regression equation indicates that when the degree of ventricular expansion is less than 40%, the predicted values are highly correlated with the actual values. However, when the expansion degree exceeds 40%, the actual values may deviate from the predicted values, posing challenges in assessing the degree of hydrocephalus expansion and its impact on clinical symptoms. Furthermore, while ventricular expansion and hydrocephalus should be differentiated, evidence suggests that ventricular expansion may serve as a high-risk precursor to hydrocephalus. Simultaneously, a reduction in ventricular size can improve clinical symptoms, indicating a potential connection between the two conditions. Therefore, we recommend increasing the sample size in future studies to more accurately classify and understand these diseases. With advancements in MRI technology—such as the 3D-SPACE sequence—and the application of neuro-endoscopy, the misdiagnosis rate of hydrocephalus is expected to decrease [28,29].

This study employed three-dimensional (3D) modeling technology to conduct a quantitative analysis of 137 cases of canine hydrocephalus, examining the relationship between the height, area, and volume of the lateral ventricles at the site of thalamic adhesion. The derived regression equation offers a preliminary clinical screening tool for hydrocephalus diagnosis, particularly when combined with assessments of third and fourth ventricular enlargement and extra-ventricular conditions. The Spearman's rank correlation coefficient found no significant correlation between clinical symptoms and height index (not present). While all data were clinically sourced, potential limitations include imbalances in breed, age, and gender distribution, as well as morphological variations in canine skull anatomy, which may influence the model's generalizability. Furthermore, the study's reliability is constrained by the hydrocephalus group's sample size and the limited control cohort, inherent to clinical data collection challenges. Future applications of this model should account for these variables and integrate case-specific clinical contexts to optimize diagnostic accuracy.

5. Conclusions

We determined that the maximum percentage of ventricular height to brain height (H-max%) at the thalamic adhesion level can serve as an important indicator of the volume and area associated with hydrocephalus.

Author Contributions: Conceptualization: H.Z. and C.H.; funding acquisition: C.H.; investigation: H.Z., Q.Y., L.Z., X.X., D.G., Q.X. and C.H.; methodology: H.Z. and C.H.; software: H.Z. and C.H.; validation: H.Z. and C.H.; formal analysis: H.Z. and C.H.; resources: Q.Y., L.Z., X.X., D.G., Q.X. and C.H.; data curation: H.Z., Q.Y., L.Z., X.X., D.G., Q.X. and C.H.; writing—original draft preparation: H.Z.; writing—review and editing: H.Z. and C.H.; visualization: H.Z.; supervision: C.H.; project administration: H.Z. and C.H. All authors have read and agreed to the published version of the manuscript.

Funding: This research was funded by the Distinguished Teacher Workshop of Huazhong Agricultural University (2023).

Institutional Review Board Statement: The animal use protocol was reviewed and approved by the Huazhong Agricultural University Animal Ethical and Welfare Committee with the approval number HZAUDO-2024-0028.

Informed Consent Statement: All owners of the dogs included in the study gave informed consent for the use of the clinical data gathered during the examinations.

Data Availability Statement: The data presented in this study are available upon request from the corresponding author.

Acknowledgments: The authors are grateful for the support of Xinhuai Shi, Haojun Xu, Xue Qi, and Yinhang Xu.

Conflicts of Interest: The authors declare no conflicts of interest.

References

1. Al Aiyan, A.; Balan, R.; Ghebrehiwot, E.; Mihreteab, Y.; Zerom, S.; Gebreigziabiher, S.; AlDarwich, A.; Willingham, A.L.; Kishore, U. Mapping of the exterior architecture of the mesocephalic canine brain. *Sci. Rep.* **2024**, *14*, 17147. [CrossRef] [PubMed]
2. Gillespie, S.; Gilbert, Z.; De Decker, S. Results of oral prednisolone administration or ventriculoperitoneal shunt placement in dogs with congenital hydrocephalus: 40 cases (2005–2016). *J. Am. Vet. Med. Assoc.* **2019**, *254*, 835–842. [CrossRef] [PubMed]
3. Schmidt, M.; Ondreka, N. Hydrocephalus in animals. *Pediatr. Hydrocephalus* **2019**, 53–95. [CrossRef]
4. Hochstetler, A.; Raskin, J.; Blazer-Yost, B.L. Hydrocephalus: Historical analysis and considerations for treatment. *Eur. J. Med. Res.* **2022**, *27*, 168. [CrossRef]
5. Farke, D.; Olszewska, A.; Büttner, K.; Schmidt, M.J. Association among raised intraventricular pressure, clinical signs, and magnetic resonance imaging findings in dogs with congenital internal hydrocephalus. *J. Vet. Intern. Med.* **2024**, *38*, 3119–3128. [CrossRef]
6. Farke, D.; Siwicka, A.K.; Olszewska, A.; Czerwik, A.; Büttner, K.; Schmidt, M.J. Risk factors, treatment, and outcome in dogs and cats with subdural hematoma and hemispheric collapse after ventriculoperitoneal shunting of congenital internal hydrocephalus. *J. Vet. Intern. Med.* **2023**, *37*, 2269–2277. [CrossRef]
7. Coates, J.R.; Axlund, T.W.; Dewey, C.; Smith, J. Hydrocephalus in dogs and cats. *Compend. Contin. Educ. Pract. Vet.* **2006**, *28*, 136–146.
8. Czerwik, A.; Schmidt, M.J.; Olszewska, A.; Hinz, S.; Büttner, K.; Farke, D. Reliability and interobserver variability of a grading system of ventricular distension in dogs. *Front. Vet. Sci.* **2023**, *10*, 1271545. [CrossRef]
9. Bongers, J.; Bernardini, M.; Rusbridge, C.; Boudreau, E.; Santifort, K.; Glass, E.; Shores, A.; Green, L.; Ortega, M.; Dutil, G. Clinical Findings, MRI Features, and Short-Term Outcome of Medium-to-Large Breed Dogs with Hydrocephalus. In Proceedings of the 34th Symposium of the European Society of Veterinary Neurology—Spinal Cord Injury, Palma de Mallorca, Spain, 23–24 September 2022.
10. Ishikawa, C.; Tanaka, N.; Sekiguchi, N.; Kitagawa, M.; Ito, D. Cerebrospinal fluid flow in small-breed dogs with idiopathic epilepsy observed using time-spatial labeling inversion pulse images: A preliminary study. *J. Vet. Med. Sci.* **2024**, *86*, 456–462. [CrossRef]
11. Spaulding, K.A.; Sharp, N.J. Ultrasonographic imaging of the lateral cerebral ventricles in the dog. *Vet. Radiol. Ultrasound* **1990**, *31*, 59–64. [CrossRef]
12. Esteve-Ratsch, B.; Kneissl, S.; Gabler, C. Comparative evaluation of the ventricles in the Yorkshire Terrier and the German Shepherd dog using low-field MRI. *Vet. Radiol. Ultrasound* **2001**, *42*, 410–413. [CrossRef] [PubMed]
13. Woo, D.; Choi, C.; Nam, J.; Ryu, K.; Jahng, G.; Lee, S.; Lee, D.; Kim, S.; Kim, H.; Ahn, K. Quantitative analysis of hydrocephalic ventricular alterations in Yorkshire terriers using magnetic resonance imaging. *Vet. Med.* **2010**, *55*, 125–132. [CrossRef]
14. Laubner, S.; Ondreka, N.; Failing, K.; Kramer, M.; Schmidt, M.J. Magnetic resonance imaging signs of high intraventricular pressure-comparison of findings in dogs with clinically relevant internal hydrocephalus and asymptomatic dogs with ventriculomegaly. *BMC Vet. Res.* **2015**, *11*, 181. [CrossRef]
15. Toma, A.K.; Holl, E.; Kitchen, N.D.; Watkins, L.D. Evans' Index Revisited: The Need for an Alternative in Normal Pressure Hydrocephalus. *Neurosurgery* **2011**, *68*, 939–944. [CrossRef]
16. Estey, C.M. Congenital hydrocephalus. *Vet. Clin. Small Anim. Pract.* **2016**, *46*, 217–229. [CrossRef] [PubMed]
17. Bramall, A.N.; Anton, E.S.; Kahle, K.T.; Fecci, P.E. Navigating the ventricles: Novel insights into the pathogenesis of hydrocephalus. *EBioMedicine* **2022**, *78*, 103931. [CrossRef]
18. Rekate, H.L. A consensus on the classification of hydrocephalus: Its utility in the assessment of abnormalities of cerebrospinal fluid dynamics. *Childs Nerv. Syst.* **2011**, *27*, 1535–1541. [CrossRef]
19. Thomas, W.B. Hydrocephalus in Dogs and Cats. *Vet. Clin. N. Am. Small Anim. Pract.* **2010**, *40*, 143–159. [CrossRef]
20. Langner, S.; Fleck, S.; Baldauf, J.; Mensel, B.; Kühn, J.P.; Kirsch, M. Diagnosis and Differential Diagnosis of Hydrocephalus in Adults. *Rofo* **2017**, *189*, 728–739. [CrossRef]
21. Missori, P.; Rughetti, A.; Peschillo, S.; Gualdi, G.; Di Biasi, C.; Nofroni, I.; Marinelli, L.; Fattapposta, F.; Currà, A. In normal aging ventricular system never attains pathological values of Evans' index. *Oncotarget* **2016**, *7*, 11860. [CrossRef]
22. Nakajima, M.; Yamada, S.; Miyajima, M.; Ishii, K.; Kuriyama, N.; Kazui, H.; Kanemoto, H.; Suehiro, T.; Yoshiyama, K.; Kameda, M.; et al. Guidelines for Management of Idiopathic Normal Pressure Hydrocephalus (Third Edition): Endorsed by the Japanese Society of Normal Pressure Hydrocephalus. *Neurol. Med. Chir.* **2021**, *61*, 63–97. [CrossRef] [PubMed]
23. Sarı, E.; Sarı, S.; Akgün, V.; Özcan, E.; İnce, S.; Babacan, O.; Saldır, M.; Açıkel, C.; Başbozkurt, G.; Yeşilkaya, Ş.; et al. Measures of Ventricles and Evans' Index: From Neonate to Adolescent. *Pediatr. Neurosurg.* **2015**, *50*, 12–17. [CrossRef]
24. Damasceno, B.P. Neuroimaging in normal pressure hydrocephalus. *Dement. Neuropsychol.* **2015**, *9*, 350–355. [CrossRef]
25. Przyborowska, P.; Adamiak, Z.; Zhalniarovich, Y. Quantification of cerebral lateral ventricular volume in cats by low- and high-field MRI. *J. Feline Med. Surg.* **2017**, *19*, 1080–1086. [CrossRef] [PubMed]

26. Choi, H.j.; Lee, K.J.; Ahn, S.J.; Kwon, Y.H.; Jung, K.Y.; Lee, H.C.; Lee, Y.W. Comparative evaluation of the lateral ventricles with computed tomography in Yorkshire Terrier, Maltese, and Shih-Tzu dogs. *J. Vet. Clin.* **2011**, *28*, 7–12.
27. Kii, S.; Uzuka, Y.; Taura, Y.; Nakaichi, M.; Takeuchi, A.; Inokuma, H.; Onishi, T. Magnetic resonance imaging of the lateral ventricles in beagle-type dogs. *Vet. Radiol. Ultrasound* **1997**, *38*, 430–433. [CrossRef]
28. Chellathurai, A.; Subbiah, K.; Abdul Ajis, B.N.; Balasubramaniam, S.; Gnanasigamani, S. Role of 3D SPACE sequence and susceptibility weighted imaging in the evaluation of hydrocephalus and treatment-oriented refined classification of hydrocephalus. *Indian J. Radiol. Imaging* **2018**, *28*, 385–394. [CrossRef]
29. Guil-Ibáñez, J.J.; García-Pérez, F.; Gomar-Alba, M.; Huete-Allut, A.; Narro-Donate, J.M.; Masegosa-González, J. ETV as treatment for obstructive hydrocephalus in an aneurysmal malformation of the vein of Galen in infants: Case report and review of literature. *Childs Nerv. Syst.* **2023**, *39*, 1667–1672. [CrossRef]

veterinary sciences

MDPI

Review

Impact of Antimicrobial Resistance of *Pseudomonas aeruginosa* in Urine of Small Companion Animals in Global Context: Comprehensive Analysis

Ana Pereira [1,2,†], Telma de Sousa [2,3,4,5,†], Catarina Silva [2], Gilberto Igrejas [3,4,5] and Patrícia Poeta [1,2,5,6,*]

1 CECAV—Veterinary and Animal Research Centre, University of Trás-os-Montes and Alto Douro, 5000-801 Vila Real, Portugal; anafilipalp@gmail.com
2 MicroART-Antibiotic Resistance Team, Department of Veterinary Sciences, University of Trás-os Montes and Alto Douro, 5000-801 Vila Real, Portugal; telmaslsousa@hotmail.com (T.d.S.); xusilva2002@gmail.com (C.S.)
3 Department of Genetics and Biotechnology, University of Trás-os-Montes and Alto Douro, 5000-801 Vila Real, Portugal; gigrejas@utad.pt
4 Functional Genomics and Proteomics Unit, University of Trás-os-Montes and Alto Douro, 5000-801 Vila Real, Portugal
5 Associated Laboratory for Green Chemistry, University NOVA of Lisbon, 1099-085 Caparica, Portugal
6 Veterinary and Animal Research Centre, Associate Laboratory for Animal and Veterinary Science (AL4AnimalS), University of Trás-os-Montes and Alto Douro, 5000-801 Vila Real, Portugal
* Correspondence: ppoeta@utad.pt
† These authors contributed equally to this work.

Simple Summary: Pets and humans are increasingly being infected with multidrug-resistant bacteria. *Pseudomonas aeruginosa* is a significant challenge in clinic practice due to its resistance to multiple antibiotics. This study aims to review the global prevalence and resistance patterns of *Pseudomonas aeruginosa* in small animals with urinary tract infection in order to improve management strategies for emerging MDR bacteria.

Abstract: The isolation of multidrug-resistant (MDR) bacteria from the urinary tracts of pets is increasingly common, particularly in animals with concurrent health conditions. *Pseudomonas aeruginosa* (PA) is one of the most significant antimicrobial-resistant bacteria affecting cats and dogs within the European Union (EU). This study aims to review the prevalence and antimicrobial resistance patterns of PA isolated from urine samples of small animals globally. This pathogen is known for its opportunistic infections and is a significant concern in veterinary medicine due to its inherent resistance to multiple antibiotics and its ability to acquire additional resistance mechanisms. This review seeks to enhance educational initiatives regarding the management of emerging MDR bacteria.

Keywords: *Pseudomonas aeruginosa*; urinary tract infections; small animals; resistance patterns; veterinary medicine

1. Introduction

Bacterial urinary tract infections (UTIs) are commonly diagnosed in veterinary medicine and are among the leading causes of antibiotic prescriptions [1,2]. Over the past two decades, there has been a reported increase in MDR bacteria associated with UTIs in companion animals [3]. This trend poses therapeutic challenges and public health concerns, as direct contact between humans and pets facilitates the transmission of MDR bacteria [3,4]. A wide variety of bacterial agents can cause UTIs, with *Escherichia coli* (*E. coli*) being the most frequently isolated pathogen in canines, felines, and even humans [5,6].

Other organisms, such as *Pseudomonas aeruginosa*, are less commonly found in uncomplicated UTIs, but become prominent in animals with recurrent infections and often exhibit multidrug resistance [7,8].

The prevalence and the percentage of resistance of PA to certain antibiotics associated with UTIs in cats and dogs have been poorly characterized in veterinary medicine [9].

Understanding which pathogens are prevalent in pet populations and their resistance patterns allows for more effective treatment options, reduces the risk of zoonotic infections, and facilitates the development of targeted interventions. This knowledge also supports responsible antibiotic use in veterinary medicine, helping to mitigate the growing concern of antibiotic resistance that can impact both animal and human health. By fostering a collaborative approach between veterinarians, pet owners, and public health officials, we can enhance overall community health and ensure safer environments for both pets and their human companions. Thus, the objectives of this review are to summarize the prevalence of PA in UTIs in small animals and to provide a clinical approach to managing these patients.

The classification of antibiotic resistance for PA will be carried out in accordance with Magiorakos et al. [10]. Thus, MDR (multidrug-resistant) bacteria refers to bacteria that demonstrate nonsusceptibility to at least one agent in three or more categories of antimicrobial agents. Bacteria classified as XDR are nonsusceptible to at least one agent in all categories of antimicrobial agents but remain susceptible to two or fewer categories. PDR (pan-drug-resistant) bacteria exhibit nonsusceptibility to all agents in all categories of antimicrobial agents. The term ESKAPE pathogens refers to a group of six bacterial pathogens—*Enterococcus faecium, Staphylococcus aureus, Klebsiella pneumoniae, Acinetobacter baumannii, Pseudomonas aeruginosa*, and *Enterobacter* spp.—which pose a significant threat due to their ability to escape the effects of antibiotics. The therapeutic options for infections caused by ESKAPE pathogens are increasingly limited, necessitating ongoing surveillance and the development of new antimicrobial strategies [11].

2. Methods

The research documents analyzed in this work were extracted from the Elsevier Scopus and PubMed databases. The search queries (TITLE-ABS-KEY({*Pseudomonas aeruginosa*}), (TITLE-ABS-KEY({antibiotic resistance}), and (TITLE-ABS-KEY({nosocomial infection }) were used in February 2020 for collecting academic documents and patents including "*Pseudomonas aeruginosa*" and/or "antibiotic resistance" and/or "nosocomial infection" terms in the title, abstract, and/or keywords, from the year 2000. Only publications in English were included. The articles from the search were assessed according to document type, language, and inclusion of subject categories. They were further analyzed, and the results were used to write this review. The article selection process was carried out manually, based on information extracted from the databases. No specific software was used to include the articles, although the management of the search and references was facilitated by tools such as Mendeley.

3. *Pseudomonas aeruginosa*

PA are opportunistic and ubiquitous Gram-negative bacteria that play a significant role in nosocomial infections [11,12]. This pathogen has the capacity to infect multiple anatomical sites across various animal species, including humans [12,13]. In recent years, there has been an alarming rise in MDR strains of PA [14] attributed to their ability to resist a variety of antibiotics, including aminoglycosides, quinolones, and β-lactams, as well as the emergence of new carbapenemase-producing strains [15]. The urgent need for the development of new antibiotics against these pathogens led to PA being included as an

"ESKAPE" pathogen and classified as one of the "high-priority pathogens" by the World Health Organization (WHO) [16–18].

3.1. Pathogenesis and Virulence Factors

PA is most associated with otitis in dogs; however, due to its opportunistic nature, it can also lead to a wide variety of other infections. Hattab et al. [19], Gómez et al. [13] and Dègi et al. [20] reported higher incidences of this pathogen in ear infections, skin infections, and, to a lesser extent, urinary tract infections. In cats, although infections caused by PA are less common, they predominantly affect the respiratory tract and rarely the urinary tract [21,22].

Understanding host–pathogen interaction is critical for the development of effective therapeutic strategies to control damage to the bladder [23,24]. The pathogenesis of this bacteria has not yet been clarified, but recent molecular reports have shown some advance in understanding it. Host colonization by uropathogens occurs in the lumen of the bladder within the stratified epithelium [24]. PA has an arsenal of extracellular products, such as pyocyanin, exotoxin A, and elastase, that damage the host tissue and impair the immune system response [25]. PA is a versatile pathogen known for its cytotoxic and invasive capabilities. Recent scientific evidence has shown that it can exhibit both intracellular and extracellular behaviors [24,26]. However, there is extensive evidence that PA preferentially binds to, invades, and injures wounded epithelium [27–29]. When uroepithelium is infected or traumatized, the superficial layers undergo exfoliation. As a result, the deeper layers are exposed, and may subsequently become colonized; it is within these underlying epithelial layers that uropathogens may persist in a quiescent state [24,30]. The ability of PA to survive intracellularly may play a significant role in contributing to the chronicity and recurrence of urinary tract infection [24]. Interestingly, in humans, it has been shown that *PA* UTIs result in higher readmission rates compared to UTIs caused by other uropathogens, underscoring the clinical importance of this problem [31].

PA can adapt to hostile environments within hosts by secreting a variety of virulence factors that contribute to successful infection and disease manifestation. For instance, it contains lipopolysaccharide (LPS), which protects the external leaflet and damages host cells [32]. It can promote tissue damage, facilitate attachment, and enhance recognition by host receptors. LPS has also been implicated in antibiotic tolerance and biofilm formation [33]. The outer membrane proteins (OMPs) play a significant role in nutrient exchange, adhesion, and antibiotic resistance [34].

The production of flagella and type IV pili facilitates adherence to the host and is responsible for motility through the urinary tract, enabling its colonization of the bladder and kidneys [25]. A possesses secretion systems that are vital for colonization of the host, adhesion, swimming, and swarming in response to chemotactic signals [35]. Hemolysin production enhances the ability to cause kidney infection [36].

Biofilm, a structured community of bacteria encased in a protective extracellular matrix, is a crucial virulence factor in PA. Biofilm formation by PA on catheters or within the bladder can enhance resistance to host immune responses and antibiotic treatment, making infections difficult to eradicate. Numerous studies in both human and veterinary medicine have reported that PA with biofilms exhibit resistance to various biocides, such as quaternary ammonium compounds, chlorine-based disinfectants, aldehydes such as glutaraldehyde, triclosan, benzalkonium chloride, chlorhexidine, hydrogen peroxide-based disinfectants, and 2-phenoxyethanol [15,37]. Quorum sensing systems are also crucial, as they enable PA to form biofilms and coordinate its behavior based on population density [38]. The pathogenesis of PA consists in its ability to adhere and colonize, form biofilms, produce virulence factors, evade immune responses, and remain in the host tissue [35].

3.2. The Opportunistic Behavior

The urinary tracts of dogs and cats exhibit anatomical and physiological differences that influence their susceptibility to PA infection. In dogs, the longer and narrower male urethra, along with the presence of the prostate gland, provides additional barriers to bacterial colonization, whereas the shorter female urethra facilitates bacterial ascension, increasing the risk of UTI. In cats, the male urethra is also narrow, but the prevalence of UTIs is lower due to the unique composition of the feline urobiome and the lower incidence of underlying predisposing conditions compared to dogs [39]. Significant physiological differences include urine concentration. Cats have an extraordinary ability to concentrate urine, an adaptive trait that reduces urine volume and may limit bacterial proliferation. However, this also increases the risk of crystalluria and urethral obstruction [40].

In humans, the composition of the urobiome and personal hygiene practices play a critical role in preventing urinary infections. Additionally, close contact with pets may act as a vector for MDR pathogens [41]. Studies indicate that the pathogenesis of PA in the urinary tract is closely associated with its ability to form biofilms, a characteristic observed in both humans and animals [42]. However, cats appear to mount a more effective immune response against initial PA infections, which may account for their lower incidence compared to dogs. In contrast, the predisposition of dogs to chronic conditions such as diabetes and kidney disease may contribute to their higher rates of PA infection [41].

Dogs and cats can acquire PA infections through environmental exposure (contaminated water, soil, surfaces), inhalation, or animal bites [25,43]. However, nosocomial infections are more common, reflecting their opportunistic nature. PA requires compromised epithelial barriers to invade and proliferate [44]. This explains the increased risk in immunocompromised animals, those with trauma, tumors, catheters, prolonged immobilization, and recurrent UTIs [45–47].

In humans, the presence of patients in hospitals for more than 30 days increases the risk of acquiring a UTI by almost 100% [48]. Although, in a veterinary oncology study, PA was isolated in 11% of patients, but only 3.6% were implicated in UTIs [49]. Broad-spectrum antibiotic use (cephalosporins, aminoglycosides) is a risk factor for MDR PA [30]. While PA accounts for 10–15% of human nosocomial infection, comparable data in veterinary medicine are lacking [18,50,51]. On the other hand, a study involving 32 canines with renal disease found that PA was the most frequently isolated microorganism in culture [52].

PA is a significant uropathogen in catheterized patients due to its ability to adhere to catheter surfaces and form robust biofilms which facilitate its establishment and persistence in the urinary tract [18,53]. On the other hand, urinary catheters disrupt the bladder's natural defenses, increasing susceptibility to bacterial colonization and ascending infection [5,18,54]. Bacteriuria in catheterized animals is notably high (10% to 55%), but the prevalence of PA in these cases is unknown [55,56]. A recent study by Ataya et al. [46] demonstrated a low incidence of PA in these patients but a high prevalence of PA MDR strains.

3.3. Antimicrobial Resistance

Once established on host tissue, PA employs a combination of intrinsic, acquired, and adaptive resistance mechanisms to survive and resist antibiotics (see Table 1) [15,57].

Intrinsic mechanisms include outer membrane permeability, which can be reduced by modulating porins, and the production of antibiotic-inactivating enzymes such as β-lactamases and aminoglycoside-modifying enzymes. Additionally, the efflux system plays a key role in removing harmful agents from within the bacterial cell, with the overexpression of efflux pumps being associated with MDR.

Acquired mechanisms involve genetic mutations or the acquisition of new genetic material through processes such as transduction, conjugation, and transformation, often influenced by exposure to antibiotics.

Finally, adaptive mechanisms allow PA to increase its resistance to antibiotics, contributing to persistent infections. Among these mechanisms, persistent cells, which are metabolically inactive and have a reversible phenotype, and the formation of biofilms, which facilitate the exchange of genetic material and protect bacteria from antibiotic treatment and disinfectants, stand out. Drug resistance associated with biofilm formation is linked to structures such as flagella, pili, and other adhesins [58]. Furthermore, the quorum sensing system enables PA to coordinate behaviors and form biofilms efficiently, especially in chronic and biofilm-associated infections. These mechanisms contribute to the high resistance of PA to a wide range of antibiotics, including β-lactams, aminoglycosides, fluoroquinolones, and colistin.

Recent studies have shown that prior use of antimicrobials promotes the emergence of resistant pathogens. This happens because the antimicrobial used causes selective pressure in all susceptible bacteria that are killed or inhibited, while resistant ones survive. As a result, the composition of the host's microbiome is altered, and the proportion of resistant bacteria may increase and become the majority in the host. This explains why the emergence of multidrug-resistant strains is more common in hospital environments or communities with uncontrolled use of these drugs [59]. Thus, Lin et al. [60] found resistance levels of PA in otitis and pyoderma in dogs to be lower than in other studies, attributing this to the fact that the animals studied had no history of prior antimicrobial use. On the other hand, Vingopoulou et al. [61] observed that resistance to fluoroquinolones was significantly higher in dogs that had been treated systemically with antimicrobials up to 6 months before the study.

Table 1. Resistance mechanisms of PA.

PA Resistance Mechanisms			
	Mechanisms	**Associated Antibiotic**	**References**
Intrinsic Natural mechanisms to eliminate antimicrobials	• Outer membrane (OM) permeability: decreases OM permeability by managing the number of non-specific porins present in the membrane. • Antibiotic-inactivating enzymes: produce enzymes capable of selectively inactivating or modifying antibiotics (ex:β-lactamases and aminoglycoside-modifying enzymes). • Efflux system: Remove harmful agents from the interior of the bacterial cell. The overexpression of multiple efflux pumps by this species has been associated with multidrug resistance.	β-lactams, Aminoglycosides, Fluoroquinolones	[15,17,62,63]
Acquired Implies a change in the microorganism's genetic material.	• Gene mutation • Acquisition of new DNA (transduction (by bacteriophage); conjugation (cell–cell contact); transformation (transfer of naked DNA). (It is influenced by external stressors, such as exposure to antibiotics.)	β-lactams, Aminoglycosides, Fluoroquinolones, Colistin	[15,62,64]
Adaptative Increases their ability to resist the effects of antibiotics and contributes to persistent infections.	• Persistent cells: metabolically inactive cells that exhibit tolerance to antibiotics and possess a transient, reversible phenotype. They arise under stressful conditions and are implicated in chronic bacterial infections as well as the recurrence of biofilm-associated infections • Biofilm: A structured community of bacteria embedded within a self-produced extracellular polymeric substance matrix. It can colonize natural tissues or inert materials. Facilitates the exchange of genetic material among bacteria. • Quorum sensing: Enables PA to form biofilms and coordinate behavior based on population density.	Increased tolerance to antibiotic and disinfectant treatments(>1000-fold).	[15,54,65–68]

4. Transmission

The transmission of antimicrobial-resistant bacteria has become a One Health issue as MDR bacteria can be transmitted between humans, animals, and the environment [59,69].

It is a fact that the use of antimicrobials in animals and humans promotes the development of MDR bacteria in both commensal and pathogenic bacteria [70]. In the environment, soil itself acts as a natural reservoir for antimicrobial-resistant bacteria; additionally, manure-treated soil and animal and human fecal waste contribute to the maintenance and the spread of MDR bacteria [71]. Furthermore, soils and irrigation water are sources of contamination for vegetables and fruits, where resistant bacteria have been identified [70]. Transmission by food consumption or wildlife vectors can also occur [59]. Thus, the impact of MDR bacteria (including MDR PA) is a problem that should be addressed with a One Health approach (Figure 1). In hospital environments, where antimicrobial pressure is higher, there is an increased likelihood of encountering MDR bacteria. In this context, the transmission of these pathogens can easily occur between hospital settings and the community [59]. The impact of this transmission is becoming increasingly relevant in clinical practice, especially as more people bring their pets closer to them.

Because of this approximation, MDR bacteria can be transmitted directly through close interactions between pets and humans; or indirectly through environmental transmission, where bacteria can spread when humans and pets coexist in the same household [21,52,72,73]. A study conducted by Jin et al. [69] demonstrated that AMR can be transmitted from animals to humans and vice versa, with evidence indicating that this transmission has risen over the past two decades. However, the authors note that most studies offer a low level of evidence regarding this transmission, highlighting the need for more robust studies.

Regarding PA, the information is lacking, but some studies and case reports deserve some consideration. Evidence indicates that some species, such as sheep, chinchillas, rabbits, and snakes, can be naturally infected by this microorganism [74–77]. Additionally, wildlife species also exhibit some sensitivity to PA [78–80]. The capacity of these species to act as reservoirs for PA increases the potential for the transmission of this pathogen [81,82]. According to Płókarz D. and Rypuła K. [41] dogs and cats are significant reservoirs for PA and can transmit it to humans through saliva, aerosols, urine, feces, and close contact. On the other hand, molecular studies have shown that clones associated with human infections have also been identified in animal infections; however, this does not serve as definitive evidence of transmission between species. Consequently, information regarding this transmission is still limited, but reports indicate the need for careful practices when PA is present [83,84].

Although there is no evidence of animal-to-animal transmission, the transmission between susceptible individuals (with cystic fibrosis) has been documented in humans [12,85]. Mohan et al. [21] described a case of PA respiratory infection transmitted from a human to a cat. Close contact between the individuals likely facilitated transmission, potentially via droplet spread, aerosolization, or indirect contact with contaminated surfaces [12]. The transmission of PA from animals to humans is not yet clearly understood [86], but a possible case of transmission from a dog to a young child with CF has been reported by Michl et al. [87]. Morris et al. also [88] described cross-contamination between the environment and owners and dogs with PA otitis. These studies were case reports or case series studies, suggesting that more rigorous primary studies, such as longitudinal, cohort, or case–control studies are needed to provide higher levels of evidence [69]. However, it is important to emphasize that the veterinary niche represents a potential reservoir of resistance genes for pathogenic bacteria in both animals and humans [84].

Most PA infections are primarily acquired from environmental sources rather than through direct transmission from infected individuals [12]. A study in human medicine demonstrated that environmental contamination, rather than cross-transmission, was the predominant source of infection in most cases [89]. The ability of PA to thrive in various potential niches and adapt to different hosts is concerning. This adaptability may enhance their capacity to acquire new traits related to both virulence and resistance [90]. These findings highlight the impact of this pathogen on human, animal, and environmental health. Because of this, a One Health approach could be used to address this pathogen.

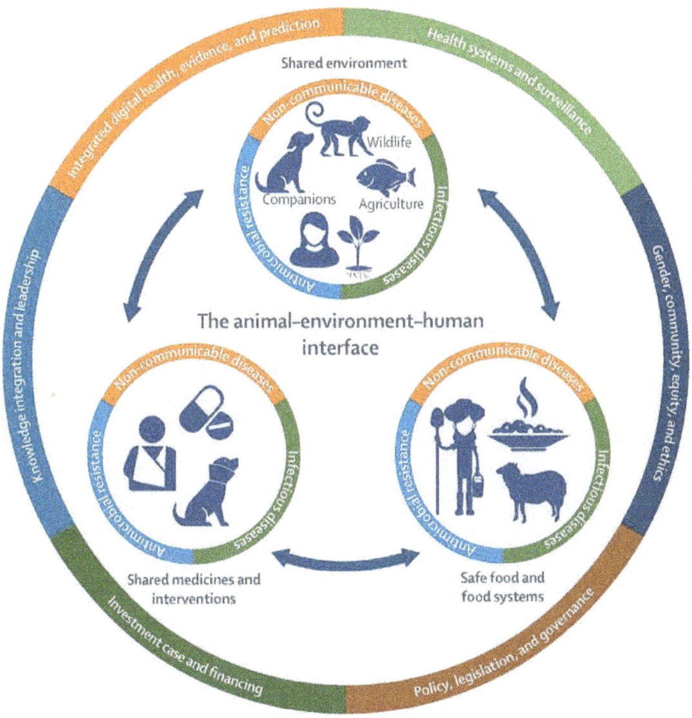

Figure 1. The importance of the One Health approach [91].

The spread of MDR bacteria can occur not only within but also between hospitals [47,92,93]. Because of this, when a PA infection occurs, it is crucial to consider all potential environments, including veterinary hospitals and homes [43,47,93]. In a Swiss study, the percentage of MDR bacteria found in clinics for companion animals was detected at 8.2% of the sampling sites [93].

To prevent the transmission of these bacteria, efforts should focus not only on reducing the use of antimicrobials, but also on establishing barriers to effectively prevent environmental contamination [90]. This can be achieved through the implementation of improved sanitation measures to contain resistant strains within healthcare settings and animal production facilities [90]. For instance, veterinary clinics with poor infection and prevention control standards showed an extensive environmental contamination [69]. The WHO has published guidelines for the "Prevention and Control of Carbapenem-Resistant Enterobacteriaceae, *Acinetobacter baumannii*, and PA in Health Care Facilities", which can be easily consulted, and which provide valuable guidance for clinical implementation [94].

5. Prevalence/Epidemiological Data

5.1. Europe

In the EU, antibiotic resistance among PA is a pressing concern, with significant implications for public health. According to The European Antimicrobial Resistance Surveillance Network (EARS-Net), the incidence of MDR PA is lower compared to other Gram-negative bacteria; however, it remains a leading cause of healthcare-associated infections across Europe [95] (Figure 2). This issue can be partly attributed to shared risk factors, including high consumption rates of broad-spectrum antimicrobials and varied infection prevention and control (IPC) practices in healthcare settings.

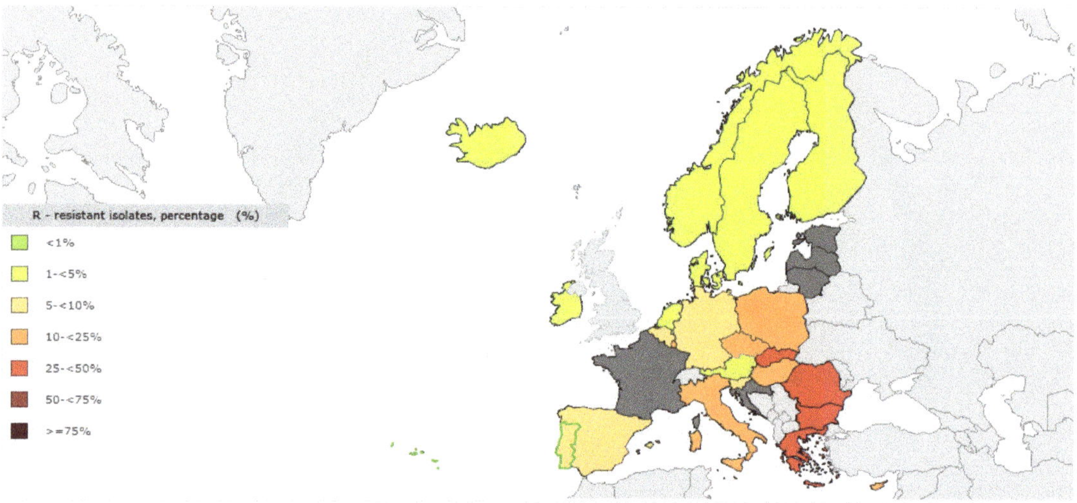

Figure 2. The distribution of MRD PA across various countries in the European Union. https://atlas.ecdc.europa.eu/public/index.aspx accessed on 30 January 2025.

A notable resistance gradient is observed across the EU, with a clear north-to-south and west-to-east pattern. The northern and western regions generally report lower resistance rates, while the eastern and southern regions show higher rates of antimicrobial resistance (Figure 1). In particular, carbapenem resistance in PA is notably more prevalent in these regions [95].

Between 2019 and 2023, the estimated incidence of PA infections resistant to piperacillin-tazobactam, ceftazidime, and carbapenems has shown significant increases, reflecting a troubling upward trend in resistance. By 2023, the highest estimated incidence of PA infections in the EU was attributed to carbapenem resistance, followed by resistance to piperacillin-tazobactam, fluoroquinolones, ceftazidime, and aminoglycosides [95].

The WHO classified carbapenem-resistant PA as a high-priority pathogen, as these agents are often utilized as last-line therapies in humans [96]. In the EU, the distribution of carbapenem resistance in PA is variable; some countries report resistance rates exceeding 50%, while others have rates below 5% [96].

In 2022, data from the European Surveillance of Antimicrobial Consumption Network (ESAC-Net) showed that 17 out of 28 countries met or exceeded the WHO's target of 60% for total antibiotic use. However, only 4 out of 15 countries that reported reached this target [97].

The slow progress towards EU targets on antimicrobial consumption, coupled with the continued rise in the use of 'Reserve' and broad-spectrum antibiotics, underscores the urgent need to strengthen efforts aimed at curbing unnecessary and inappropriate antimicrobial use across all sectors of healthcare.

5.2. Portugal

Portugal has reported resistance rates to the main classes of antimicrobials exceeding 10%, although there has been a declining trend over the years (Figure 3). However, this decline has not been observed in the group of multidrug-resistant bacteria.

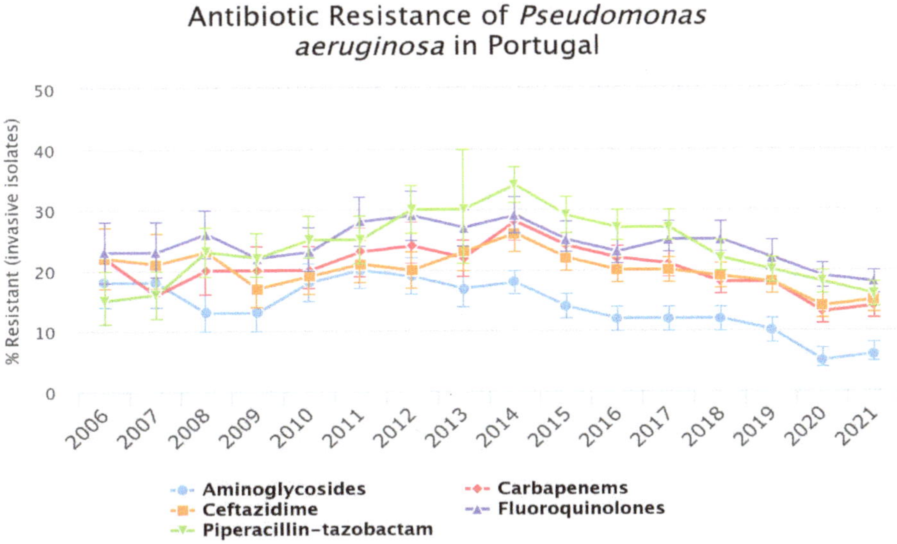

Figure 3. Antibiotic resistance of Pseudomonas in Portugal. OneHealthTrust. ResistanceMap: antibiotic resistance of *Pseudomonas aeruginosa* in Portugal. 2025. OneHealthTrust. Date accessed: 28 January 2025.

Resistance to PA in Portugal gradually increased until around 2014–2016, after which there has been a steady decrease in resistance levels (Figure 3). Interestingly, data from the European Medicines Agency (EMA) indicated that antibiotic sales in Portugal fluctuated during this period, peaking in 2016 [98]. The veterinary sales percentage ranged from 70% in 2011 to 55% in 2021. Among the antibiotics, fluoroquinolones and piperacillin-tazobactam showed the highest levels of resistance by PA in Portugal. Since 2011, the EMA has reported a significant overall decrease of 51.5% in annual antibiotics sales, including a 25% reduction in third- and fourth-generation cephalosporins, a 36.9% decline in fluoroquinolone sales, and a remarkable 76.5% reduction in polymyxin sales [98].

The EMA attributes these declines in antibiotic sales to various factors, including changes in legislation, the implementation of electronic prescriptions in January 2022, and workshops and training sessions for veterinarians and producers. It is possible that the reduction in resistance rates of PA may be connected to these decreases in antibiotic sales [98].

5.3. Animals with UTI

The prevalence of PA in UTIs in cats and dogs is unclear as of yet, and varies with geographic location, the presence of underlying conditions, and prior antibiotic use [46]. In contrast, in humans, it is well documented and is responsible for 2.4–9% of UTIs [38]. Table 2 shows the various published studies on PA in dogs with UTIs around the world.

In the United States, studies conducted across various states have reported prevalence rates ranging from 1.3% in Louisiana (2012–2014) to 3.6% in California (1969–1995), with a higher prevalence observed in female dogs and the Rottweiler breed [45,99]. However, in Illinois (2019–2020), the prevalence was 2.99%, with low resistance rates to gentamicin, piperacillin-tazobactam, amikacin, and imipenem [100].

In Europe, the prevalence of PA also varies significantly. In Italy (2013–2015), the rate was 13.4%, with gentamicin identified as the most effective antimicrobial [101]. In Portugal (1999–2016), the prevalence was only 2% [102] while in Germany (2019–2020), it was 2.1%, with 33% of isolates classified as MDR. Larger European studies conducted between 2013 and 2018 reported prevalences ranging from 3% to 4%, with high resistance levels to fluoroquinolones such as enrofloxacin and marbofloxacin [103].

In Asia, a study conducted in Chiang Mai, Thailand (2012–2016), reported a prevalence of 13.1% [4]. In Hong Kong (2018–2020), the prevalence was 4.9%, with high resistance levels to antimicrobials such as amoxicillin, ampicillin, and trimethoprim-sulfamethoxazole, while susceptibility to ciprofloxacin and ofloxacin was below 25% [9].

In Brazil (2021–2022), the prevalence was 2%, with high resistance rates to penicillin, ceftriaxone, and trimethoprim-sulfamethoxazole [104]. Conversely, in Egypt (2024), prevalence reached 12.5%, with all feline isolates classified as MDR, showing susceptibility only to fluoroquinolones and nitrofurantoin [105].

These variations can be attributed to factors such as the geographic differences, study periods, data collection methodologies, and canine populations evaluated. Furthermore, the antimicrobial resistance of PA is a growing concern in veterinary medicine. Studies indicate that this bacterium is often resistant to multiple antibiotics, including fluoroquinolones and aminoglycosides, complicating the treatment of infections in dogs [106].

In cats, the data also reveal significant variations between countries. Table 3 shows the various published studies on PA from catswith UTIs around the world.

In Australia, among the 107 samples analyzed, the prevalence was 1.6%, with resistance documented in 5 of the 14 antimicrobials tested. Susceptibility was observed for marbofloxacin and ciprofloxacin, while enrofloxacin showed intermediate results [107].

In Italy (2013–2015), the prevalence was 5.5%, with gentamicin being the most effective antimicrobial, and a documented increase in resistance over time [101]. In Spain (2016–2018), 5.2% of cats tested were positive for PA, with 4.6% being multidrug-resistant, and 0.2% XDR. Furthermore, one isolate was classified as pan-drug-resistant (PDR), with high levels of resistance ($\geq$50%) to antimicrobials such as amoxicillin-clavulanate, ampicillin, and first- to third-generation cephalosporins [1].

In Southeast Asia, in Chiang Mai, Thailand (2012–2016), the prevalence was significantly higher (25%), with all isolates classified as MDR [4]. These data are alarming, as they demonstrate a growing global trend of resistance in regions where the control of antimicrobial use and hygiene practices may be less rigorous, exacerbating the problem of bacterial resistance. In Portugal, data indicate a prevalence of 4% (1999–2016) and 1.6% (2017–2021), reinforcing the importance of local surveillance [108]. In Hong Kong (2018–2020), the prevalence was 4.8%, with 90% of isolates showing resistance to cefovecin and reduced susceptibility to fluoroquinolones and nitrofurantoin [9].

In Germany (2019–2020), 4.1% of the analyzed cats presented PA, with 41.7% MDR and four strains classified as XDR [3]. In Brazil (2021–2022), the prevalence was 2%, with high rates of resistance to trimethoprim-sulfamethoxazole, penicillin, and ceftriaxone [104]. Finally, European studies from 2013 to 2018 indicated prevalences ranging from 0% to 4.2%, with high rates of resistance to enrofloxacin and marbofloxacin [103]. Resistance to these classes of antibiotics is of particular concern because these drugs are frequently used to treat bacterial infections in animals and humans.

Table 2. Published studies on PA in dogs with UTIs around the world.

Country and Year	No. of Dogs and %PA Isolates	Resistance	Notes	References
California (EUA) 1969–1995	N = 2165 %PA = 3.6%		>Frequency of PA in dog females and Rottweiler breed	[45]
Italy 2013–2015	N = 253 %PA = 13.4%	GEN is the most effective	An increase in antimicrobial resistance over time	[101]
Louisiana (EUA) 2012–2014	N = 208 %PA = 1.3%			[99]
Spain 2016–2018	N = 52 %PA = 3.8%	≥50%—AMC, AMP, LEX, CEF, CXM, CTX, CVN, ENR, PRA DOX, FOF, NIT, SXT	In total, 4% of isolates were MDR	[1]
Chiang Mai (Thailand) 2012–2016	N = 203 %PA = 13.1%			[4]
Portugal 1999–2016	N = 700 %PA = 2%			[102]
Ilinois (EUA) 2019–2020	N = 803 %PA = 2.99%	Low resistance to GENT, TZP, AMK, IPM		[100]
Central Italy 2020	N = 450 %PA = 3.55% 93.7% MDR	AMP, NIT, TET, SXT, CPP, FQs	Susceptibility: AMK, DOX; GENT, IMI	[109]
Germany 2019–2020	N = 697 %PA = 2.1%	ENR PRA	33% MDR	[3]
Hong Kong 2018–2020	N = 2011 %PA = 4.9%	AMX, AMC, AMP, CFE, CVN, CPD, CRO, CET, DOX, NIT, SXT <25%: CIP, ENR, MAR, OFX		[9]
Brazil 2021–2022	N = 90 %PA = 2%	SXT, PEN, CRO, CTX		[104]
Romania 2022–2023	N = 83 %PA = 4%	CXM NIT		[110]
Europe 2013–2014	N = 606 %PA = 3%	1GC, TET, CVN, SXT.	Extremely high resistance levels against ENR and MAR	[103]
Europe 2017–2018	N = 773 %PA = 4%			
Cairo (Egypt) 2024	N = 146 %PA = 12.5%	AMX, 3GC, 4GC, DOX, SXT	Susceptibility: FQS and NIT 100% PA isolated from feline were MDR strains	[105]

1GC: first-generation cephalosporins; 3GC: third-generation cephalosporins; 4GC: fourth-generation cephalosporins; AMC: amoxicillin clavulanic acid; AMP: ampicillin; AMK: amikacin; AMX: amoxicillin; CEF: ceftiofur; CPP: cephalosporins; CET: cefalotin; CIP: ciprofloxacin; CPD: cefpodoxime; CRO: ceftriaxone; CTX: cefotaxime; CVN: cefovecin; CXM: cefuroxime (axetil or sodium); DOX: doxycycline; ENR: enrofloxacin; FOF: fosfomycin; FQS: fluoroquinolones; GEN: gentamicin;IMI: imipenem; LEX: cephalexin; MAR: marbofloxacin; NIT: nitrofurantoin; OFX: ofloxacin; PEN: penicillin; PRA: pradofloxacin; SXT: trimethoprim–sulfamethoxazole; TET: tetracycline; TZP: piperacillin–tazobactam.

Table 3. Published studies on PA in cats with UTIs around the world.

Country and Year	No. of Cats and %PA Isolates		Resistance	Notes	References
Austrália	N = 107	%PA = 1.6%	In total, 5/14 antimicrobials tested	Susceptible: MAR and CIP Intermediate: ENR	[107]
Italy 2013–2015	N = 81	%PA = 5.5%	GEN is the most effective	An increase in antimicrobial resistance over time	[101]
Spain 2016–2018	N = 24 4.6% MDR 0.2% XDR	%PA = 5.2%	≥50%—AMC, AMP, LEX, CEF, CXM, CTX, CVN, ENR, PRA DOX, FOF, NIT, SXT	In total, 4% of isolates were MDR and one isolate from a cat was PDR	[1]
Chiang Mai (Thailand) 2012–2016	N = 49	%PA = 25%		All isolates were MDR	[4]
Portugal 1999–2016	N = 240	%PA = 4%			[102]
Central Italy 2020	N = 185 100%MD	%PA = 6.48%	AMP, NIT, TET, SXT, CPP, FQs	Susceptibility: AMK, DOX; GENT, IMI	[109]
Germany 2019–2020	N = 313	%PA = 4.1%	ENR PRA	41.7% MDR 4 XDR e 1PDR	[3]
Hong Kong 2018–2020	N = 3708	%PA = 4.8%	AMX, AMC, AMP, CFE, CVN, CPD, CRO, CET, DOX, NIT, SXT <25%: CIP, ENR, MAR, OFX	In cats, 90% were resistant to CVN	[9]
Portugal 2017–2021	N = 5306	%PA = 1.6%			[108]
Brazil 2021–2022	N = 90	%PA = 2%	SXT, PEN, CRO, CTX		[104]
Romania 2022–2023	N = 83	%PA = 0%			[110]
European 2013–2014	N = 263	%PA = 0%	1GC, TET, CVN, SXT	Extremely high resistance levels against ENR and MAR	[103]
European 2017–2018	N = 379	%PA = 4.2%			

1GC: first-generation cephalosporins;; AMC: amoxicillin clavulanic acid; AMP: ampicillin; AMK: amikacin; AMX: amoxicillin; CFE: ceftiofur; CPP: cephalosporins; CET: cefalotin; CIP: ciprofloxacin; CPD: cefpodoxime; CRO: ceftriaxone; CTX: cefotaxime; CVN: cefovecin; CXM: cefuroxime (axetil or sodium); DOX: doxycycline; ENR: enrofloxacin; FOF: fosfomycin; FQS: fluoroquinolones; GEN: gentamicin;IMI: imipenem; LEX: cephalexin; MAR: marbofloxacin; NIT: nitrofurantoin; OFX: ofloxacin; PEN: penicillin; PRA: pradofloxacin; SXT: trimethoprim–sulfamethoxazole; TET: tetracycline.

6. Diagnosis

Diagnosis is based on a combination of clinical signs and standard bacterial culture (gold standard) to identify the pathogen [111]. Furthermore, while traditional culture takes 24–72 h, rapid methods like matrix-assisted laser desorption/ionization time-of-flight mass spectrometry (MALDI-TOF MS) and genomic analysis provide results within minutes to an hour, enabling timely treatment [112,113]. Antimicrobial susceptibility testing (AST), including MIC and disk diffusion methods, is crucial for choosing the best treatment [114]. Biofilm-associated infections, particularly those caused by PA, can yield negative culture results due to bacterial aggregation within the biofilm matrix [115,116]. Newer molecular techniques like 16S rRNA PCR, LAMP, FISH, and CLSM offer promising alternatives for detecting biofilm-associated pathogens [116,117].

During cystotomy for urolithiasis, bladder mucosal biopsies should be cultured alongside urolith analysis to identify potential uropathogens missed by standard urine cultures [118]. PCR testing can characterize bacteria and detect resistance genes, aiding antimicrobial selection [119,120]. However, this approach is not routinely used for PA.

7. Treatment

The treatment should always be guided by culture and susceptibility testing; however, initial empirical treatment may be necessary. For uncomplicated UTIs, the International Society for Companion Animal Infectious Diseases (ISCAID) guidelines recommend amoxicillin/amoxicillin-clavulanic acid or trimethoprim-sulfamethoxazole as first-line options [5]. In PA infections, due to its intrinsic resistance mechanisms, only third-generation cephalosporins, fluoroquinolones, or aminoglycosides are considered suitable for UTI treatment [3]. While carbapenems are effective against severe nosocomial PA infections, their use should be reserved for human medicine [121]. Aminoglycoside use in UTIs is limited by nephrotoxicity, parenteral administration requirements, reduced effectiveness in acidic urine, and inactivation by purulent debris [122]. For complicated UTIs, nitrofurantoin, fluoroquinolones, and third-generation cephalosporins may be reasonable first-line choices. In pyelonephritis, veterinary fluoroquinolones or cefpodoxime are considered appropriate [5]. In catheterized animals, systemic antibiotics are recommended, and catheter removal or replacement should be considered prior to initiating therapy. Inadequate management of UTIs, especially in catheterized animals, increases the risk of bacteremia [50].

Some studies have investigated antimicrobial combinations or novel agents to address the challenges in treating PA, particularly those involving biofilms. Saini et al. [66] demonstrated the potential efficacy of azithromycin combined with ciprofloxacin in treating biofilm-associated urinary tract infections. Novel antibiotics such as doripenem (a carbapenem) and plazomicin (a next-generation aminoglycoside) may offer future therapeutic options [123].

Treatment success can be challenging, and recurrent infections following treatment cessation have been reported [124]. However, data on the specific rate of treatment failure in dogs and cats is limited [12].

Alternative Therapeutic Strategies

Excessive antibiotic use contributes to the emergence and dissemination of MDR organisms, including PA [123]. This necessitates the exploration and development of alternative therapeutic strategies. Several promising approaches are summarized in Table 4. However, the translation of many of these innovative strategies into clinical practice remains limited due to concerns regarding cost, potential side effects, and overall safety [123].

The analysis of alternative therapeutic approaches for the treatment of PA reveals a diverse array of strategies, each with varying levels of efficacy according to the reviewed studies. The AMP2041 peptide has shown promising results against MDR strains in both humans and animals, with inhibition rates exceeding 80% in some studies [125]. Fruit extracts, such as those from *Cornus mas* L. and *Sorbus aucuparia*, have demonstrated a significant reduction in bacterial growth, with an efficacy of approximately 60–70% in UTIs in companion animals. Nanoparticles have excelled in reducing biofilm formation, achieving success rates above 70% [110,126].

Phillyrin, a quorum sensing inhibitor, has exhibited up to an 80% reduction in pathogen virulence in experimental models [127]. Phage therapy has shown exceptionally high efficacy, with up to 90% success in eradicating skin infections in dogs models [60,128]. Protein Epitope Mimics, while promising, still require further evaluation, showing an efficacy of around 60–70% in inhibiting resistant strains. Although vaccines against PA are not yet widely available, experimental vaccines have demonstrated up to 70% efficacy in preventing infections [129,130]. Additionally, the use of electrochemical structures to combat biofilms and enhance antibiotic penetration has proven effective, achieving success rates of 75–80% [131].

Alternative strategies for the treatment of PA reveal promising advances in combating infections caused by this MDR pathogen. Despite differences in reported efficacy rates, which range between 60% and 90% depending on the approach and experimental context, these results highlight the importance of exploring alternatives to traditional antibiotic-based therapies. However, there is still a need for additional studies to validate the clinical applicability of these strategies on a large scale and in different epidemiological scenarios.

Table 4. Published research on alternative strategies for antimicrobial treatment.

Alternative Strategy	Description	References
Peptide Amp 2041	Positive results in human and animal research.	[125]
Fruit extracts (*Cornus mas* L. and *Sorbus aucuparia*)	Might be used as substitutes or adjuvants for antibiotics in UTI isolates of companion animals.	[110]
Nanoparticles	Promising in reducing PA growth as well as in the formation of its biofilm.	[126]
Phillyrin	It is an effective inhibitor of quorum sensing with potential for PA infection therapy.	[127]
Phages	In a study on dogs with PA skin infection, the application of phages was greatly effective in the control of this pathogen.	[60,128]
PEM (protein epitope mimetic)	It has emerged as a novel class of antibiotics against PA. They inhibit the transport of LPS to the outer membrane.	[123]
Vaccines	There is not a PA vaccine currently available on the market, although several different vaccines and several monoclonal antibodies have been developed in recent decades.	[129,130]
Electrochemical scaffold	Disruption of bacterial biofilms and increase in antibiotic penetration	[131]

8. Discussion

UTIs are among the most prevalent medical conditions affecting domestic cats and dogs. These infections not only pose significant health risks, but also contribute to the emergence and proliferation of antibiotic resistance [46].

While there is extensive literature on the antimicrobial resistance patterns of PA in human medicine, data for this in veterinary medicine have been poorly characterized, particularly in UTIs [9]. These findings may be explained by the greater impact of PA on human health compared to that on small animals. For instance, this pathogen has been associated with high morbidity and a substantial mortality rate in human medicine, accounting for 3.57 million deaths in 2019 [16].

PA is frequently isolated from ear and wound swabs in dogs and from the nasal cavity in cats [22]; this may explain the limited number of studies focusing on the prevalence of PA in the urinary tract in these species.

With this review, we aim to provide insights into the prevalence of urinary tract infections in dogs and cats, as well as how this has changed over the years. However, due to the aforementioned reasons, the data available for clarification are still limited. Nonetheless, we believe that, given its impact on public health, further investigation into this topic should not be overlooked. Tables 2 and 3 present a compilation of studies from around the world that document the isolation of PA in UTIs in dogs and cats, respectively.

The prevalence of PA varies over the years and across different countries in these studies. However, when compared to other pathogen isolates, the prevalence of PA remains low in both species (varies between 1.3% and 13.4% in dogs and 0 and 25% in cats). These findings may be attributed to the opportunistic nature of PA, which only affects a small group of patients (immunocompromised animals). Furthermore, the bacterium's ability to form biofilms may lead to an underestimation of its presence in culture results, potentially resulting in false negatives [116]. This underscores the need for further research to enhance our understanding of the epidemiology of PA in UTIs among small animals. We believe that understanding the prevalence of PA among pets is crucial from both veterinary and human medicine perspectives [83].

European studies from 2013 to 2018 indicated prevalences ranging from 0% to 4.2% in cats and 3–4% in dogs (Tables 2 and 3). According to Figure 2, in 2023 the percentage of MDR PA varies greatly from country to country, ranging from 1% to 50%.

E. coli is the most isolated pathogen in UTIs [5,6,132] in contrast, a study conducted in Chiang Mai reported PA as the predominant isolate in feline UTIs (25% of the isolates) [4]. This discrepancy between the microbial profiles observed in Thailand and those in other countries may be attributable to variations in empirical antimicrobial prescription practices, which can create selective pressure on bacterial populations, or can be justified by natural geographical differences [7].

The number of dogs with suspected UTIs was higher than that of cats (Tables 2 and 3). This observation may be attributed to a greater number of dogs visiting veterinary clinics or the ease of obtaining samples from canines [108]. These findings are also consistent with numerous studies that report a higher incidence of non-infection urinary tract diseases (idiopathic and inflammatory) in cats [133,134]. In fact, according to Rodriguez [135] and Heseltine [134], less than 10% of cats with urinary clinical signs have a positive urine culture. In a study by to Płókarz D. and Rypuła K. [41] dogs presented higher frequencies of PA (92%) than cats (72%), but these data considered samples from the nasal cavity, ear, and wounds. Perhaps cats are more resistant to PA than dogs, which could explain the differences observed between the two species regarding this bacterium in these reports. But more studies are needed to determine this.

In the context of UTI, sufficient data are not yet available. The data regarding UTIS in dogs and cats cannot be compared based on the information provided in Tables 2 and 3. This is primarily due to the significant differences in the population sizes of cats and dogs included in the respective studies, which may influence the observed prevalence rates. Therefore, without more balanced data or additional studies specifically targeting PA prevalence in UTIs among both species, it remains unclear whether the trend observed in other samples would hold true for UTIs [30]. On the other hand, different methods were employed to assess antibiotic susceptibility, and the years of the studies also varied. These factors are crucial for the interpretation of results, as the guidelines for interpretation PA sensitivity have undergone changes. In humans, the elderly and women appear to be more susceptible to PA UTI. In companion animals, bacterial UTIs are also more common in females, and age is a significant risk factor as well. In a previous study, Norris et al. [45] found a higher frequency of PA in female dogs and in the Rottweiler breed (Table 2) [136].

Some studies [103,137] in Tables 2 and 3 indicated a gradual increase in the prevalence of MDR pathogens over the years [3,14]. If we compare with EARS-NET data, we can observe a similar trend. A significant isolation of MDR PA strains was observed across all the studies. Included in this, Gómez et al. [13] and Darwich et al. [1] identified PA as one of the bacterial species with the highest levels of MDR strains. These findings underscore the versatility of PA, attributed to its diverse resistance mechanisms, the presence of numerous virulence factors, and its capacity to form biofilms [14]. At same time, these findings have enhanced the overuse and misuse of antibiotics over the years and the need to improve antimicrobial prescription procedures [138,139]. Likewise, the EMA and ESAC-Net report a high consumption of antibiotics in recent years in Europe.

The prevalence of AMR strains was found to be higher in cats compared to dogs [13]. Similar results were also obtained by Gómez-Beltránthe authors of [13]. Although there are currently no publications addressing this disparity, it is plausible that differences in the urobiome composition between these species may contribute to this observation. Additionally, the nature of urinary diseases in cats, which are often inflammatory, may facilitate the growth of MDR PA in the urinary tract. Further studies are necessary to elucidate these relationships and to better understand the underlying factors contributing to the observed resistance patterns in this bacterium. Likewise, PDR strains were isolated exclusively from feline species in Spain and in Germany. These results are frightening, as in these cases clinicians did not have any effective antimicrobials to eliminate this pathogen [3].

Damborg et al. [43] propose that pet-associated bacterial zoonoses represent a relatively neglected area of study compared with food borne zoonoses, and we support this assertion due to the lack of information available on the topic of PA in companion animals. While the precise transmission routes of PA are not fully understood, some studies suggest potential transmission between humans, animals, and the environment. Therefore, a One Health approach (Figure 1) is crucial for effectively addressing these infections [14]. Robust biosecurity and infection control measures are essential for combating the spread of PA, particularly in hospital environments [140]. However, controlling PA presents significant challenges due to its inherent resistance to certain biocides and its ability to adapt and survive in challenging environments [15].

The innate resistance of PA contributes to its reduced susceptibility to beta-lactams, cephalosporins, trimethoprim, and sulfonamides [1,103]. As shown in Tables 2 and 3, the levels of resistance to these antimicrobials were consistently high across all studies. Additionally, elevated resistance levels were also observed for tetracyclines and cefovecin [1,103]. The resistance patterns to fluoroquinolones and aminoglycosides displayed significant variability among the studies. This variation may be attributed to factors such as geographical location, differences in prescription practices, or the specific populations studied [105].

According to ISCAID guidelines, nitrofurantoin, fluoroquinolones, and third-generation cephalosporins may be reasonable first-line choices for complicated or resistant UTIs [5]. However, these studies revealed a wide range of resistance patterns for these antibiotics. For instance, fluoroquinolones demonstrated efficacy in populations in Europe, Illinois, and Egypt, but were ineffective in a feline population in Germany. Gentamicin exhibited consistent efficacy across species and locations in most studies, likely due to careful dosing practices in veterinary settings to minimize toxicity [109]. These findings suggest geographical consistency in PA susceptibility patterns [103]. Notably, high resistance rates to marbofloxacin and enrofloxacin were observed in European studies, potentially attributed to the widespread use of fluoroquinolones in these species, particularly for UTIs [141–143]. Likewise, the data reported by EARS-Net show the highest incidence of PA infections in the EU which was attributed to carbapenem resistance, followed by piperacillin-tazobactam, and fluoroquinolones. Although PA resistance to carbapenems was generally low in companion animal studies, they were not evaluated in all studies. Because of this, the antimicrobial profiles of *PA* isolates are difficult to compare between different studies [83]. But in the Illinois study, some isolates were found to be resistant to imipenem [83,100]. In Europe, the eastern and southern regions show higher rates of carbapenem antimicrobial resistance. It may be interesting to introduce this drug in future studies in these countries.

It is important to note that the in vitro susceptibility results may not always accurately predict the clinical outcomes. Some antibiotics achieve significantly higher concentrations in urine compared to plasma, which can enhance their efficacy in treating UTIs despite in vitro resistance [111]. Understanding in vivo sensitivity could clarify treatment failure rates for this versatile pathogen. In human medicine, the relationship between multidrug resistance and clinical outcomes also remains unclear [144].

The optimal treatment approach should be guided by culture and AST results [139,141]. However, in clinical practice, constraints such as time, financial limitations, and physical resources can impede the implementation of this standard [139]. Consequently, clinicians may resort to empirical antimicrobial therapy. While necessary in some cases, this practice can contribute to the emergence and spread of antimicrobial resistance [101]. A study conducted in an Italian veterinary teaching hospital found that only 5% of antimicrobial prescriptions were based on microbiological culture and susceptibility testing [138]. Additionally, Robbins et al. [141] and Mouiche et al. [142] reported a substantial number of cases where antimicrobials were prescribed unnecessarily. Additionally, data from ESAC-NET reveal that antimicrobial consumption in most EU countries exceeds the target set by the WHO. These results highlight the need for an antimicrobial control system.

Inadequate regulatory oversight and ineffective surveillance of antibiotic prescriptions contribute to the emergence and spread of MDR in pets with UTIs [145]. For example, the implementation of an online antimicrobial stewardship program in Switzerland significantly reduced antibiotic prescriptions for UTIs in dogs and cats [146,147]. Urgent action is needed to improve the surveillance and monitoring of antimicrobial resistance in veterinary medicine [145,148]. We believe that effective antimicrobial stewardship strategies include educational initiatives, clinical guidelines, and computer-based decision support systems [141,145].

Due to this paradigm, future treatments may focus less on eradicating pathogens and more on maintaining the integrity of the urinary tract barrier [30]. Concurrently, research is actively exploring alternative strategies for eliminating pathogens, as summarized in Table 4. While these approaches offer promising potential, further research is necessary to validate their efficacy and safety. We hope that this review serves as a foundation for future studies aimed at investigating the prevalence of PA in UTIs, improving monitoring techniques, and developing effective strategies for its elimination.

9. Conclusions

Current understanding of PA prevalence and its resistance patterns in canine and feline UTIs is limited. Existing research suggests a low overall prevalence of PA in UTIs, but the incidence of MDR strains appears to be alarmingly high. Further investigation is crucial, particularly regarding feline UTIs. PA antimicrobial resistance patterns varied geographically and temporally. Therefore, the empirical antimicrobial agents suggested by the ISCAID guidelines are not always effective against isolated strains. This highlights the importance of bacterial culture. which should be performed whenever UTIs are suspected to minimize antimicrobial overuse. Future research can include the correlation of MDR strains with clinical outcomes in order to clarify their impact on treatment success. Future studies should also consider more epidemiological data on animals with PA infection. Genomics has proven to be a powerful tool for understanding PA and guiding targeted antibiotic therapy. The exploration of alternative therapies to antibiotics can also be a useful tool to minimize their impact. In any case, a collaborative effort (human, animal, and environment) is essential to eliminate this pathogen (a One Health approach).

Implementing antimicrobial stewardship programs is increasingly urgent, and research like this can contribute to enhanced monitoring efforts.

Author Contributions: Conceptualization, A.P.; investigation, A.P.; formal analysis, A.P. and T.d.S.; data curation, A.P., T.d.S. and C.S.; writing—original draft preparation, A.P., T.d.S. and C.S.; supervision, G.I. and P.P. All authors have read and agreed to the published version of the manuscript.

Funding: This research was funded by FCT (Fundação para a Ciência e a Tecnologia), associated with a Ph.D. grant with the reference 2020.05332.BD, and the projects UI/00772 and LA/P/0059/2020.

Institutional Review Board Statement: Not applicable.

Informed Consent Statement: Not applicable.

Data Availability Statement: Not applicable.

Acknowledgments: This work received support and help from FCT/MCTES (LA/P/0008/2020 DOI 10.54499/LA/P/0008/2020, UIDP/50006/2020 DOI 10.54499/UIDP/50006/2020, and UIDB/50006/2020 DOI 10.54499/UIDB/50006/2020) through national funds.

Conflicts of Interest: The authors declare no conflicts of interest.

References

1. Darwich, L.; Seminati, C.; Burballa, A.; Nieto, A.; Durán, I.; Tarradas, N.; Molina-López, R.A. Antimicrobial Susceptibility of Bacterial Isolates from Urinary Tract Infections in Companion Animals in Spain. *Vet. Rec.* **2021**, *188*, e60. [CrossRef] [PubMed]
2. Hernando, E.; Vila, A.; D'Ippolito, P.; Rico, A.J.; Rodon, J.; Roura, X. Prevalence and Characterization of Urinary Tract Infection in Owned Dogs and Cats from Spain. *Top. Companion Anim. Med.* **2021**, *43*, 100512. [CrossRef] [PubMed]
3. Aurich, S.; Prenger-Berninghoff, E.; Ewers, C. Prevalence and Antimicrobial Resistance of Bacterial Uropathogens Isolated from Dogs and Cats. *Antibiotics* **2022**, *11*, 1730. [CrossRef] [PubMed]
4. Amphaiphan, C.; Yano, T.; Som-in, M.; Kungwong, P.; Wongsawan, K.; Pusoonthornthum, R.; Salman, M.D.; Tangtrongsup, S. Antimicrobial Drug Resistance Profile of Isolated Bacteria in Dogs and Cats with Urologic Problems at Chiang Mai University Veterinary Teaching Hospital, Thailand (2012–2016). *Zoonoses Public Health* **2021**, *68*, 452–463. [CrossRef] [PubMed]
5. Weese, J.S.; Blondeau, J.; Boothe, D.; Guardabassi, L.G.; Gumley, N.; Papich, M.; Jessen, L.R.; Lappin, M.; Rankin, S.; Westropp, J.L.; et al. International Society for Companion Animal Infectious Diseases (ISCAID) Guidelines for the Diagnosis and Management of Bacterial Urinary Tract Infections in Dogs and Cats. *Vet. J.* **2019**, *247*, 8–25. [CrossRef]
6. Kocúreková, T.; Koščová, J.; Hajdučková, V. Infections of the Urinary Tract of Bacterial Origin in Dogs and Cats. *Folia Vet.* **2021**, *65*, 59–66. [CrossRef]
7. Wong, C.; Epstein, S.E.; Westropp, J.L. Antimicrobial Susceptibility Patterns in Urinary Tract Infections in Dogs (2010–2013). *J. Vet. Intern. Med.* **2015**, *29*, 1045–1052. [CrossRef]
8. Johnstone, T. A Clinical Approach to Multidrug-Resistant Urinary Tract Infection and Subclinical Bacteriuria in Dogs and Cats. *N. Z. Vet. J.* **2020**, *68*, 69–83. [CrossRef]
9. Chan, O.S.K.; Baranger-Ete, M.; Lam, W.W.T.; Wu, P.; Yeung, M.; Lee, E.; Bond, H.; Swan, O.; Tun, H.M. A Retrospective Study of Antimicrobial Resistant Bacteria Associated with Feline and Canine Urinary Tract Infection in Hong Kong SAR, China—A Case Study on Implication of First-Line Antibiotics Use. *Antibiotics* **2022**, *11*, 1140. [CrossRef]
10. Magiorakos, A.P.; Srinivasan, A.; Carey, R.B.; Carmeli, Y.; Falagas, M.E.; Giske, C.G.; Harbarth, S.; Hindler, J.F.; Kahlmeter, G.; Olsson-Liljequist, B.; et al. Multidrug-Resistant, Extensively Drug-Resistant and Pandrug-Resistant Bacteria: An International Expert Proposal for Interim Standard Definitions for Acquired Resistance. *Clin. Microbiol. Infect.* **2012**, *18*, 268–281. [CrossRef]
11. Bernal-Rosas, Y.; Osorio-Muñoz, K.; Torres-García, O. *Pseudomonas aeruginosa*: An Emerging Nosocomial Trouble in Veterinary. *Rev. MVZ Cordoba* **2015**, *20*, 4937–4946. [CrossRef]
12. Nielsen, S.S.; Bicout, D.J.; Calistri, P.; Canali, E.; Drewe, J.A.; Garin-Bastuji, B.; Gonzales Rojas, J.L.; Gortázar, C.; Herskin, M.; Michel, V.; et al. Assessment of Listing and Categorisation of Animal Diseases within the Framework of the Animal Health Law (Regulation (EU) No 2016/429): Antimicrobial-Resistant *Pseudomonas aeruginosa* in Dogs and Cats. *EFSA J.* **2022**, *20*, e07310. [CrossRef] [PubMed]
13. Gómez-Beltrán, D.A.; Villar, D.; López-Osorio, S.; Ferguson, D.; Monsalve, L.K.; Chaparro-Gutiérrez, J.J. Prevalence of Antimicrobial Resistance in Bacterial Isolates from Dogs and Cats in a Veterinary Diagnostic Laboratory in Colombia from 2016–2019. *Vet. Sci.* **2020**, *7*, 173. [CrossRef] [PubMed]
14. De Sousa, T.; Hébraud, M.; Enes Dapkevicius, M.L.N.; Maltez, L.; Pereira, J.E.; Capita, R.; Alonso-Calleja, C.; Igrejas, G.; Poeta, P. Genomic and Metabolic Characteristics of the Pathogenicity in *Pseudomonas aeruginosa*. *Int. J. Mol. Sci.* **2021**, *22*, 12892. [CrossRef] [PubMed]
15. Verdial, C.; Serrano, I.; Tavares, L.; Gil, S.; Oliveira, M. Mechanisms of Antibiotic and Biocide Resistance That Contribute to *Pseudomonas aeruginosa* Persistence in the Hospital Environment. *Biomedicines* **2023**, *11*, 1221. [CrossRef]

16. WHO. *WHO Bacterial Priority Pathogens List, 2024: Bacterial Pathogens of Public Health Importance to Guide Research, Development and Strategies to Prevent and Control Antimicrobial Resistance*; WHO: Geneva, Switzerland, 2024; Volume 72.
17. Barbu, I.C.; Gheorghe-Barbu, I.; Grigore, G.A.; Vrancianu, C.O.; Chifiriuc, M.C. Antimicrobial Resistance in Romania: Updates on Gram-Negative ESCAPE Pathogens in the Clinical, Veterinary, and Aquatic Sectors. *Int. J. Mol. Sci.* **2023**, *24*, 7892. [CrossRef]
18. Elfadadny, A.; Ragab, R.F.; AlHarbi, M.; Badshah, F.; Ibáñez-Arancibia, E.; Farag, A.; Hendawy, A.O.; De los Ríos-Escalante, P.R.; Aboubakr, M.; Zakai, S.A.; et al. Antimicrobial Resistance of *Pseudomonas aeruginosa*: Navigating Clinical Impacts, Current Resistance Trends, and Innovations in Breaking Therapies. *Front. Microbiol.* **2024**, *15*, 1374466. [CrossRef]
19. Hattab, J.; Mosca, F.; Di Francesco, C.E.; Aste, G.; Marruchella, G.; Guardiani, P.; Giorgio Tiscar, P. Occurrence, Antimicrobial Susceptibility, and Pathogenic Factors of *Pseudomonas aeruginosa* in Canine Clinical Samples. *Vet. World* **2021**, *14*, 978. [CrossRef]
20. Dégi, J.; Moțco, O.A.; Dégi, D.M.; Suici, T.; Mareș, M.; Imre, K.; Cristina, R.T. Antibiotic Susceptibility Profile of *Pseudomonas aeruginosa* Canine Isolates from a Multicentric Study in Romania. *Antibiotics* **2021**, *10*, 846. [CrossRef]
21. Mohan, K.; Fothergill, J.L.; Storrar, J.; Ledson, M.J.; Winstanley, C.; Walshaw, M.J. Transmission of *Pseudomonas aeruginosa* Epidemic Strain from a Patient with Cystic Fibrosis to a Pet Cat. *Thorax* **2008**, *63*, 839–840. [CrossRef]
22. Sharma, D.; Pakravan, N.; Pritchard, J.C.; Hartmann, F.A.; Young, K.M. Mucoid *Pseudomonas aeruginosa* Infection in a Cat with Severe Chronic Rhinosinusitis. *Vet. Clin. Pathol.* **2019**, *48*, 300–304. [CrossRef] [PubMed]
23. Alhazmi, A. *Pseudomonas aeruginosa*—Pathogenesis and Pathogenic Mechanisms. *Int. J. Biol.* **2015**, *7*, 44. [CrossRef]
24. Penaranda, C.; Chumbler, N.M.; Hung, D.T. Dual Transcriptional Analysis Reveals Adaptation of Host and Pathogen to Intracellular Survival of *Pseudomonas aeruginosa* Associated with Urinary Tract Infection. *PLoS Pathog.* **2021**, *17*, e1009534. [CrossRef] [PubMed]
25. De Sousa, T.; Garcês, A.; Silva, A.; Lopes, R.; Alegria, N.; Hébraud, M.; Igrejas, G.; Poeta, P. The Impact of the Virulence of *Pseudomonas aeruginosa* Isolated from Dogs. *Vet. Sci.* **2023**, *10*, 343. [CrossRef] [PubMed]
26. Sana, T.G.; Baumann, C.; Merdes, A.; Soscia, C.; Rattei, T.; Hachani, A.; Jones, C.; Bennett, K.L.; Filloux, A.; Superti-Furga, G.; et al. Internalization of *Pseudomonas aeruginosa* Strain PAO1 into Epithelial Cells Is Promoted by Interaction of a T6SS Effector with the Microtubule Network. *Am. Soc. Microbiol.* **2015**, *6*, e00712-15. [CrossRef]
27. De Bentzmann, S.; Plotkowski, C.; Puchelle, E. Receptors in the *Pseudomonas aeruginosa* Adherence to Injured and Repairing Airway Epithelium. *Am. J. Respir. Crit. Care Med.* **2012**, *154*, S155–S162. [CrossRef]
28. Zahm, J.M.; Chevillard, M.; Puchelle, E. Wound Repair of Human Surface Respiratory Epithelium. *Am. J. Respir. Cell Mol. Biol.* **2012**, *5*, 242–248. [CrossRef]
29. Yamaguchi, T.; Yamada, H. Role of Mechanical Injury on Airway Surface in the Pathogenesis of *Pseudomonas aeruginosa*. *Am. Rev. Respir. Dis.* **2012**, *144*, 1147–1152. [CrossRef]
30. Newman, J.N.; Floyd, R.V.; Fothergill, J.L. Invasion and Diversity in *Pseudomonas aeruginosa* Urinary Tract Infections. *J. Med. Microbiol.* **2022**, *71*, 1458. [CrossRef]
31. Gomila, A.; Carratalà, J.; Eliakim-Raz, N.; Shaw, E.; Wiegand, I.; Vallejo-Torres, L.; Gorostiza, A.; Vigo, J.M.; Morris, S.; Stoddart, M.; et al. Risk Factors and Prognosis of Complicated Urinary Tract Infections Caused by *Pseudomonas aeruginosa* in Hospitalized Patients: A Retrospective Multicenter Cohort Study. *Infect. Drug Resist.* **2018**, *11*, 2571–2581. [CrossRef]
32. Park, W.S.; Lee, J.; Na, G.; Park, S.; Seo, S.K.; Choi, J.S.; Jung, W.K.; Choi, I.W. Benzyl Isothiocyanate Attenuates Inflammasome Activation in *Pseudomonas aeruginosa* LPS-Stimulated THP-1 Cells and Exerts Regulation through the MAPKs/NF-KB Pathway. *Int. J. Mol. Sci.* **2022**, *23*, 1228. [CrossRef] [PubMed]
33. Chambers, J.R.; Cherny, K.E.; Sauer, K. Susceptibility of *Pseudomonas aeruginosa* Dispersed Cells to Antimicrobial Agents Is Dependent on the Dispersion Cue and Class of the Antimicrobial Agent Used. *Antimicrob. Agents Chemother.* **2017**, *61*, e00846-17. [CrossRef] [PubMed]
34. Sabnis, A.; Hagart, K.L.H.; Klöckner, A.; Becce, M.; Evans, L.E.; Furniss, R.C.D.; Mavridou, D.A.I.; Murphy, R.; Stevens, M.M.; Davies, J.C.; et al. Colistin Kills Bacteria by Targeting Lipopolysaccharide in the Cytoplasmic Membrane. *Elife* **2021**, *10*, 65836. [CrossRef] [PubMed]
35. Qin, S.; Xiao, W.; Zhou, C.; Pu, Q.; Deng, X.; Lan, L.; Liang, H.; Song, X.; Wu, M. *Pseudomonas aeruginosa*: Pathogenesis, Virulence Factors, Antibiotic Resistance, Interaction with Host, Technology Advances and Emerging Therapeutics. *Signal Transduct. Target. Ther.* **2022**, *7*, 1–27. [CrossRef]
36. Mittal, R.; Khandwaha, R.K.; Gupta, V.; Mittal, P.K.; Harjai, K. Phenotypic Characters of Urinary Isolates of *Pseudomonas aeruginosa* & Their Association with Mouse Renal Colonization. *Indian. J. Med. Res.* **2006**, *123*, 67–72.
37. El Zowalaty, M.E.; Al Thani, A.A.; Webster, T.J.; El Zowalaty, A.E.; Schweizer, H.P.; Nasrallah, G.K.; Marei, H.E.; Ashour, H.M. *Pseudomonas aeruginosa*: Arsenal of Resistance Mechanisms, Decades of Changing Resistance Profiles, and Future Antimicrobial Therapies. *Future Microbiol.* **2015**, *10*, 1683–1706. [CrossRef]
38. Shaji, J.C.; Kumar, R. *Pseudomonas aeruginosa* as a Cause for Urinary Tract Infection in Children with No Predisposing Factors. *Pediatr. Infect. Dis.* **2022**, *3*, 119–120. [CrossRef]

39. Guardabassi, L.; Schwarz, S.; Lloyd, D.H. Pet Animals as Reservoirs of Antimicrobial-Resistant Bacteria: Review. *J. Antimicrob. Chemother.* **2004**, *54*, 321–332. [CrossRef]
40. Jangsangthong, A.; Lugsomya, K.; Apiratwarrasakul, S.; Phumthanakorn, N. Distribution of Sequence Types and Antimicrobial Resistance of Clinical *Pseudomonas aeruginosa* Isolates from Dogs and Cats Visiting a Veterinary Teaching Hospital in Thailand. *BMC Vet. Res.* **2024**, *20*, 1–6. [CrossRef]
41. Płókarz, D.; Czopowicz, M.; Bierowiec, K.; Rypuła, K. Virulence Genes as Markers for *Pseudomonas aeruginosa* Biofilm Formation in Dogs and Cats. *Animals* **2022**, *12*, 422. [CrossRef]
42. Cole, S.J.; Records, A.R.; Orr, M.W.; Linden, S.B.; Lee, V.T. Catheter-Associated Urinary Tract Infection by *Pseudomonas aeruginosa* Is Mediated by Exopolysaccharide-Independent Biofilms. *Infect. Immun.* **2014**, *82*, 2048–2058. [CrossRef] [PubMed]
43. Damborg, P.; Broens, E.M.; Chomel, B.B.; Guenther, S.; Pasmans, F.; Wagenaar, J.A.; Weese, J.S.; Wieler, L.H.; Windahl, U.; Vanrompay, D.; et al. Bacterial Zoonoses Transmitted by Household Pets: State-of-the-Art and Future Perspectives for Targeted Research and Policy Actions. *J. Comp. Pathol.* **2016**, *155*, S27–S40. [CrossRef] [PubMed]
44. Pina-Sánchez, M.; Rua, M.; Pozo, J.L. Del Present and Future of Resistance in *Pseudomonas aeruginosa*: Implications for Treatment. *Rev. Esp. Quim.* **2023**, *36* (Suppl. 1), 54–58. [CrossRef] [PubMed]
45. Norris, C.R.; Williams, B.J.; Ling, G.V.; Franti, C.E.; Johnson, D.L.; Ruby, A.L. Recurrent and Persistent Urinary Tract Infections in Dogs: 383 Cases (1969–1995). *J. Am. Anim. Hosp. Assoc.* **2000**, *36*, 484–492. [CrossRef]
46. Ataya, H.A.S.; Soliman, S.M.; Kayaf, K.A.H.; Marouf, S.; Al-Amry, K. Incidence, Bacterial Causes and Antibiotic Resistance Patterns of Urinary Tract Infection in Pet Animals. *J. Appl. Vet. Sci.* **2023**, *8*, 35–43. [CrossRef]
47. Zendri, F.; Isgren, C.M.; Devaney, J.; Schmidt, V.; Rankin, R.; Timofte, D. Resistome-Based Surveillance Identifies ESKAPE Pathogens as the Predominant Gram-Negative Organisms Circulating in Veterinary Hospitals. *Front. Microbiol.* **2023**, *14*, 1252216. [CrossRef]
48. Xu, Z.G.; Gao, Y.; He, J.G.; Xu, W.F.; Jiang, M.; Jin, H.S. Effects of Azithromycin on *Pseudomonas aeruginosa* Isolates from Catheter-Associated Urinary Tract Infection. *Exp. Ther. Med.* **2014**, *9*, 569. [CrossRef]
49. Curran, K.; Leeper, H.; O'Reilly, K.; Jacob, J.; Bermudez, L.E. An Analysis of the Infections and Determination of Empiric Antibiotic Therapy in Cats and Dogs with Cancer-Associated Infections. *Antibiotics* **2021**, *10*, 700. [CrossRef]
50. Flores-Mireles, A.; Hreha, T.N.; Hunstad, D.A. Pathophysiology, Treatment, and Prevention of Catheter-Associated Urinary Tract Infection. *Top. Spinal. Cord. Inj. Rehabil.* **2019**, *25*, 228–240. [CrossRef]
51. Farzin, A.; Mizanur Rahman, M.; Ara Mollika, F. *Pseudomonas aeruginosa*: The Alarming Pathogen of Hospital Acquired Infection. In *Pseudomonas aeruginosa—New Perspectives and Applications*; InTech: Rijeka, Croatia, 2024.
52. Fernandes, M.R.; Sellera, F.P.; Moura, Q.; Carvalho, M.P.N.; Rosato, P.N.; Cerdeira, L.; Lincopan, N. Zooanthroponotic Transmission of Drug-Resistant *Pseudomonas aeruginosa*, Brazil. *Emerg. Infect. Dis. J.* **2018**, *24*, 1160–1162. [CrossRef]
53. Mekonnen, S.A.; El Husseini, N.; Turdiev, A.; Carter, J.A.; Belew, A.T.; El-Sayed, N.M.; Lee, V.T. Catheter-Associated Urinary Tract Infection by *Pseudomonas aeruginosa* Progresses through Acute and Chronic Phases of Infection. *Proc. Natl. Acad. Sci. USA* **2022**, *119*, e2209383119. [CrossRef] [PubMed]
54. De Sousa, T.; Hébraud, M.; Alves, O.; Costa, E.; Maltez, L.; Pereira, J.E.; Martins, Â.; Igrejas, G.; Poeta, P. Study of Antimicrobial Resistance, Biofilm Formation, and Motility of *Pseudomonas aeruginosa* Derived from Urine Samples. *Microorganisms* **2023**, *11*, 1345. [CrossRef] [PubMed]
55. Hugonnard, M.; Chalvet-Monfray, K.; Dernis, J.; Pouzot-Nevoret, C.; Barthélémy, A.; Vialard, J.; Goy-Thollot, I. Occurrence of Bacteriuria in 18 Catheterised Cats with Obstructive Lower Urinary Tract Disease: A Pilot Study. *J. Feline Med. Surg.* **2013**, *15*, 843–848. [CrossRef]
56. Ruple-Czerniak, A.; Aceto, H.W.; Bender, J.B.; Paradis, M.R.; Shaw, S.P.; Van Metre, D.C.; Weese, J.S.; Wilson, D.A.; Wilson, J.H.; Morley, P.S. Using Syndromic Surveillance to Estimate Baseline Rates for Healthcare-Associated Infections in Critical Care Units of Small Animal Referral Hospitals. *J. Vet. Intern. Med.* **2013**, *27*, 1392–1399. [CrossRef]
57. Sawa, T.; Shimizu, M.; Moriyama, K.; Wiener-Kronish, J.P. Association between *Pseudomonas aeruginosa* Type III Secretion, Antibiotic Resistance, and Clinical Outcome: A Review. *Crit. Care* **2014**, *18*, 668. [CrossRef]
58. Ozer, E.; Yaniv, K.; Chetrit, E.; Boyarski, A.; Meijler, M.M.; Berkovich, R.; Kushmaro, A.; Alfonta, L. An inside Look at a Biofilm: *Pseudomonas aeruginosa* Flagella Biotracking. *Sci. Adv.* **2021**, *7*, eabg8581. [CrossRef]
59. Argudín, M.A.; Deplano, A.; Meghraoui, A.; Dodémont, M.; Heinrichs, A.; Denis, O.; Nonhoff, C.; Roisin, S. Bacteria from Animals as a Pool of Antimicrobial Resistance Genes. *Antibiotics* **2017**, *6*, 12. [CrossRef]
60. Lin, D.M.; Koskella, B.; Lin, H.C. Phage Therapy: An Alternative to Antibiotics in the Age of Multi-Drug Resistance. *World J. Gastrointest. Pharmacol. Ther.* **2017**, *8*, 162. [CrossRef]
61. Vingopoulou, E.I.; Delis, G.A.; Batzias, G.C.; Kaltsogianni, F.; Koutinas, A.; Kristo, I.; Pournaras, S.; Saridomichelakis, M.N.; Siarkou, V.I. Prevalence and Mechanisms of Resistance to Fluoroquinolones in *Pseudomonas aeruginosa* and *Escherichia coli* Isolates Recovered from Dogs Suffering from Otitis in Greece. *Vet. Microbiol.* **2018**, *213*, 102–107. [CrossRef]

62. Pelegrin, A.C.; Palmieri, M.; Mirande, C.; Oliver, A.; Moons, P.; Goossens, H.; Van Belkum, A. *Pseudomonas aeruginosa*: A Clinical and Genomics Update. *FEMS Microbiol. Rev.* **2021**, *45*, fuab026. [CrossRef]
63. Lister, P.D.; Wolter, D.J.; Hanson, N.D. Antibacterial-Resistant *Pseudomonas aeruginosa*: Clinical Impact and Complex Regulation of Chromosomally Encoded Resistance Mechanisms. *Clin. Microbiol. Rev.* **2009**, *22*, 582–610. [CrossRef] [PubMed]
64. Moradali, M.F.; Ghods, S.; Rehm, B.H.A. *Pseudomonas aeruginosa* Lifestyle: A Paradigm for Adaptation, Survival, and Persistence. *Front. Cell Infect. Microbiol.* **2017**, *7*, 249785. [CrossRef] [PubMed]
65. Lupo, A.; Haenni, M.; Madec, J.-Y. Antimicrobial Resistance in *Acinetobacter* spp. and *Pseudomonas* spp. *Microbiol. Spectr.* **2018**, *6*, 377–393. [CrossRef] [PubMed]
66. Saini, H.; Chhibber, S.; Harjai, K. Azithromycin and Ciprofloxacin: A Possible Synergistic Combination against *Pseudomonas aeruginosa* Biofilm-Associated Urinary Tract Infections. *Int. J. Antimicrob. Agents* **2015**, *45*, 359–367. [CrossRef]
67. Ocak, F.; Turkyilmaz, S. Investigation of Antimicrobial Resistance, Biofilm Production, Biofilm Associated Virulence Genes and Integron Genes of *Pseudomonas Aeroginosa* Isolates Obtained from Animal Clinical Samples. *Isr. J. Vet. Med.* **2022**, *77*, 15–26.
68. Patel, H.; Buchad, H.; Gajjar, D. *Pseudomonas aeruginosa* Persister Cell Formation upon Antibiotic Exposure in Planktonic and Biofilm State. *Sci. Rep.* **2022**, *12*, 1–12. [CrossRef]
69. Jin, M.; Osman, M.; Green, B.A.; Yang, Y.; Ahuja, A.; Lu, Z.; Cazer, C.L. Evidence for the Transmission of Antimicrobial Resistant Bacteria between Humans and Companion Animals: A Scoping Review. *One Health* **2023**, *17*, 100593. [CrossRef]
70. McEwen, S.A.; Fedorka-Cray, P.J. Antimicrobial Use and Resistance in Animals. *Clin. Infect. Dis.* **2002**, *34*, S93–S106. [CrossRef]
71. Thanner, S.; Drissner, D.; Walsh, F. Antimicrobial Resistance in Agriculture. *mBio* **2016**, *7*, e02227-15. [CrossRef]
72. Panagea, S.; Winstanley, C.; Walshaw, M.J.; Ledson, M.J.; Hart, C.A. Environmental Contamination with an Epidemic Strain of *Pseudomonas aeruginosa* in a Liverpool Cystic Fibrosis Centre, and Study of Its Survival on Dry Surfaces. *J. Hosp. Infect.* **2005**, *59*, 102–107. [CrossRef]
73. Walther, B.; Hermes, J.; Cuny, C.; Wieler, L.H.; Vincze, S.; Elnaga, Y.A.; Stamm, I.; Kopp, P.A.; Kohn, B.; Witte, W.; et al. Sharing More than Friendship—Nasal Colonization with Coagulase-Positive *Staphylococci* (CPS) and Co-Habitation Aspects of Dogs and Their Owners. *PLoS ONE* **2012**, *7*, e35197. [CrossRef] [PubMed]
74. Varriale, L.; Dipineto, L.; Russo, T.P.; Borrelli, L.; Romano, V.; D'orazio, S.; Pace, A.; Menna, L.F.; Fioretti, A.; Santaniello, A. Antimicrobial Resistance of *Escherichia coli* and *Pseudomonas aeruginosa* from Companion Birds. *Antibiotics* **2020**, *9*, 780. [CrossRef] [PubMed]
75. Wright, E.A.; Di Lorenzo, V.; Trappetti, C.; Liciardi, M.; Orru, G.; Viti, C.; Bronowski, C.; Hall, A.J.; Darby, A.C.; Oggioni, M.R.; et al. Divergence of a Strain of *Pseudomonas aeruginosa* during an Outbreak of Ovine Mastitis. *Vet. Microbiol.* **2015**, *175*, 105–113. [CrossRef]
76. Goldstein, E.J.C.; Agyare, E.O.; Vagvolgyi, A.E.; Halpern, M. Aerobic Bacterial Oral Flora of Garter Snakes: Development of Normal Flora and Pathogenic Potential for Snakes and Humans. *J. Clin. Microbiol.* **1981**, *13*, 954–956. [CrossRef]
77. Von Degerfeld, M.M.; Banchi, P.; Quaranta, G. Successful Treatment of Pyometra Caused by *Pseudomonas aeruginosa* Infection in a Rabbit. *Top. Companion Anim. Med.* **2020**, *41*, 100473. [CrossRef]
78. Martins, W.M.B.S.; Narciso, A.C.; Cayô, R.; Santos, S.V.; Fehlberg, L.C.C.; Ramos, P.L.; da Cruz, J.B.; Gales, A.C. SPM-1-Producing *Pseudomonas aeruginosa* ST277 Clone Recovered from Microbiota of Migratory Birds. *Diagn. Microbiol. Infect. Dis.* **2018**, *90*, 221–227. [CrossRef]
79. Zhang, R.; Liu, Z.; Li, J.; Lei, L.; Yin, W.; Li, M.; Wu, C.; Walsh, T.R.; Wang, Y.; Wang, S.; et al. Presence of VIM-Positive *Pseudomonas* Species in Chickens and Their Surrounding Environment. *Antimicrob. Agents Chemother.* **2017**, *61*. [CrossRef]
80. Oliveira, M.; Serrano, I.; Santos, J.P.; Bilocq, F.; Pereira, N.; de Santos Loureiro, N.; Tavares, L.; Pirnay, J.P.; De Vos, D. *Pseudomonads* from Wild Free-Living Sea Turtles in Príncipe Island, Gulf of Guinea. *Ecol. Indic.* **2017**, *81*, 260–264. [CrossRef]
81. Ruiz-Roldán, L.; Rojo-Bezares, B.; de Toro, M.; López, M.; Toledano, P.; Lozano, C.; Chichón, G.; Alvarez-Erviti, L.; Torres, C.; Sáenz, Y. Antimicrobial Resistance and Virulence of *Pseudomonas* Spp. among Healthy Animals: Concern about Exolysin ExlA Detection. *Sci. Rep.* **2020**, *10*, 1–11. [CrossRef]
82. Colinon, C.; Jocktane, D.; Brothier, E.; Rossolini, G.M.; Cournoyer, B.; Nazaret, S. Genetic Analyses of *Pseudomonas aeruginosa* Isolated from Healthy Captive Snakes: Evidence of High Inter- and Intrasite Dissemination and Occurrence of Antibiotic Resistance Genes. *Environ. Microbiol.* **2010**, *12*, 716–729. [CrossRef]
83. Płókarz, D.; Rypuła, K. A One Health Perspective on the Human-Pets *Pseudomonas aeruginosa* Transmission. *Appl. Microbiol.* **2022**, *8*, 1–3.
84. Scott, A.; Pottenger, S.; Timofte, D.; Moore, M.; Wright, L.; Kukavica-Ibrulj, I.; Jeukens, J.; Levesque, R.C.; Freschi, L.; Pinchbeck, G.L.; et al. Reservoirs of Resistance: Polymyxin Resistance in Veterinary-Associated Companion Animal Isolates of *Pseudomonas aeruginosa*. *Vet. Rec.* **2019**, *185*, 206. [CrossRef] [PubMed]
85. Fothergill, J.L.; Walshaw, M.J.; Winstanley, C. Transmissible Strains of *Pseudomonas aeruginosa* in Cystic Fibrosis Lung Infections. *Eur. Respir. J.* **2012**, *40*, 227–238. [CrossRef] [PubMed]

86. Pomba, C.; Rantala, M.; Greko, C.; Baptiste, K.E.; Catry, B.; van Duijkeren, E.; Mateus, A.; Moreno, M.A.; Pyörälä, S.; Ružauskas, M.; et al. Public Health Risk of Antimicrobial Resistance Transfer from Companion Animals. *J. Antimicrob. Chemother.* **2017**, *72*, 957–968. [CrossRef] [PubMed]
87. Michl, R.K.; Beck, J.F.; Mainz, J.G. Cystic Fibrosis Patient's Best Friend? Potential Transmission of *Pseudomonas aeruginosa* from a Dog. *Klin. Padiatr.* **2017**, *229*, 245–246. [CrossRef]
88. Morris, D.O.; Davis, M.F.; Palmeiro, B.S.; O'Shea, K.; Rankin, S.C. Molecular and Epidemiological Characterization of Canine *Pseudomonas* Otitis Using a Prospective Case-Control Study Design. *Vet. Dermatol.* **2017**, *28*, 118-e25. [CrossRef]
89. Pham, T.M.; Büchler, A.C.; Voor in 't holt, A.F.; Severin, J.A.; Bootsma, M.C.J.; Gommers, D.; Kretzschmar, M.E.; Vos, M.C. Routes of Transmission of VIM-Positive *Pseudomonas aeruginosa* in the Adult Intensive Care Unit-Analysis of 9 Years of Surveillance at a University Hospital Using a Mathematical Model. *Antimicrob. Resist. Infect. Control* **2022**, *11*, 1–9. [CrossRef]
90. Da Costa, P.M.; Loureiro, L.; Matos, A.J.F. Transfer of Multidrug-Resistant Bacteria Between Intermingled Ecological Niches: The Interface Between Humans, Animals and the Environment. *Int. J. Environ. Res. Public Health* **2013**, *10*, 278–294. [CrossRef]
91. ResistanceMap. Available online: https://resistancemap.onehealthtrust.org/CountryPage.php?countryId=29&country=Portugal (accessed on 10 February 2025).
92. Sebola, D.C.; Oguttu, J.W.; Malahlela, M.N.; Kock, M.M.; Qekwana, D.N. Occurrence and Characterization of ESKAPE Organisms on the Hands of Veterinary Students before Patient Contact at a Veterinary Academic Hospital, South Africa. *BMC Vet. Res.* **2024**, *20*, 475. [CrossRef]
93. Schmidt, J.S.; Kuster, S.P.; Nigg, A.; Dazio, V.; Brilhante, M.; Rohrbach, H.; Bernasconi, O.J.; Büdel, T.; Campos-Madueno, E.I.; Gobeli Brawand, S.; et al. Poor Infection Prevention and Control Standards Are Associated with Environmental Contamination with Carbapenemase-Producing Enterobacterales and Other Multidrug-Resistant Bacteria in Swiss Companion Animal Clinics. *Antimicrob. Resist. Infect. Control* **2020**, *9*, 1–13. [CrossRef]
94. WHO. *Guidelines for the Prevention and Control of Carbapenem-Resistant Enterobacteriaceae, Acinetobacter Baumannii and* Pseudomonas aeruginosa *in Health Care Facilities*; WHO: Geneva, Switzerland, 2017.
95. European Antimicrobial Resistance Surveillance Network (EARS-Net). Available online: https://www.ecdc.europa.eu/en/about-us/networks/disease-networks-and-laboratory-networks/ears-net-data (accessed on 30 January 2025).
96. Search | European Centre for Disease Prevention and Control. Available online: https://www.ecdc.europa.eu/en/search?s=pseudomonas+aeruginosa (accessed on 30 January 2025).
97. Antimicrobial Consumption Dashboard (ESAC-Net). Available online: https://www.ecdc.europa.eu/en/antimicrobial-consumption/surveillance-and-disease-data/database (accessed on 30 January 2025).
98. Antimicrobial Resistance in Veterinary Medicine | European Medicines Agency (EMA). Available online: https://www.ema.europa.eu/en/veterinary-regulatory-overview/antimicrobial-resistance-veterinary-medicine (accessed on 30 January 2025).
99. McGovern, D.A.; Gaschen, F.; Roy, A. Antimicrobial Susceptibility Patterns and Clinical Parameters in 208 Dogs with Positive Urine Cultures (2012-2014). *J. Am. Anim. Hosp. Assoc.* **2019**, *55*, 306–313. [CrossRef] [PubMed]
100. Yudhanto, S.; Hung, C.C.; Maddox, C.W.; Varga, C. Antimicrobial Resistance in Bacteria Isolated from Canine Urine Samples Submitted to a Veterinary Diagnostic Laboratory, Illinois, United States. *Front. Vet. Sci.* **2022**, *9*, 867784. [CrossRef] [PubMed]
101. Rampacci, E.; Bottinelli, M.; Stefanetti, V.; Hyatt, D.R.; Sgariglia, E.; Coletti, M.; Passamonti, F. Antimicrobial Susceptibility Survey on Bacterial Agents of Canine and Feline Urinary Tract Infections: Weight of the Empirical Treatment. *J. Glob. Antimicrob. Resist.* **2018**, *13*, 192–196. [CrossRef] [PubMed]
102. Salas, C.S. *Caracterização da Infeção do Trato Urinário em cães e gatos e Utilização de "Formula for Rational Antimicrobial Therapy" na Avaliação da Antibioterapia Empírica em Portugal*; Universidade de Lisboa: Lisboa, Portugal, 2021.
103. Temmerman, R.; Berlamont, H.; El Garch, F.; Rose, M.; Simjee, S.; Meschi, S.; de Jong, A. Antimicrobial Susceptibility of Canine and Feline Urinary Tract. Infection Pathogens Isolated from Animals with Clinical Signs in European Veterinary Practices during the Period 2013–2018. *Antibiotics* **2024**, *13*, 500. [CrossRef]
104. Medeiros de Rezende, G.; Eduarda Monteiro Silva, M.; Bistritschan Israel, C.; Ferreira Bernardo, V.; Junio Moreira Fernandes, A.; Ferreira Bastos, B. Infecções Do Trato Urinário De Cães E Gatos: Avaliação "In Vitro" Da Sensibilidade Bacteriana A Antibióticos. *Rev. De Med. Vet. Do UNIFESO* **2023**, *3*, 61–67.
105. Farag, H.S.; Ali, M.E.; Abdel Masseih, E.S.; Bakry, N.M. Diagnostic Ultrasonography and Antimicrobial Resistance of Different Pathogens Associated with Canine and Feline Lower Urinary Tract Disorders. *Comp. Immunol. Microbiol. Infect. Dis.* **2024**, *112*, 102216. [CrossRef]
106. Elfadadny, A.; Uchiyama, J.; Goto, K.; Imanishi, I.; Ragab, R.F.; Nageeb, W.M.; Iyori, K.; Toyoda, Y.; Tsukui, T.; Ide, K.; et al. Antimicrobial Resistance and Genotyping of *Pseudomonas aeruginosa* Isolated from the Ear Canals of Dogs in Japan. *Front. Vet. Sci.* **2023**, *10*, 1074127. [CrossRef]
107. Litster, A.; Moss, S.M.; Honnery, M.; Rees, B.; Trott, D.J. Prevalence of Bacterial Species in Cats with Clinical Signs of Lower Urinary Tract Disease: Recognition of *Staphylococcus felis* as a Possible Feline Urinary Tract Pathogen. *Vet. Microbiol.* **2007**, *121*, 182–188. [CrossRef]

108. Garcês, A.; Lopes, R.; Silva, A.; Sampaio, F.; Duque, D.; Brilhante-Simões, P. Bacterial Isolates from Urinary Tract Infection in Dogs and Cats in Portugal, and Their Antibiotic Susceptibility Pattern: A Retrospective Study of 5 Years (2017–2021). *Antibiotics* **2022**, *11*, 1520. [CrossRef]
109. Smoglica, C.; Evangelisti, G.; Fani, C.; Marsilio, F.; Trotta, M.; Messina, F.; Di Francesco, C.E. Antimicrobial Resistance Profile of Bacterial Isolates from Urinary Tract Infections in Companion Animals in Central Italy. *Antibiotics* **2022**, *11*, 1363. [CrossRef]
110. Aurori, M.; Novac, C.Ș.; Nadăș, G.C.; Crăciun, S.; Fiț, N.; Andrei, S. The Antimicrobial Effect of *Cornus mas* L. and *Sorbus aucuparia* L. Fruit Extracts against Resistant Uropathogens in Correlation with the Prevalence of Urinary Tract Infections in Companion Animals. *Pharmaceuticals* **2024**, *17*, 814. [CrossRef] [PubMed]
111. Foster, J.D. Today's Veterinary Practice. Use of Antibiotics for Treating UTIs in Dogs and Cats. *Focus. Pharmacol.* **2017**, *7*, 119–127.
112. Pinthanon, A.; Nithitarnwat, C.; Pintapin, C.; Siripanee, C.; Yindee, J.; Am-in, N.; Kesdangsakonwut, S.; Surachetpong, S.; Prapasarakul, N. Rapid Identification of Canine Uropathogens by Matrix-Assisted Laser Desorption/Ionization-Time-of-Flight Mass Spectrometry and the Clinical Factors That Correlated Bacterial Species and Antimicrobial Resistance. *Vet. Res. Commun.* **2023**, *47*, 1457–1469. [CrossRef] [PubMed]
113. Maeda, H.; Hayashi, K.; Ishige, T.; Sunagawa, T.; Tanigawa, S.; Mishina, M.; Watanabe, T.; Sogawa, K. Use of the Maldi Biotyper System and Rapid Bacpro with Maldi-Tof MS for Rapid Identification of Microorganisms Causing Bacterial Urinary Tract Infection in Feline Urine Samples. *J. Vet. Med. Sci.* **2018**, *80*, 1490–1494. [CrossRef]
114. Feßler, A.T.; Wang, Y.; Burbick, C.R.; Diaz-Campos, D.; Fajt, V.R.; Lawhon, S.D.; Li, X.-Z.; Lubbers, B.V.; Maddock, K.; Miller, R.A.; et al. Antimicrobial Susceptibility Testing in Veterinary Medicine: Performance, Interpretation of Results, Best Practices and Pitfalls. *One Health Adv.* **2023**, *1*, 1–16. [CrossRef]
115. Hall-Stoodley, L.; Stoodley, P.; Kathju, S.; Høiby, N.; Moser, C.; William Costerton, J.; Moter, A.; Bjarnsholt, T. Towards Diagnostic Guidelines for Biofilm-Associated Infections. *FEMS Immunol. Med. Microbiol.* **2012**, *65*, 127–145. [CrossRef]
116. Abdullahi, U.F.; Igwenagu, E.; Mu'azu, A.; Aliyu, S.; Umar, M.I. Intrigues of Biofilm: A Perspective in Veterinary Medicine. *Vet. World* **2016**, *9*, 12–18. [CrossRef]
117. Wu, H.; Moser, C.; Wang, H.Z.; Høiby, N.; Song, Z.J. Strategies for Combating Bacterial Biofilm Infections. *Int. J. Oral. Sci.* **2015**, *7*, 1–7. [CrossRef]
118. Gatoria, I.S.; Saini, N.S.; Rai, T.S.; Dwivedi, P.N. Comparison of Three Techniques for the Diagnosis of Urinary Tract Infections in Dogs with Urolithiasis. *J. Small Anim. Pract.* **2006**, *47*, 727–732. [CrossRef]
119. Evans, S.R.; Tran, T.T.T.; Hujer, A.M.; Hill, C.B.; Hujer, K.M.; Mediavilla, J.R.; Manca, C.; Domitrovic, T.N.; Perez, F.; Farmer, M.; et al. Rapid Molecular Diagnostics to Inform Empiric Use of Ceftazidime/Avibactam and Ceftolozane/Tazobactam Against *Pseudomonas aeruginosa*: PRIMERS IV. *Clin. Infect. Dis.* **2019**, *68*, 1823–1830. [CrossRef]
120. Voulgari, E.; Miliotis, G.; Siatravani, E.; Tzouvelekis, L.S.; Tzelepi, E.; Miriagou, V. Evaluation of the Performance of Acuitas® Resistome Test and the Acuitas Lighthouse® Software for the Detection of β-Lactamase-Producing Microorganisms. *J. Glob. Antimicrob. Resist.* **2020**, *22*, 184–189. [CrossRef] [PubMed]
121. Hu, Y.Y.; Cao, J.M.; Yang, Q.; Chen, S.; Lv, H.Y.; Zhou, H.W.; Wu, Z.; Zhang, R. Risk Factors for Carbapenem-Resistant *Pseudomonas aeruginosa*, Zhejiang Province, China. *Emerg. Infect. Dis.* **2019**, *25*, 1861. [CrossRef] [PubMed]
122. Giguère, S.; Prescott, J.F.; Dowling, P.M. *Antimicrobial Therapy in Veterinary Medicine*, 5th ed.; Giguère, S., Prescott, J.F., Dowling, P.M., Eds.; Wiley-Blackwell: Hoboken, NJ, USA, 2013.
123. Pang, Z.; Raudonis, R.; Glick, B.R.; Lin, T.J.; Cheng, Z. Antibiotic Resistance in *Pseudomonas aeruginosa*: Mechanisms and Alternative Therapeutic Strategies. *Biotechnol. Adv.* **2019**, *37*, 177–192. [CrossRef]
124. Barnard, N.; Foster, A. How to Treat *Pseudomonas* Otitis in Dogs. *Vet. Rec.* **2018**, *182*, 109–110. [CrossRef]
125. Cabassi, C.S.; Sala, A.; Santospirito, D.; Alborali, G.L.; Carretto, E.; Ghibaudo, G.; Taddei, S. Activity of AMP2041 against Human and Animal Multidrug Resistant *Pseudomonas aeruginosa* Clinical Isolates. *Ann. Clin. Microbiol. Antimicrob.* **2017**, *16*, 17. [CrossRef]
126. Ghasemian, E.; Naghoni, A.; Rahvar, H.; Kialha, M.; Tabaraie, B. Evaluating the Effect of Copper Nanoparticles in Inhibiting *Pseudomonas aeruginosa* and *Listeria monocytogenes* Biofilm Formation. *Jundishapur J. Microbiol.* **2015**, *8*, e17430. [CrossRef]
127. Zhou, S.; Zhang, A.; Chu, W. Phillyrin Is an Effective Inhibitor of Quorum Sensing with Potential as an Anti-*Pseudomonas aeruginosa* Infection Therapy. *J. Vet. Med. Sci.* **2019**, *81*, 473. [CrossRef]
128. Furusawa, T.; Iwano, H.; Higuchi, H.; Yokota, H.; Usui, M.; Iwasaki, T.; Tamura, Y. Bacteriophage Can Lyse Antibiotic-Resistant *Pseudomonas aeruginosa* Isolated from Canine Diseases. *J. Vet. Med. Sci.* **2016**, *78*, 1035–1038. [CrossRef]
129. Döring, G.; Pier, G.B. Vaccines and Immunotherapy against *Pseudomonas aeruginosa*. *Vaccine* **2008**, *26*, 1011–1024. [CrossRef]
130. Guo, Z.; Zhu, Y.; Du, G.; Qin, M.; He, C.; He, P.; Song, Y.; Chen, W.; Bai, S.; Wu, F.; et al. Rapid Development of a Subunit Nano-Vaccine against Drug-Resistant *Pseudomonas aeruginosa* with Effective Cross-Protection. *Nano Today* **2022**, *43*, 101398. [CrossRef]
131. Sultana, S.T.; Atci, E.; Babauta, J.T.; Mohamed Falghoush, A.; Snekvik, K.R.; Call, D.R.; Beyenal, H. Electrochemical Scaffold Generates Localized, Low Concentration of Hydrogen Peroxide That Inhibits Bacterial Pathogens and Biofilms. *Sci. Rep.* **2015**, *5*, 14908. [CrossRef] [PubMed]

132. DiBartola, S.P.; Westropp, J.D. *Small Animall Internal Medicine*, 6th ed.; Couto, G., Richard, N., Eds.; Elsevier: St Louis, MI, USA, 2023.
133. Kovarikova, S. Urinary Biomarkers of Renal Function in Dogs and Cats: A Review. *Vet. Med.* **2015**, *60*, 589–602. [CrossRef]
134. Diagnosing and Managing Feline Lower Urinary Tract Disease. Available online: https://todaysveterinarypractice.com/urology-renal-medicine/diagnosing-and-managing-feline-lower-urinary-tract-disease/ (accessed on 7 January 2025).
135. Rodriguez, M.D. *Enfermedades Del Tracto Urinario*; Manual Clínico de Medicina Interna En Pequeños Animales, Ed.; 5M Publishing Ltd., Benchmark House: London, UK, 2016; Volume II.
136. Pereira, A.; Jota Baptista, C.; Oliveira, P.A.; Coelho, A.C. Urinary Tract Bacterial Infections in Small Animal Practice: Clinical and Epidemiological Aspects. *Vet. Stanica* **2024**, *55*, 703–720. [CrossRef]
137. Marques, C.; Belas, A.; Franco, A.; Aboim, C.; Gama, L.T.; Pomba, C. Increase in Antimicrobial Resistance and Emergence of Major International High-Risk Clonal Lineages in Dogs and Cats with Urinary Tract Infection: 16 Year Retrospective Study. *J. Antimicrob. Chemother.* **2018**, *73*, 377–384. [CrossRef]
138. Escher, M.; Vanni, M.; Intorre, L.; Caprioli, A.; Tognetti, R.; Scavia, G. Use of Antimicrobials in Companion Animal Practice: A Retrospective Study in a Veterinary Teaching Hospital in Italy. *J. Antimicrob. Chemother.* **2011**, *66*, 920–927. [CrossRef] [PubMed]
139. Caneschi, A.; Bardhi, A.; Barbarossa, A.; Zaghini, A. The Use of Antibiotics and Antimicrobial Resistance in Veterinary Medicine, a Complex Phenomenon: A Narrative Review. *Antibiotics* **2023**, *12*, 487. [CrossRef] [PubMed]
140. Mesaros, N.; Nordmann, P.; Plésiat, P.; Roussel-Delvallez, M.; Van Eldere, J.; Glupczynski, Y.; Van Laethem, Y.; Jacobs, F.; Lebecque, P.; Malfroot, A.; et al. *Pseudomonas aeruginosa*: Resistance and Therapeutic Options at the Turn of the New Millennium. *Clin. Microbiol. Infect.* **2007**, *13*, S27–S40. [CrossRef]
141. Robbins, S.N.; Goggs, R.; Lhermie, G.; Lalonde-Paul, D.F.; Menard, J. Antimicrobial Prescribing Practices in Small Animal Emergency and Critical Care. *Front. Vet. Sci.* **2020**, *7*, 110. [CrossRef]
142. Mouiche, M.M.M.; Mpouam, S.E.; Moffo, F.; Nkassa, C.M.N.; Mbah, C.K.; Mapiefou, N.P.; Awah-Ndukum, J. Prescription Pattern of Antimicrobial Use in Small Animal Veterinary Practice in Cameroon. *Top. Companion Anim. Med.* **2021**, *44*, 100540. [CrossRef]
143. Foglia Manzillo, V.; Peruzy, M.F.; Gizzarelli, M.; Izzo, B.; Sarnelli, P.; Carrella, A.; Vinciguerra, G.; Chirollo, C.; Ben Fayala, N.E.H.; Balestrino, I.; et al. Examining the Veterinary Electronic Antimicrobial Prescriptions for Dogs and Cats in the Campania Region, Italy: Corrective Strategies Are Imperative. *Animals* **2023**, *13*, 2869. [CrossRef]
144. Kothari, A.; Kherdekar, R.; Mago, V.; Uniyal, M.; Mamgain, G.; Kalia, R.B.; Kumar, S.; Jain, N.; Pandey, A.; Omar, B.J. Age of Antibiotic Resistance in MDR/XDR Clinical Pathogen of *Pseudomonas aeruginosa*. *Pharmaceuticals* **2023**, *16*, 1230. [CrossRef] [PubMed]
145. Vercelli, C.; Della Ricca, M.; Re, M.; Gambino, G.; Re, G. Antibiotic Stewardship for Canine and Feline Acute Urinary Tract Infection: An Observational Study in a Small Animal Hospital in Northwest Italy. *Antibiotics* **2021**, *10*, 562. [CrossRef] [PubMed]
146. Hubbuch, A.; Schmitt, K.; Lehner, C.; Hartnack, S.; Schuller, S.; Schüpbach-Regula, G.; Mevissen, M.; Peter, R.; Müntener, C.; Naegeli, H.; et al. Antimicrobial Prescriptions in Cats in Switzerland before and after the Introduction of an Online Antimicrobial Stewardship Tool. *BMC Vet. Res.* **2020**, *16*, 229. [CrossRef] [PubMed]
147. Lehner, C.; Hubbuch, A.; Schmitt, K.; Schuepbach-Regula, G.; Willi, B.; Mevissen, M.; Peter, R.; Muentener, C.R.; Naegeli, H.; Schuller, S. Effect of Antimicrobial Stewardship on Antimicrobial Prescriptions for Selected Diseases of Dogs in Switzerland. *J. Vet. Intern. Med.* **2020**, *34*, 2418–2431. [CrossRef]
148. Ghadaksaz, A.; Imani Fooladi, A.A.; Hosseini, H.M.; Amin, M. The Prevalence of Some *Pseudomonas* Virulence Genes Related to Biofilm Formation and Alginate Production among Clinical Isolates. *J. Appl. Biomed.* **2015**, *13*, 61–68. [CrossRef]

veterinary sciences

MDPI

Article

Immunohistochemical Analysis of Inter-Alpha-Trypsin Inhibitor Heavy Chain 2 and Enolase 1 in Canine Mammary Tumors: Associations with Tumor Aggressiveness and Prognostic Significance

Luadna dos Santos e Silva [1], Pedro Henrique Fogaça Jordão [2], Beatriz Castilho Balieiro [1], Laura de Souza Baracioli [1], Daniela Farias de Nóbrega [3], Adriana Alonso Novais [4], Luiz Gustavo de Almeida Chuffa [5] and Debora Aparecida Pires de Campos Zuccari [1,*]

1 Molecular Investigation of Cancer Laboratory (MICL), Department of Molecular Biology, Faculty of Medicine of São José do Rio Preto (FAMERP), São José do Rio Preto 15090-000, Brazil; lua_dna@outlook.com (L.d.S.e.S.); bia-balieiro@hotmail.com (B.C.B.); laura.baracioli@hotmail.com (L.d.S.B.)
2 Biochemistry and Molecular Biology Research Center (NPBIM), Faculty of Medicine of São José do Rio Preto (FAMERP), São José do Rio Preto 15090-000, Brazil; pedro.hf.jordao@hotmail.com
3 Pat Animal Laboratory, São José do Rio Preto 15070-000, Brazil; danielanobrega@patanimal.com.br
4 Institute of Health Science (ICS), Universidade Federal de Mato Grosso (UFMT), Sinop 78550-728, Brazil; adriana.novais@ufmt.br
5 Department of Structural and Functional Biology, Institute of Biosciences, UNESP, São Paulo State University, Botucatu 18618-689, Brazil; luiz-gustavo.chuffa@unesp.br
* Correspondence: debora.zuccari@famerp.br

Simple Summary: Mammary tumors in dogs are commonly seen in middle-aged and older females, particularly in those that are not spayed. These tumors share many similarities with human breast cancer, making dogs a useful model for studying the disease. This study focuses on two potential biomarkers—Inter-Alpha-Trypsin Inhibitor Heavy Chain 2 (ITIH2) and Enolase 1 (ENO1)—which have been identified in previous research as relevant to canine mammary tumors. We used immunohistochemistry to analyze tissue samples from healthy dogs and those with benign or malignant tumors. The results showed specific patterns of expression for both ITIH2 and ENO1, suggesting that these proteins may play important roles in tumor development. This study highlights their potential as biomarkers for diagnosing and treating mammary tumors, both in dogs and, possibly, in human breast cancer.

Abstract: Mammary neoplasms in dogs are a common clinical concern, especially in middle-aged and older intact females. These tumors share similarities with human breast cancer in terms of histology, disease progression, and risk factors, making dogs a relevant model for breast cancer research. The search for biomarkers in canine mammary tumors is essential to understand tumor progression and identify potential therapeutic targets. This study investigated the expression of two potential biomarkers—Inter-Alpha-Trypsin Inhibitor Heavy Chain 2 (ITIH2) and Enolase 1 (ENO1)—in the mammary glands of healthy and tumor-bearing dogs using immunohistochemistry. Both proteins were identified in previous proteomic analyses of extracellular vesicles derived from the plasma of healthy and tumor-bearing dogs. A total of fifty-one canine mammary tissue samples were analyzed and categorized into three groups: (i) the control group, composed of five samples of normal mammary tissue without neoplasia; (ii) benign tumors, composed of nineteen samples of benign mixed tumors; and (iii) malignant tumors, which included six carcinomas in grade 1 mixed tumors, five carcinomas in grade 2 mixed tumors, thirteen solid carcinomas of grade 3, one papillary carcinoma, and two tubular carcinomas. Regarding the intensity of staining, quantified by histoscore, there were no significant differences in the comparison between the groups; for ITIH2, the p-value was 0.33, and for ENO1, the p-value was 0.57. Regarding the predictive potential of their respective ROC curves, the proteins demonstrated low predictive power in canine mammary tumors. These findings indicate that neither ITIH2 nor ENO1 demonstrated strong prognostic value in this setting, as demonstrated by their moderate AUC values, wide confidence intervals, and lack of statistical significance. However, this study found distinct tissue localization patterns for ITIH2 and subcellular localization for ENO1. As an additional way to examine possible associations of these proteins with epithelial–mesenchymal transition, the ZEB1 antibody was tested by both single and double immunohistochemistry, demonstrating a tendency to be more intensely expressed in the malignant group and tending to be associated with ENO1 in canine mammary tumors.

Keywords: bitches; mammary tumors; extracellular vesicles (EVs); biomarkers; prognosis

1. Introduction

Mammary neoplasms are the most common group of tumors encountered in veterinary practice, exhibiting a biological behavior that ranges from benign to highly malignant. Metastases are reported in 25 to 50% of cases, emphasizing the aggressive potential of these tumors. Mammary cancer in dogs serves as an exceptional model for studying breast cancer in women due to striking similarities in tumor histology, disease progression, behavior, associated risk factors, and hormone receptor profiles [1].

These neoplasms often originate as benign lesions. However, under certain conditions, they may accumulate genetic and epigenetic changes that drive their progression to malignancy. On the other hand, malignant mammary tumors typically exhibit an aggressive phenotype from the moment of diagnosis, including invasiveness and metastatic potential. Studies report that approximately 25% of dogs experience the development of new nodules following surgical removal of initial tumors. Interestingly, these subsequent nodules may display malignant characteristics, even when the primary lesion was histologically classified as benign [1]. As previously reported [2], hormonal stimulation—often associated with recurrent pseudocyesis—can exacerbate this process. In cases where tumors are not surgically removed promptly, hormonal influence may lead to tumor disorganization, the accumulation of alterations and mutations, and ultimately, a worse prognosis. Delays in surgical excision increase the likelihood that these tumors will undergo malignant transformation, further complicating treatment and prognosis [3].

Mammary neoplasms mainly affect middle-aged and elderly female dogs that have not been spayed. Breeds such as Poodles, Dachshunds, Yorkshire Terriers, Cocker Spaniels, German Shepherds, Boxers, Fox Terriers, and mixed-breed dogs are predisposed to developing these tumors. Malignant mammary neoplasms, particularly carcinomas, are characterized by an aggressive inflammatory process, rapid growth, regional lymphatic infiltration, and metastasis to distant tissues [3]. The lymphatic system is the primary route of dissemination, with the lungs being the most commonly affected site in metastatic cases. Consequently, thorough evaluation using complementary diagnostic exams is essential for accurate staging and treatment planning [4].

Cytological criteria, such as variation in nuclear size, prominent nuclei, altered nuclear-to-cytoplasmic ratios, and variations in the number, shape, and size of the nucleoli are important predictive factors of malignancy. However, one of the main challenges in obtaining an accurate cytological diagnosis in mammary neoplasms in dogs is the significant heterogeneity of tumor tissue. This heterogeneity arises from the presence of necrotic and inflamed regions, which can distort cellular morphology, as well as from distinct niches of tumor cells displaying varying levels of genetic and phenotypic alterations. Such variability complicates the interpretation of cytological features and may obscure the distinction between benign and malignant characteristics [5].

Dogs have five pairs of mammary glands, and the incidence of multiple mammary tumors has been reported to reach as high as 60%. Despite the relatively high occurrence of synchronous mammary tumors, the relationship between these tumors remains poorly understood. Most reports suggest that they are independent primary tumors with distinct histopathological features, leading to the recommendation that all tumors be biopsied to assess malignancy. Therefore, benign and malignant tumors are generally considered separate entities, with the former not acting as precursors to the latter. However, recent studies challenge this assumption. One publication reported an increased risk of developing new malignant primary tumors in other mammary glands in dogs with a previous diagnosis of mammary carcinoma compared to those with benign tumors. Additionally, another study reported a high incidence of concurrent carcinoma in situ in dogs diagnosed with carcinomas [6].

The search for biomarkers in canine mammary tumors is particularly relevant for predicting tumor progression in both species, providing insights into shared mechanisms of carcinogenesis between dogs and humans. These similarities emphasize the importance of identifying early diagnostic biomarkers for metastasis and prognostic markers that can guide clinical practice, ultimately improving patient outcomes in both species.

2. Inter-Alpha-Trypsin Inhibitor Heavy Chain 2 (ITIH2)

ITIH2 belongs to a family of structurally related plasma serine protease inhibitors. These proteins contribute to matrix stability by covalent binding to hyaluronan and play significant roles in inflammation and cancer metastasis prevention [7,8].

ZEB1 is a hyaluronan-binding protein, and ITIH2 is involved in constructing a hyaluronan network. It is secreted by mesenchymal lung cancer cells when co-cultured with cancer-associated fibroblasts. The protein transcriptionally regulates 1, an epithelial–mesenchymal transition (EMT)-inducing transcription factor. Together, the ZEB1-regulated EMT signaling contributes to establishing a pro-metastatic hyaluronan network in the tumor microenvironment, presenting a novel strategy to target this network and suppress lung cancer progression and metastasis [9].

Overall, underexpression of ITIH2 is associated with longer overall survival of colorectal cancer patients; the overexpression of ITIH2 was associated with poor overall survival. ITIH2 is enriched in normal liver tissues and liver cancer tissues. However, the ITIH2 gene was not detected in tumor cells in colon cancer [10].

2.1. ITIH2 and Breast Cancer in Women

Normal breast epithelial cells clearly express ITIH2, whereas its expression is consistently lost or strongly decreased in invasive breast cancer. Patients with abundant ITIH5 expression have better clinical outcomes compared with those with reduced expression in node-negative invasive breast cancer. The strong correlation between ITIH2 and ITIH5, especially since they are located on the same chromosome, suggests that these molecules interact in their metastasis-suppressive and metastasis-repressive properties [11].

2.2. ITIH2 and Mammary Tumors in Bitches

The role of ITIH2 in canine mammary neoplasms remains unexplored, representing a promising opportunity to expand knowledge on comparative biomarkers and tumor mechanisms.

3. Enolase 1 (ENO1)

Enolase 1 is a key enzyme in glycolysis, responsible for converting 2-phosphoglycerate into phosphoenolpyruvate, a fundamental step in glucose breakdown for energy production. This enzyme is widely expressed in various tissues, primarily located in the cytoplasm of cells [12]. Beyond its metabolic role, ENO1 functions as a plasminogen receptor on the surface of cancer cells, facilitating extracellular matrix degradation, cell invasion, and metastasis. In the cytoplasm, it catalyzes the glycolytic conversion between 2-phosphoglycerate and phosphoenolpyruvate. On the cell surface, it mediates plasminogen activation and extracellular matrix degradation, promoting cell migration and metastasis. Additionally, in the nucleus, ENO1 acts as a C-Myc promoter-binding protein (MBP-1) [13,14].

3.1. ENO1 and Breast Cancer in Women

ENO1 has been implicated in several cancer types, showing relevance in oncological studies. Its overexpression is associated with overall survival in patients, including those with breast cancer.

In an experimental study [15], the effect of reduced Enolase 1 expression in human umbilical vein endothelial cells (HUVECs)/MDA-MB-231 cells on the response to hypoxia and the possible mechanisms involved were investigated. After Enolase 1-transfected breast cancer cells were injected into nude mice, tumor growth significantly decreased, and tumor volume and weight both reduced. After radiotherapy treatment, tumor size decreased significantly in both groups, and the greatest reduction was observed in the transfected group. Reduction in Enolase 1 expression significantly decreases the response to hypoxia and increases the sensitivity of cells to radiotherapy. Other studies have also demonstrated that tumor-conditioned medium from the breast carcinoma cell line, MDA-MB-231 cells, led to upregulated expression of Enolase 1 in HUVECs, elevated reproductive and antiapoptotic capacity of endothelial cells, increased cell cycle progression, and improved angiogenesis in vitro [16,17].

Overexpression of ENO 1 has also been associated with tamoxifen resistance in breast cancer, in addition to reduced overall survival in cancer patients. The proteins E6AP (ubiquitin ligase E6-associated protein) and ENO1 interact predominantly in the cytoplasmic periphery of breast cancer cells. E6AP has been shown to target ENO1 for degradation, highlighting the importance of this interaction in oncological contexts, especially in breast cancer [18].

The ENO1 supports anaerobic proliferation via the Warburg effect and promotes cancer invasion through its surface expression as a plasminogen receptor. Its dual roles in metabolism and metastasis make it a promising target for cancer therapy [19].

3.2. ENO1 and Mammary Tumors in Bitches

There is only a single published study on ENO1 and canine mammary tumors, and that study showed that ENO1 overexpression was detected only in canine mammary carcinoma tumor cells and significantly correlated with shorter 5-year cause-specific survival. The results of age-adjusted Cox regression analysis also indicated that ENO1 overexpression was significantly and independently associated with shorter cause-specific survival. In contrast to the results of the human breast cancer study, the findings suggested that ER positivity was not associated with ENO1 overexpression in canine mammary carcinoma [20].

Although there has been a single published study on Enolase 1 in canine mammary tumors that used the immunohistochemical technique, it did not focus on the various prognostic regulations that this protein can influence depending on the labeling site, as ours did.

The proteins ITIH2 and Enolase1 (ENO1) were found to be overexpressed in extracellular vesicles (EVs) extracted from the plasma of bitches with malignant mammary tumors in a previous proteomic study carried out by our research group. Therefore, the aim of our study is to validate these proteins in the mammary gland tissue of healthy bitches and those with benign and malignant mammary tumors to investigate whether they have prognostic potential for these tumors.

4. Materials and Methods

4.1. Samples and Histological Processing

The tissue blocks selected for this study were previously submitted to histopathological evaluation (hematoxylin–eosin stain) by the clinic from which they were acquired, focusing on cellular morphology, lymphatic invasion, and metastasis to identify what type of tumor each sample was, whether benign or malignant; this was carried out by adhering to the criteria established by the Consensus on Mammary Tumors in Dogs and Cats [21]. A total of fifty-one canine mammary tissue samples were analyzed and categorized into three groups: (i) the control group, consisting of five samples of normal mammary tissue without neoplasia; (ii) benign tumors, consisting of nineteen samples of benign mixed tumors; and (iii) malignant tumors, which included six grade 1 carcinomas in mixed tumors, five grade 2 carcinomas in mixed tumors, thirteen grade 3 solid carcinomas, one papillary carcinoma, and two tubular carcinomas. All samples were obtained from a veterinary pathology laboratory tumor biobank, ensuring standardized preservation and accessibility for this study. Regarding the test with ZEB1 performed, only 10 tumor samples were used (4 benign and 6 malignant, with the malignant ones subdivided into good and reserved prognosis).

4.2. Positive and Negative Controls

Positive and negative controls for each antibody used (Figure 1) were obtained according to The Atlas Human Protein website.

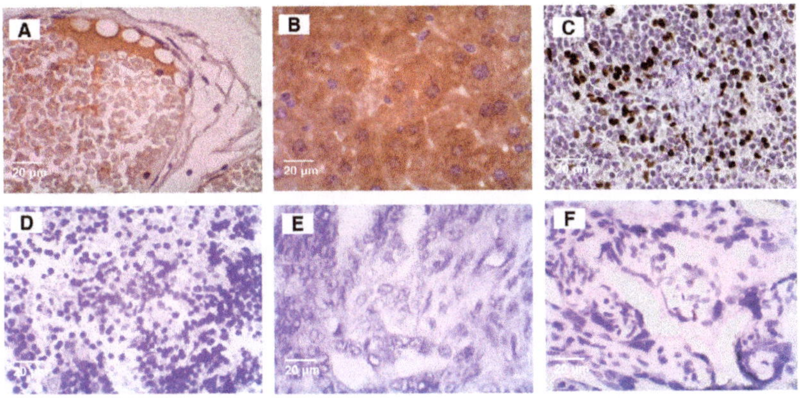

Figure 1. Photomicrographs of the positive and negative controls of ITIH2, ENO1, and ZEB1. (**A**)—Positive control of ITIH2 (fallopian tube) and (**D**)—negative control (tonsil); (**B**)—Positive control of ENO1 (liver) and (**E**)—negative control (antibody suppression in dog mammary gland tissue); (**C**)—Positive control of ZEB1 (lymph node) and (**F**)—negative control (placenta).

4.3. Single Immunohistochemical Labeling

For the single immunohistochemical labeling, 4 μm thick tissue sections were prepared using a microtome from the manufacturer Leica (Model 2255, Wetzlar, Germany). The slides were deparaffinized in xylene and rehydrated through a graded alcohol series, followed by incubation in an oven at 60 °C for 3 h. Antigen retrieval was performed using citrate buffer (pH 6.0) for ITIH2 or EDTA buffer (pH 7.4) for ENO1, both heated to 100 °C in a steam cooker (Histo Bath, from the manufacturer Easy Path, São Paulo-SP, Brazil) for 30 min. Endogenous peroxidase activity was blocked using a DAKO peroxide solution for 15 min, and non-specific protein binding was minimized with a 3% BSA solution in PBS for 10 min.

Primary antibodies (anti-ITIH2 at 1:300, anti-ENO1 at 1:3000, and anti-ZEB1 at 1:500) were applied and incubated overnight at 4 °C to ensure specific binding. Afterward, horseradish peroxidase (HRP)-conjugated secondary antibodies were applied for 25 min. Immunoreactivity was visualized using diaminobenzidine tetrahydrochloride (DAB) as the chromogen (1 drop per 1000 μL of DAKO substrate) for 2 min. Finally, the slides were counterstained with hematoxylin for 2 min to enhance cellular structure visualization and mounted for analysis.

4.4. Double Immunohistochemical Labeling

The double labeling procedure followed the same initial steps as single labeling, including tissue section preparation, deparaffinization, antigen retrieval, and blocking. The distinguishing feature was the sequential application of two primary antibodies. Following the first immunolabeling, which utilized DAB as the chromogen, an acid block was applied to inactivate residual HRP activity and prevent cross-reactivity with the second antibody.

Next, the second primary antibody was introduced, followed by incubation with a distinct HRP-conjugated secondary antibody and visualization using a contrasting chromogen to differentiate it from the first labeling. Stringent washing steps were performed between each stage to remove residual reagents or minimize nonspecific binding.

This method ensures precise localization and differentiation of multiple target antigens within the same tissue section, facilitating a deeper understanding of co-expressed biomarkers.

4.5. Microscopy and Quantification of Immunohistochemical Labeling Intensity

The photomicrographs were captured at 40× magnification on a NIKON ECLIPSE LEICA EC3 microscope, model E200, and using the LAZ EZ Imaging software for Windows, software, version 3.4.0. These images were then analyzed by the IMAGE J [22], using the IHC Profiler plugin, version 1.46r [23] that quantified the DAB and hematoxylin staining, generating three numerical results for each sample. First, the software determined the staining intensity by dividing the pixels into three categories: high positive, low positive, and negative. The values obtained were recorded in an Excel spreadsheet, where the following formula was applied to calculate the staining histoscore (HS) 16, 17, which ranges from 0 to 300: HS = (1 × % low positive cells) + (2 × % positive cells) + (3 × % high positive cells).

4.6. Statistical Analyses

Statistical analyses were performed using R [24,25] and GraphPad Prism (v. 5.0). The normality of residuals of quantitative variables was analyzed by the Shapiro–Wilk test, and Levene's test assessed variance homogeneity. Quantitative data were described as mean ± standard error or median with minimum and maximum values, depending on data distribution. Group comparisons were performed using ANOVA followed by Bonferroni post hoc tests, *t*-tests with Welch's correction, or Kruskal–Wallis with Dunn's post hoc, as appropriate. For categorical variables, Fisher's exact test was applied. Residuals from the chi-square test were assessed to identify significance in the observed distributions. Count data were modeled using a quasi-Poisson regression. All graphs were generated using the ggplot2 package [26] and GraphPad Prism (v. 10.3.1).

4.7. Prognostic Classification

The terms "good prognosis" and "reserved prognosis" refer to tumors that, in most cases, tend not to progress to death and those that progress rapidly to death, respectively. For this classification, criteria such as lymphatic invasion, tumor size, morphological aspects of the cells and metastasis were considered.

5. Results

5.1. Age and Breed

This study revealed significant age differences among the groups. The mean age was significantly higher in the malignant group (M = 10.82 ± 0.70) compared to the benign group (M = 7.40 ± 0.68; p = 0.003), while no difference was observed between the malignant and control groups (M = 8.5 ± 0.86; p = 0.49) or the benign and control groups (p = 1.0) (Figure 2A). Regarding breeds, the malignant group included a wide variety of breeds such as Border Collie (n = 01), Dachshund (n = 01), Fox Paulistinha (n = 01), Golden Retriever (n = 01), Lhasa Apso (n = 01), German Shepherd (n = 01), Pinscher (n = 02), Poodle (n = 06), Shih Tzu (n = 03), Teckel (n = 01), and undefined breed (NDB) (n = 08). Similarly, the benign group included Border Collie (n = 02), Boxer (n = 01), Cocker Spaniel (n = 01), Lhasa Apso (n = 01), Maltese (n = 02), German Shepherd (n = 01), Pinscher (n = 01), Poodle (n = 02), Shih Tzu (n = 03), and NDB (n = 02). The control group, however, was composed only of German Shepherd (n = 01), Shih Tzu (n = 03), and NDB (n = 01) (Figure 2B).

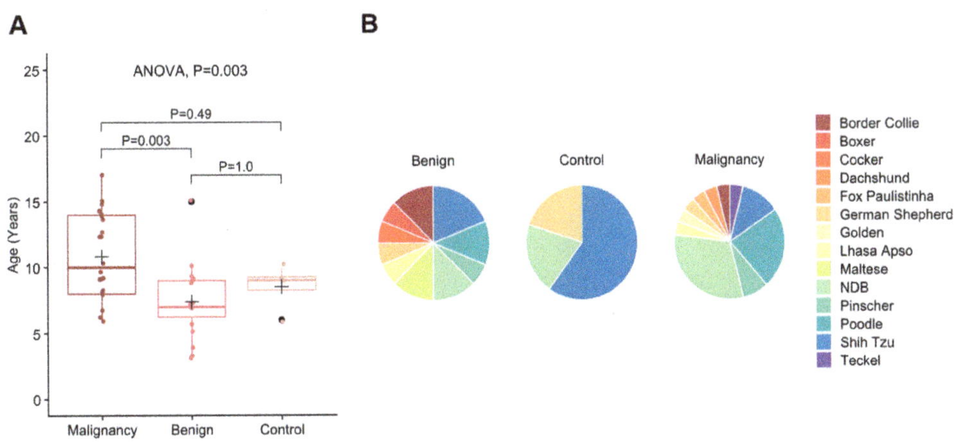

Figure 2. (**A**) Box plot illustrating age distribution across the malignant, benign, and control groups. p-values were calculated using ANOVA with Bonferroni post hoc tests. The black cross represents the mean age within each group. (**B**) Proportion representation of breeds within the malignant, benign, and control groups.

5.2. Anatomopathological Profile

Frequent necrosis was significantly more common in malignant tumors (r_{ij} = 1.83), while it was less prevalent in the benign group (r_{ij} = −1.73; p = 0.027). On the other hand, mild anisokaryosis was observed more frequently in benign tumors (r_{ij} = 2.3), whereas moderate anisokaryosis was predominant in malignant tumors (r_{ij} = 1.71; p = 0.005). Furthermore, marked nuclear pleomorphism was significantly associated with malignant tumors (r_{ij} = 1.71) and was notably less frequent in the benign group (r_{ij} = −1.61; $p < 0.001$). However, mild pleomorphism was more common in benign tumors (r_{ij} = 1.66) and less prevalent in the malignant group (r_{ij} = −0.89; p = 0.001). Tumor diameter did not differ significantly between the malignant and benign groups (p = 0.063), and the presence of vascular invasion and metastases was similar between these groups (Table 1). All sample data are available in the Supplementary Materials.

Table 1. Anatomopathological profile of the malignant, benign, and control groups.

	Malignant	Benign	Control	*p*
Necrosis n (%)				
Absent	14 (37)	19 (50)	05 (13)	
Moderate	01 (100)	0 (0)	0 (0)	
Central	2 (100)	0 (0)	0 (0)	0.027 *
Frequent	08 (100)	0 (0)	0 (0)	
Abundant	01 (100)	0 (0)	0 (0)	
Accentuated	01 (100)	0 (0)	0 (0)	
Anisokaryosis n (%)				
Absent	17 (47)	14 (39)	05 (14)	
Discret	0 (0)	05 (100)	0 (0)	0.005 *
Moderate	07 (100)	0 (0)	0 (0)	
Frequent	03 (100)	0 (0)	0 (0)	
Nuclear pleomorphism n (%)				
Absent	02 (17)	05 (42)	05 (42)	
Discret	05 (36)	09 (64)	0 (0)	<0.001 *
Moderate	13 (72)	05 (28)	0 (0)	
accentuated	07 (100)	0 (0)	0 (0)	
Tumor diameter (cm), mean ± SE (n)	1.34 ± 0.87 (11)	0.76 ± 0.40 (17)	–	0.063 †
Lymphatic invasion, n (%)				
Absent	22 (54)	19 (46)	–	0.067 *
Present	05 (100)	0 (0)	–	
Metastasis, n (%)				
Absent	23 (55)	19 (45)	–	0.131 *
Present	19 (100)	0 (0)	–	

SEM = standard error of the mean. * p-values for Fisher's exact test. † p-value for a t-test with Welch's correction.

To analyze the mitotic rate, a regression model with a quasi-Poisson distribution was applied. The mean mitotic rate in the malignant group was approximately 25.33 mitoses per sample ($\beta = 3.23 \pm 0.21$; $p < 0.001$). In contrast, the benign group exhibited a mean mitotic rate of 0.92 mitoses per sample ($\beta = -3.31 \pm 0.75$; $p < 0.001$). This indicates a substantial difference, with the mitotic rate between the two groups being approximately 0.04% of that observed in the malignant group. These findings underscore the significantly higher proliferative activity in malignant tumors compared to benign ones.

5.3. Protein Expression

The protein expression of ITIH2 (Figure 3) and ENO1 (Figure 4) was evaluated, and histoscore quantifications were performed. The expression levels were compared between the malignant, benign, and control groups (Figure 5). However, the immunohistochemical staining patterns did not differ between the groups. Regarding the ENO1 staining sites, it was observed that it was present in several cellular compartments (Figure 6). In the control group, 100% of the stainings were of the cytoplasm and nucleus simultaneously; in the benign group, 84.21% of the stainings were of the cytoplasm and nucleus simultaneously, and 15.79% were of the cytoplasm only; as for the malignant group, there was 18.52% of cytoplasm staining, 55% of simultaneous cytoplasm and nucleus staining, and 25.92% of plasma membrane staining.

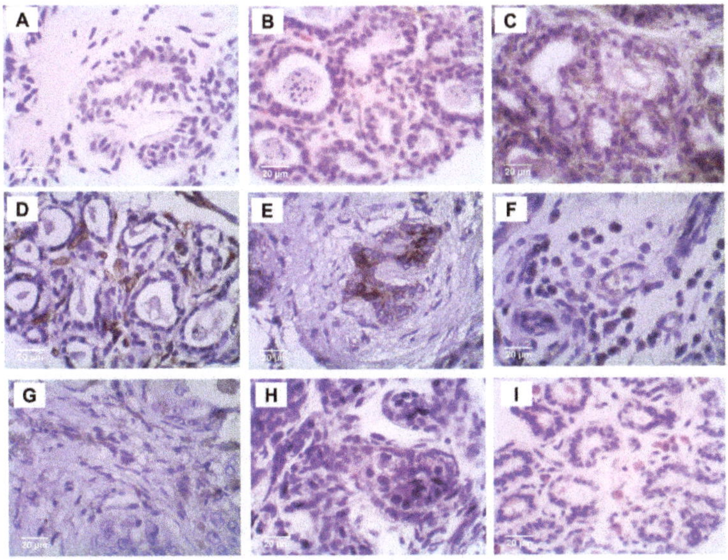

Figure 3. Immunohistochemical staining of ITIH2. Control group (**A–C**); benign tumors (**D–F**); and malignant tumors (**G–I**). Photomicrographs at 20 µm showing weak or absent immunostaining in the cytoplasm.

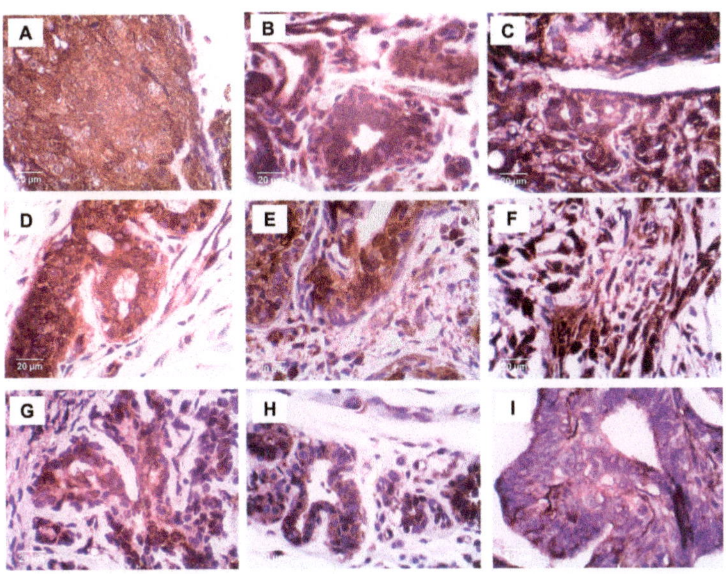

Figure 4. Immunohistochemical staining of ENO1. Control group (**A–C**); benign tumors (**D–F**); and malignant tumors (**G–I**). Photomicrographs at 20 µm showing strong immunostaining in nucleus, cytoplasm, and membrane (**I**).

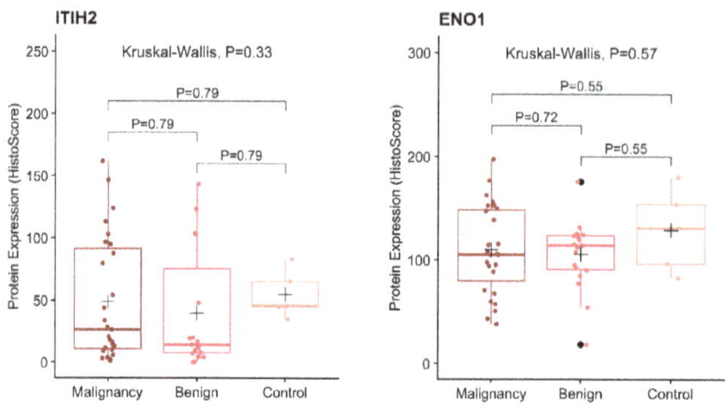

Figure 5. Protein expression (HistoScore) of ITIH2 (Inter-Alpha-Trypsin Inhibitor Heavy Chain 2) and ENO1 (Enolase 1). *p*-values for the Kruskal–Wallis test and Dunn's post hoc test. The black cross represents the mean.

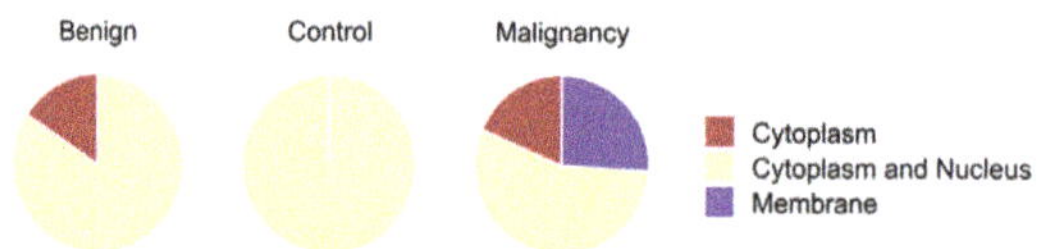

Figure 6. Immunohistochemical localization of ENO1 across cellular compartments.

Immunohistochemistry for the mesenchymal–epithelial transformation marker ZEB1 was performed to analyze its presence or absence in the samples. In cases where ZEB1 was present, the staining intensity was assessed observationally (Figure 7(1A,1B)). Double immunohistochemistry staining for ENO1 and ZEB1 (Figure 8) was performed to evaluate potential similarities or differences in their expression patterns. This analysis aimed to identify shared molecular pathways, mechanisms, or relevant distinctions between tumors with different prognoses.

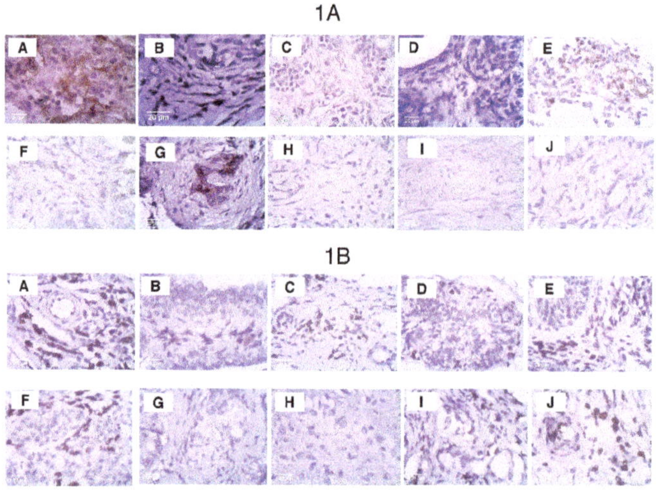

Figure 7. Immunohistochemical comparison of ITIH2 (**1A**) and ZEB1 (**1B**) staining. The tissues in (**1A**) and (**1B**) are identical, differing only by the marker used. Tumor types and grades include carcinoma in mixed tumor grade 2 (**A**); carcinoma in mixed tumor grade 1 (**H–J**); tubular carcinoma (**C**); solid carcinoma grade 3 (**E**,**F**); and benign mixed tumor (**B**,**D**,**G**). Metastatic tumors are shown in (**A**,**C**,**E**).

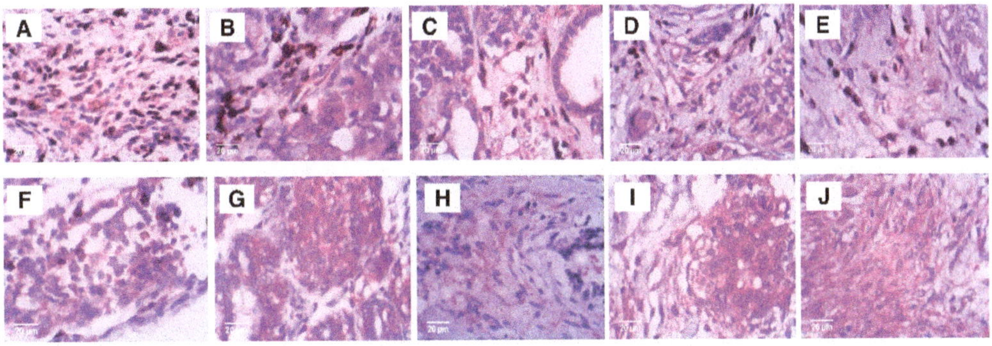

Figure 8. Double immunohistochemical staining showing ENO1 (magenta) and ZEB1 (brown) expression. Tumor types and grades include carcinoma in mixed tumor grade 2 (**A**); carcinoma in mixed tumor grade 1 (**H–J**); tubular carcinoma (**C**); solid carcinoma grade 3 (**E**,**F**); and benign mixed tumor (**B**,**D**,**G**).

5.4. Test to Evaluate the Association of ITIH2 and ENO1 with ZEB1 in Female Dog Mammary Tumors

Immunohistochemical analysis using the ZEB1 antibody was performed (Figure 7(1A,1B)) on samples (n = 6) of benign and malignant tumors with good prognosis (benign mixed tumor and carcinoma in mixed tumor, grade 1) and on samples (n = 4) of malignant tumors with reserved prognosis (carcinoma in mixed tumor, grade 2; tubular carcinoma; and solid carcinoma, grade 3). This particular analysis involved two observers who individually analyzed each of these 10 samples so that we could then proceed with considerations on staining intensities. When comparing the staining intensities of ITIH2 with those of ZEB1 in the samples with good prognosis, it was observed that in approximately 33% of the samples, ZEB1 was absent (G and H), with ITIH2 being moderate in one of these samples (G) and absent in the other (H); and in approximately 66% of the samples, ZEB1 showed strong staining (B, D, I, J), whereas ITIH2 also stained strongly in one (approximately 16.66%) of these samples (B) and was absent in 50% of them (D, I, J). In contrast, in malignant tumors with poor prognosis, n = 4, ITIH2 staining ranged from moderate in 50% of the samples (A, E) to absent in 50% (C, F), whereas ZEB1 showed strong intensity in all of these samples. Double staining with ENO1 and ZEB1 (Figure 7) revealed distinct patterns. In tumors with good prognosis, ENO1 was present in all samples, ranging from moderate in approximately 16.6% of samples (H) to strong in approximately 83% of samples (B, D, I, J, G), accompanied by strong ZEB1 staining in approximately 33% of samples (B, D) and absence of ZEB1 in approximately 66.6% of these samples (I, J, G, H). However, in malignant tumors with reserved prognosis, ENO1 showed moderate staining in approximately 50% of samples (E, F) and strong in the other 50% (A, C), while ZEB1 showed consistently strong intensity.

5.5. Predictive Potential

The prognostic potential of ITIH2 and ENO1 (Figure 9) for mammary tumors in bitches was assessed using ROC curve analysis, which revealed limited predictive power. For ITIH2, the area under the curve (AUC) was 0.67 ± 0.07 (95% confidence interval (CI) of 0.54–0.80; p = 0.217). The optimal cutoff value was (<44.50), providing a sensitivity of 65.22% (CI: 49.75–78.65%) and a specificity of 80% (CI: 28.36–99.49%). ENO1 exhibited an AUC of 0.64 ± 0.14 (95% CI: 0.37–0.92; p = 0.296), with a cutoff value of <127.9 yielding a sensitivity of 76.09% (CI: 61.23–87.41%) and a specificity of 60% (CI: 14.66–94.73%).

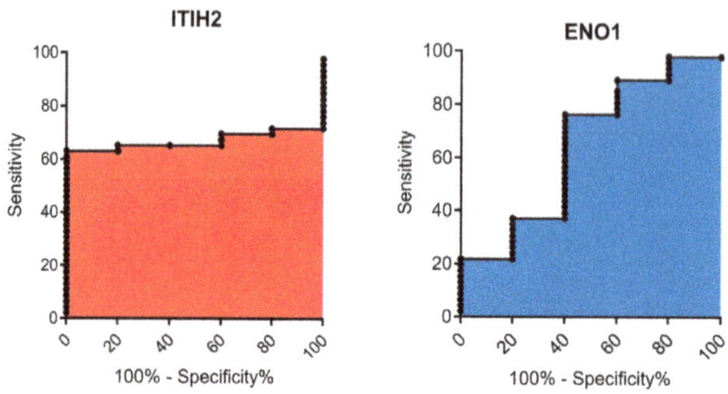

Figure 9. Roc curves of ITIH2 (Inter-Alpha-Trypsin Inhibitor Heavy Chain 2) and ENO1 (Enolase 1).

6. Discussion

The malignant group exhibited a higher mean age compared to the benign group, consistent with findings reported in the literature [27]. Conversely, tumor size did not differ significantly between the two groups, a result that contrasts with previous studies [28]. Anatomopathological characteristics such as necrosis, anisoacaryosis, nuclear pleomorphism, and increased mitotic activity were more pronounced in the malignant group, in alignment with established literature [29].

As a secreted protein, ITIH2 is expected to be present in both normal and tumor tissue, as well as in various locations, including plasma within vessels and in the extracellular medium. This widespread distribution complicates the establishment of a direct relationship between ITIH2 and breast tumors, as it is observed in all samples across groups. To overcome this challenge, a specific staining criterion was adopted: staining was considered positive only when present within tumor cells. In our findings, no intense ITIH2 staining was detected within cells in any sample. Moderate to weak cytoplasm staining was observed in some cases, with rare nuclear staining. This weak intracellular presence suggests that there is another function of ITIH2 besides its primary one, in the extracellular matrix and plasma. However, its occasional intracellular detection warrants further investigation into what its potential additional roles are, especially in cancer biology. Despite these results highlighting a limited cellular presence of ITIH2, its role in inflammation and extracellular matrix stabilization suggests potential relevance to breast cancer prognosis [30]. ITIH2 appears to have a dual, context-dependent function in tumor biology. On the one hand, its antiproteolytic function and regulation of the extracellular matrix (ECM) may suppress tumor progression. On the other hand, interactions with molecules such as hyaluronic acid and modulation of the inflammatory microenvironment may exert ambivalent effects, depending on the tumor's stage and context. A comprehensive understanding of ITIH2's role in tumors requires considering the specific cancer type, disease stage, and the surrounding microenvironment.

The expression of ITIH2 in breast tumors can be explained by its binding with hyaluronic acid, a molecule overexpressed in malignant tumors. Interestingly, hyaluronic acid serum levels do not increase in patients with localized breast tumors but rise during advanced disease stages [31]. In contrast, studies have shown reduced hyaluronic acid levels in infiltrating breast tumors [32,33]. Hyaluronic acid cross-linking with ITIH2 via tumor necrosis factor-induced protein 6 (TSG6) reduces its degradation and alters extracellular matrix (ECM) interactions, which may offer cancer-protective effects [34]. Although our discussions on ITIH2 are primarily based on studies of human breast cancer, this is due to the lack of specific data for dogs. This gap underscores the need for future investigations exploring the role of this protein in canine mammary tumors, contributing to a better understanding of its relevance in this context.

Regarding ENO1, the fact that there were no differences between the groups draws attention to the specificity of the antibody used. A higher frequency of simultaneous cytoplasmic and nuclear staining was observed in the same cell in all groups, especially in the control group, which exclusively presented this type of staining; in the benign group, there was mostly simultaneous staining of the cytoplasm and nucleus, but there was also staining only in the cytoplasm in some samples; in the malignant group, there was a lot of simultaneous staining of the cytoplasm and nucleus, but in some samples it stained only the cytoplasm, and in others it stained the plasma membrane. Cytoplasmic staining is in line with its role as a glycolytic enzyme, while nuclear localization—reported predominantly in benign tumors—is linked to its isoform MBP-1, which binds to C-myc, inactivating it, thus acting as a tumor suppressor [35]. However, cytoplasmic staining does not always predict ENO1, and it may be MBP1, its isoform.

In malignant cells, there was simultaneous staining of the cytoplasm and nucleus, as well as staining of the cell membrane, which did not occur in the benign group or in the control group. This fact corroborates studies that report the expression of ENO1 on the cell surface exceptionally in malignant tumors, where it binds to plasminogen to promote ECM degradation and metastasis [36]. Our findings using the anti-ENO1 antibody revealed expression of both ENO1 (cytoplasm/membrane) and MBP1 (nucleus and cytoplasm). A study showed that both proteins can be expressed in normal and tumoral mammary glands [37]. Our results show that in all groups there was a majority staining of the nucleus and cytoplasm, with additional expression of the membrane, exclusive to the malignant tumor group.

This shows that concomitant staining of ENO1 and MBP1 may have occurred in all groups, but we were unable to differentiate them because we used an antibody that recognizes ENO1 and its isoform MBP1. We interpret this as stating that it is necessary to use antibodies specific for ENO1 and another for MBP1; only then could we be certain of the quantifications related to each one, as well as knowing which of the two isoforms would be in the cytoplasm, since both can be expressed in the cytoplasm. Corroborating this, an article [37] stated that the studies conducted so far in tumor and normal tissue samples depend on the use of commercially and non-commercially available antibodies recognizing both ENO1 and MBP-1, which has made it difficult to clearly assess the relative expression of the two proteins by immunohistochemistry. We hypothesize that if there is greater expression of ENO1 to the detriment of MBP1, this could evidence energetic favoritism for the tumor; on the other hand, if there is majority expression of MBP1, it could be inferred that the tumor cells are in hypoxia, since MBP1 does not have an enzymatic function in glycolysis like ENO1. As a conclusion regarding the expression of ENO1, we can state that we are certain of its membrane expression exclusively in malignant tumors.

The low AUC values indicate that ITIH2 and ENO1 have limited predictive capability. The wide confidence intervals and lack of statistical significance further highlight their limited utility as individual prognostic markers. These findings underscore the need for larger studies to confirm these results and suggest that combining biomarkers or exploring alternative analytical approaches may yield better predictive performance. Further research into the biological roles of ITIH2 and ENO1 is also warranted.

In canine mammary tumors, foreign tissues such as bone and cartilage are commonly observed due to epithelial–mesenchymal transition. Knowing that ZEB1 regulates the transcription of ITIH2, the additional test we performed with ZEB1 aimed to investigate its presence for two reasons: the first was to find out whether there is a relationship between ZEB1 and ITIH2 in canine mammary tumors, and the second was to confirm whether ZEB1 regulates the transcription of ITIH2 [38]. Our results regarding the immunohistochemical test with ZEB1 and visual analysis showed that this protein is indeed present in both benign and malignant mammary tumors of bitches. However, it is important to clarify that statistically the number of samples, both total and distributed in the subgroups "good prognosis and reserved prognosis", was insufficient. However, our objective was only to collaborate with future studies, verifying the absence or presence of ZEB1 and whether this coincided with the intensity of staining of ITIH2 and ENO1, as well as the absence/presence of these two proteins, all this to suggest future studies of colocalization/co-expression of these proteins in mammary tumors of bitches. It was observed that in tumors with good prognosis, ITIH2 presented absent to moderate staining, while ZEB1 presented intense staining in most samples, although absent in a small part; in tumors with reserved prognosis, ITIH2 did not differ in its staining pattern, but ZEB1 stained intensely and strongly in all samples.

In double staining with ENO1 and ZEB1, we observed that in tumors with good prognosis, while ENO1 was marked strongly in most samples, ZEB1 was absent. In tumors with reserved prognosis, it was observed that ZEB1 was present with strong intensity in all samples. Moreover, ENO1 was also present in all samples, but was ranging from moderate to strong intensity. In summary, in mammary tumors of dogs, ZEB1 was found in tumors with good prognosis and in those with poor prognosis but showed a tendency to be present and to mark more intensely in tumors with poor prognosis; in addition, its expression showed no association with ITIH2, but it did with ENO1.

CD44 is a transmembrane glycoprotein that acts as an endocytic HA receptor in human breast tumor cells [39]; therefore, future investigations should explore the interactions between ITIH2, hyaluronic acid, and CD44 in canine and human breast tumors, given the evidence of potential functional relationships. Furthermore, ZEB1 could also be included in these investigations.

Regarding ENO1, future studies with larger sample sizes are essential to elucidate its interaction with ZEB1 and its subcellular localization in tumor cells, as these factors provide significant information on tumor progression and prognosis.

7. Conclusions

This study investigated the immunohistochemical expression of ITIH2 and ENO1 in mammary tissues from healthy dogs and those with benign and malignant tumors. Although no statistically significant differences in histoscores were observed between groups, we identified two findings: ENO1 staining of the plasma membrane was exclusive to malignant tumors, and its nuclear presence was observed in both benign and malignant tumors, contradicting previous reports of its restriction to benign cases. Furthermore, we confirmed the presence of the MBP-1 (nuclear) isoform in normal mammary tissue, highlighting its relevance beyond tumor contexts.

Although this study faced limitations, including sample size, our findings open avenues for exploring the mechanisms underlying cancer in the canine species. Notably, the involvement of ITIH2 and ENO1 in extracellular vesicles (EVs) and their prevalence in the bloodstream highlight their potential as biomarkers and therapeutic targets. These results highlight the importance of further research into the role of these proteins in cancer progression, with implications for veterinary and comparative oncology.

Supplementary Materials: The following supporting information can be downloaded at: https://www.mdpi.com/article/10.3390/vetsci12020110/s1, To ensure transparency and accessibility of data, we provide a table with the Histoscore data and anatomopathological profiles of all patients and another table with the metadata.

Author Contributions: D.A.P.d.C.Z. and A.A.N. contributed substantially to conception and design. D.F.d.N. and L.d.S.e.S. were involved in acquiring biological material. D.F.d.N., B.C.B., L.d.S.B. and

L.d.S.e.S. were involved in immunohistochemical analysis. D.F.d.N. was involved in histopathological analysis. D.A.P.d.C.Z., P.H.F.J. and L.d.S.e.S. were involved in data interpretation. D.A.P.d.C.Z., L.G.d.A.C., P.H.F.J. and L.d.S.e.S. were involved in drafting the manuscript or revising it critically for important intellectual content. D.A.P.d.C.Z. gave final approval for the version to be published. Each author has participated sufficiently in the work to take public responsibility for appropriate parts of the content and agrees to be accountable for all aspects of the work in ensuring that questions related to the accuracy or integrity of any part of the work are appropriately investigated and resolved. All authors have read and agreed to the published version of the manuscript.

Funding: This research was funded by the FAPESP 2020/12970-8 CNPq, grant number 131890/2023-9/307288/2022-6.

Institutional Review Board Statement: This project has already been submitted to the Animal Use Ethics Committee (CEUA) of the Faculty of Medicine of São José do Rio Preto—SP and was duly approved, being registered under protocol number 001-004391/2019 and its approval report number 4,007,723.

Informed Consent Statement: Not applicable, since this research did not involve human subjects.

Data Availability Statement: To ensure transparency and accessibility of data, we provide a table with the Histoscore data and anatomopathological profile of all patients, and another table with the metadata.

Acknowledgments: We would like to thank CNPq for supporting this project (Proc. No. 131890/2023-9) and veterinarian Daniela de Farias Nóbrega, from the PAT ANIMAL Veterinary Pathology Laboratory, for her assistance and availability in collecting canine samples for this research.

Conflicts of Interest: The authors declare no conflicts of interest.

References

1. Sorenmo, K.U.; Kristiansen, V.M.; Cofone, M.A.; Shofer, F.S.; Breen, A.M.; Langeland, M.; Mongil, C.M.; Grondahl, A.M.; Teige, J.; Goldschmidt, M.H. Canine mammary gland tumours, a histological continuum from benign to malignant; clinical and histopathological evidence. *Veter. Comp. Oncol.* **2009**, *7*, 162–172. [CrossRef] [PubMed]
2. Kaszak, I.; Ruszczak, A.; Kanafa, S.; Kacprzak, K.; Król, M.; Jurka, P. Current biomarkers of canine mammary tumors. *Acta Vet. Scand.* **2018**, *60*, 66. [CrossRef] [PubMed] [PubMed Central]
3. Zuccari, D.A.P.D.C.; Santana, A.E.; Rocha, N.S. Correlação entre a citologia aspirativa por agulha fina e a histologia no diagnóstico de tumores mamários de cadelas. *Braz. J. Vet. Res. Anim. Sci.* **2001**, *38*, 38–41. [CrossRef]
4. Borecka, P.; Ciaputa, R.; Janus, I.; Bubak, J.; Piotrowska, A.; Ratajczak-Wielgomas, K.; Podhorska-Okolów, M.; Dzięgiel, P.; Nowak, M. Expression of Periostin in Mammary Cancer Cells of Female Dogs. *In Vivo* **2020**, *34*, 3255–3262. [CrossRef] [PubMed] [PubMed Central]
5. Vigneau, A.; Rico, C.; Boerboom, D.; Paquet, M. Statins downregulateYAPandTAZand exert anti-cancer effects in canine mammary tumour cells. *Veter. Comp. Oncol.* **2021**, *20*, 437–448. [CrossRef] [PubMed]
6. Fischer, A.H.; Zhao, C.; Li, Q.K.; Gustafson, K.S.; Eltoum, I.; Tambouret, R.; Benstein, B.; Savaloja, L.C.; Kulesza, P. The cytologic criteria of malignancy. *J. Cell. Biochem.* **2010**, *110*, 795–811. [CrossRef]
7. Gómez, B.; Ramirez, M.; Maldonado, J. Presence of lung metastases in bitches affected by malignant mammary neoplasms in Medellin (Colombia). *Rev. MVZ Córdoba* **2012**, *17*, 2983–2990. [CrossRef]
8. de Almeida, L.G.N.; Young, D.; Chow, L.; Nicholas, J.; Lee, A.; Poon, M.-C.; Dufour, A.; Agbani, E.O. Proteomics and Metabolomics Profiling of Platelets and Plasma Mediators of Thrombo-Inflammation in Gestational Hypertension and Preeclampsia. *Cells* **2022**, *11*, 1256. [CrossRef] [PubMed]
9. Bost, F.; Diarra-Mehrpour, M.; Martin, J.P. Inter-alpha-trypsin inhibitor proteoglycan family—A group of proteins binding and stabilizing the extracellular matrix. *Eur. J. Biochem.* **1998**, *252*, 339–346. [CrossRef] [PubMed]
10. Lee, S.; Park, J.; Kim, E.J.; Park, S.; Kurie, J.M.; Ahn, Y.H. ZEB1-driven reconstruction of the hyaluronan network establishes a pro-metastatic microenvironment in lung adenocarcinoma. In Proceedings of the American Association for Cancer Research Annual Meeting, Orlando, FL, USA, April 2023.
11. Kang, X.; Bai, L.; Qi, X.; Wang, J.; Kang, X.; Bai, L.; Qi, X. Screening and identification of key genes between liver hepatocellular carcinoma (LIHC) and cholangiocarcinoma (CHOL) by bioinformatic analysis. *Medicine* **2020**, *99*, e23563. [CrossRef] [PubMed] [PubMed Central]

12. Hamm, A.; Veeck, J.; Bektas, N.; Wild, P.J.; Hartmann, A.; Heindrichs, U.; Kristiansen, G.; Werbowetski-Ogilvie, T.; Del Maestro, R.; Knuechel, R.; et al. Frequent expression loss of Inter-alpha-trypsin inhibitor heavy chain (ITIH) genes in multiple human solid tumors: A systematic expression analysis. *BMC Cancer* **2008**, *8*, 25. [CrossRef] [PubMed]
13. Niu, L.; Gao, C.; Li, Y. Identification of potential core genes in colorectal carcinoma and key genes in colorectal cancer liver metastasis using bioinformatics analysis. *Sci. Rep.* **2021**, *11*, 23938. [CrossRef] [PubMed]
14. Gao, J.; Zhao, R.; Xue, Y.; Niu, Z.; Cui, K.; Yu, F.; Zhang, B.; Li, S. Role of enolase-1 in response to hypoxia in breast cancer: Exploring the mechanisms of action. *Oncol. Rep.* **2013**, *29*, 1322–1332. [CrossRef] [PubMed]
15. Li, A.; Li, H.; Jin, G.; Xiu, R. A proteomic study on cell cycle progression of endothelium exposed to tumor conditioned medium the possible role of cyclin, D.1/E. *Clin. Hemorheol. Microcirc.* **2003**, *29*, 383–390. [PubMed]
16. Li, A.; Li, H.; Zhang, J.; Jin, G.; Xiu, R. The mitogenic and anti[1]apoptotic activity of tumor conditioned medium on endothelium. *Clin. Hemorheol. Microcirc.* **2003**, *29*, 375–382. [PubMed]
17. Ma, J.; Yang, F.; Both, S.K.; Prins, H.J.; Helder, M.N.; Pan, J.; Cui, F.-Z.; Jansen, J.A.; van den Beucken, J.J. In vitro and in vivo angiogenic capacity of BM-MSCs/HUVECs and AT-MSCs/HUVECs cocultures. *Biofabrication* **2014**, *6*, 015005. [CrossRef]
18. Gou, Y.; Li, F.; Huo, X.; Hao, C.; Yang, X.; Pei, Y.; Li, N.; Liu, H.; Zhu, B. ENO1 monoclonal antibody inhibits invasion, proliferation and clone formation of cervical cancer cells. *Am. J. Cancer Res.* **2021**, *11*, 1946–1961. [PubMed]
19. Mishra, M.; Sharma, A.; Thacker, G.; Trivedi, A.K. Nano-LC based proteomic approach identifies that E6AP interacts with ENO1 and targets it for degradation in breast cancer cells. *IUBMB Life* **2019**, *71*, 1896–1905. [CrossRef] [PubMed]
20. Chu, P.Y.; Hsu, N.C.; Liao, A.T.; Shih, N.Y.; Hou, M.F.; Liu, C.H. Overexpression of α-enolase correlates with poor survival in canine mammary carcinoma. *BMC Vet. Res.* **2011**, *7*, 1. [CrossRef] [PubMed]
21. Jensen, K.; Krusenstjerna-Hafstrøm, R.; Lohse, J.; Petersen, K.H.; Derand, H. A novel quantitative immunohistochemistry method for precise protein measurements directly in formalin-fixed, paraffin-embedded specimens: Analytical performance measuring HER2. *Mod. Pathol.* **2017**, *30*, 180–193. [CrossRef]
22. Varghese, F.; Bukhari, A.B.; Malhotra, R.; De, A. IHC Profiler: An Open Source Plugin for the Quantitative Evaluation and Automated Scoring of Immunohistochemistry Images of Human Tissue Samples. *PLoS ONE* **2014**, *9*, e96801. [CrossRef] [PubMed]
23. Capello, M.; Ferri-Borgogno, S.; Cappello, P.; Novelli, F. α-Enolase: A promising therapeutic and diagnostic tumor target. *FEBS J.* **2011**, *278*, 1064–1074. [CrossRef]
24. Cassali, G.D.; Jark, P.C.; Gamba, C.; Damasceno, K.A.; Lima, A.E.; Nardi, A.B.D.; Ferreira, E.; Horta, R.S.; Firmo, B.F.; Sueiro, F.A.; et al. Consensus Regarding the Diagnosis, Prognosis and Treatment of Canine and Feline Mammary Tumors—2019. *Braz. J. Vet. Pathol.* **2020**, *13*, 555–574. [CrossRef]
25. Posit Team. *RStudio: Integrated Development Environment for R*; Posit Software: Boston, MA, USA, 2024.
26. R Core Team. R: A Language and Environment for Statistical Computing. Vienna: R Foundation for Statistical Computing. 2024. Available online: https://www.R-project.org/ (accessed on 26 January 2025).
27. Villanueva RA, M.; Chen, Z.J. Ggplot2: Elegant Graphics for Data Analysis. 2019. Available online: https://ms.mcmaster.ca/~bolker/misc/ggb.pdf (accessed on 26 January 2025).
28. Oliveira Filho, J.C.; Kommers, G.D.; Masuda, E.K.; Marques, B.M.; Fighera, R.A.; Irigoyen, L.F.; Barros, C.S. Estudo retrospectivo de 1.647 tumores mamários em cães. *Pesqui. Veterinária Bras.* **2010**, *30*, 177–185. [CrossRef]
29. Bitencourt, A.G.V.; Lima, E.N.P.; Chojniak, R.; Marques, E.F.; Souza, J.A.D.; Graziano, L.; Andrade, W.P.; Osório, C.A.B.D.T. Correlação entre resultado do PET/CT e achados histológicos e imuno-histoquímicos em carcinomas mamários. *Radiol. Bras.* **2014**, *47*, 67–73. [CrossRef] [PubMed]
30. Mendes, T.C.; Guim, T.N.; Dias, M.F.; Bonel-Raposo, J.; Fernandes, C.G. Comparação entre os sistemas histomorfológico e de graduação histológica para classificação prognóstica de tumores mamários em cadelas. *Acta Sci. Vet.* **2007**, *35*, 339–343. [CrossRef]
31. Zhang, X.-F.; Zhang, X.-L.; Guo, L.; Bai, Y.-P.; Tian, Y.; Luo, H.-Y. The function of the inter-alpha-trypsin inhibitors in the development of disease. *Front. Med.* **2024**, *11*, 1432224. [CrossRef] [PubMed]
32. Arnold, J.M.; Gu, F.; Ambati, C.R.; Rasaily, U.; Ramirez-Pena, E.; Joseph, R.; Manikkam, M.; Martin, R.S.; Charles, C.; Pan, Y.; et al. UDP-glucose 6-dehydrogenase regulates hyaluronic acid production and promotes breast cancer progression. *Oncogene* **2020**, *39*, 3089–3101, Erratum in *Oncogene* **2020**, *39*, 3226–3228. [CrossRef] [PubMed] [PubMed Central]
33. García, G.A.; Hernández, S.; Mejía, Ó.R.; Baez, S.A.; García, A. Biología y patobiología humana del ácido hialurónico en la estabilización de la matriz extracelular y la inflamación. *Rev. Med.* **2006**, *14*, 80–87.
34. Ruibal, A.; Arias, J.; Del Rio, M.C.; Resino, C.; Tejerina, A. Carcinomas ductales infiltrantes de mama en mujeres mayores de 60 años. Asociación con mayor proliferación celular y menores concentraciones de pS2 y ácido hialurónico citosólico y de membrana. *Rev. Española De Med. Nucl.* **2001**, *20*, 525–529. [CrossRef] [PubMed]
35. Xu, Y.; Benedikt, J.; Ye, L. Moléculas Interagindo com Ácido Hialurônico Mediadas Crosstalk entre Células Cancerosas e Microambiente de Tumor Primário a Metástase Distante. *Cânceres* **2024**, *16*, 1907.
36. Didiasova, M.; Schaefer, L.; Wygrecka, M. When place matters: Shuttling of enolase-1 across cellular compartments. *Front. Cell Dev. Biol.* **2019**, *7*, 61. [CrossRef]

37. Lo Presti, M.; Ferro, A.; Contino, F.; Mazzarella, C.; Sbacchi, S.; Roz, E.; Feo, S. Myc promoter-binding protein-1 (MBP-1) is a novel potential prognostic marker in invasive ductal breast carcinoma. *PLoS ONE* **2010**, *5*, e12961. [CrossRef] [PubMed]
38. Lee, S.; Park, J.; Kim, E.J.; Cho, S.; Park, S.; Kurie, J.M.; Ahn, Y.H. ZEB1-regulated hyaluronan network in the lung tumor microenvironment. *Cancer Res.* **2024**, *84* (Suppl. S6), 4265. [CrossRef]
39. de Tótaro, P.I.S. *Aplicação de Nanopartículas Multifuncionais Baseadas em Fosfato de Cálcio Como Plataforma Para Imageamento e Tratamento do Câncer de Mama In Vitro*; Universidade Federal de Minas Gerais: Belo Horizonte, Brazil, 2017.

veterinary sciences

MDPI

Article

How Does *Saccharomyces cerevisiae* DSM 34246 (Canobios-BL) *var. boulardii* Supplementation Impact the Fecal Parameters of Healthy Adult Dogs?

Nicolò Lonigro [1], Francesca Perondi [2], Natascia Bruni [3], Mauro Bigliati [3], Annalisa Costale [1], Elena Pagani [4], Ilaria Lippi [2], Alice Melocchi [5], Lucia Zema [5], Giorgia Meineri [6] and Elisa Martello [7,*]

1 Department of Drug Science and Technology, University of Turin, 10124 Turin, Italy; nicolo.lonigro@unito.it (N.L.); annalisa.costale@unito.it (A.C.)
2 Department of Veterinary Science, University of Pisa, 56122 Pisa, Italy; f.perondi87@gmail.com (F.P.); ilaria.lippi@unipi.it (I.L.)
3 Candioli Pharma S.r.l., 10092 Turin, Italy; natascia.bruni@candioli.it (N.B.); bigliativet@virgilio.it (M.B.)
4 Monge & C. S.p.A, 12030 Cuneo, Italy; elena.pagani@monge.it
5 PhormulaMI Lab, Department of Pharmaceutical Sciences, University of Milan, 20133 Milan, Italy; alice.melocchi@unimi.it (A.M.); lucia.zema@unimi.it (L.Z.)
6 Department of Veterinary Sciences, School of Agriculture and Veterinary Medicine, University of Turin, 10095 Turin, Italy; giorgia.meineri@unito.it
7 Division of Epidemiology and Public Health, School of Medicine, University of Nottingham, Nottingham NG5 1PB, UK
* Correspondence: martello.elisa@gmail.com or elisa.martello1@nottingham.ac.uk

Simple Summary: This study tested *Saccharomyces cerevisiae* DSM 34246 (Canobios-BL) *var. boulardii* on adult dogs (West Highland White Terrier (WT) and German Shepherd (GS)) to evaluate its impact on gut health. A total of 53 healthy adult dogs were randomly assigned to control (CTR: WT 14/28, GS 12/25) and treated (SACC: WT 14/28, GS 13/25) groups. Both were fed a dry diet twice daily, with the additive given to the SACC group (5×10^9 CFU/kg) and a placebo to the CTR. Over 35 days, body weight, body condition score, fecal parameters, and water intake were measured. Statistical analysis showed significant improvements in body condition, fecal parameters, and IgA, indicating *S. boulardii* positively affected gut health in dogs.

Abstract: Background: A supplement containing *Saccharomyces cerevisiae* DSM 34246 (Canobios-BL) *var. boulardii* was tested to demonstrate its efficacy as a gut health stabilizer in two breeds of dogs, West Highland White Terrier (WT) and German Shepherd (GS). Methods: A randomized double-blind controlled trial was performed on a total of 53 healthy adult dogs, 28 WT and 25 GS. The animals were randomly assigned to a control group (CTR: WT 14/28, GS 12/25) and to a treated group (SACC; WT 14/28, GS 13/25). The CTR group was fed with dry commercial food, while the SACC group was fed with the same food but supplemented with *S. boulardii* at the concentration of 5×10^9 CFU/kg. The study lasted 35 days and included six evaluation time points (T_0-T_5). Body Weight (BW), Body Condition Score (BCS), Fecal Score (FS), FS measured with a penetrometer (FSp), Fecal Dry matter (DM), Fecal Humidity (UM), and Fecal IgA (IgA) were measured at each time point. Results: A significant improvement ($p < 0.1$) in BCS, FS, FSp, DM, UM, and IgA in the SACC group compared to the CTR for both WT and GS was reported. Conclusions: We found a positive effect of *S. boulardii* supplementation on the improvement of the fecal parameters and on the maintenance of good physiological as well as biological conditions in healthy adult dogs in breeding conditions.

Keywords: pet; probiotic; feed supplements; yeast products; immunomodulation; gut health

1. Introduction

The gastrointestinal (GI) tract is populated by a variety of microbes, which were recently demonstrated to play a major role in both human and animal health [1]. In fact, "the healthy gut" is overall linked to the welfare of the host not only because it is a vital organ for digestion and absorption of nutrients, but also as it is involved in the defense against pathogens and in the maintenance of the immune system [2,3].

The gut microbiota is essential for maintaining the homeostasis of the host, especially when there is a microbial imbalance in the GI tract, which might possibly lead to a change in the microbiome, influencing gut health [1,3]. The use of supplements for maintaining gut health is also increasing in veterinary medicine, as their administration has shown them to be effective and associated with limited side effects [4]. For example, probiotics—intended to be administered as live organisms able to provide health benefits to the host [3,4]—are largely used to maintain GI health and mitigate stress-dependent GI problems in animals [5]. Specifically, they support the barrier function of intestinal tissues and their regeneration [3,6].

Several bacteria have recently been identified as probiotics [7] and, for example, beneficial effects related to the supplementation of *Saccharomyces cerevisiae var. boulardii* in both animals [3,8–10] and humans [11,12] have been preliminary reported. Indeed, the administration of *S. boulardii* in humans affected by GI diseases, such as chronic diarrhea and inflammatory bowel disease (IBD), has demonstrated clinical effectiveness, indicating that such a probiotic might positively interfere with inflammatory conditions [4,11]. Thus, *S. boulardii's* probiotic activity in humans has been linked with multiple effects, such as improvements in gut barrier function, pathogen competitive exclusion, the production of antimicrobial peptides, immune modulation, and trophic effects on the gut [4,13]. Similarly, *S. boulardii* recently started to be used as a probiotic in pets [2,3,14]. One study showed an improvement in the intestinal status and general health conditions of dogs suffering from chronic enteropathies when treated for 60 days with *S. boulardii* [14]. In addition, a different study involving healthy dogs in breeding conditions reported promising outcomes regarding intestinal microbiota and nutritional as well as stress status, resulting from the supplementation of *S. boulardii* for 42 days [3].

Based on the above-mentioned considerations, the aim of this randomized double-blind controlled trial was to demonstrate the efficacy of the feed additive *S. cerevisiae* DSM 34246 (Canobios-BL) *var. boulardii* as a "gut flora stabilizer" in healthy dogs.

2. Materials and Methods

2.1. Animals and Study Design

This randomized double-blind controlled trial included a group of 53 adult dogs (aged from 1 to 5 yr). In detail, 28 dogs were West Highland White Terriers (WTs) (10 males and 18 females), and the other 25 were German Shepherds (GS) (13 males and 12 females), selected from two ENCI-registered breeders in Italy.

Animals were randomly assigned to control (CTR: WT 14/28, GS 12/25) and treated (SACC: WT 14/28, GS 13/25) groups.

Both groups were fed with a commercial dry food diet for adult dogs (Monge Natural Superpremium Grain Free all breeds Adult Anchovies with Potatoes and Peas) two times a day, from at least 7 days before the beginning of the study.

The amount of daily food was calculated based on the following equation: (NRC, 2006): ME (kcal/day) = $110 \times (\text{kg BW})^{0.75}$.

A placebo (Maltodextrin powder) or a supplement containing *S. cerevisiae* DSM 34246 (Canobios-BL) *var. boulardii* (5.0×10^9 CFU/kg) [15] was added to the pet food given to the CTR or to the SACC group, respectively, once a day for the whole duration of the study (35 consecutive days). The study was divided into five experimental times (from T_0 to T_5), with each time point separated from the previous and/or the following one by a seven-day interval. The number of animals and the length of the trial were set according to EFSA (European Food Safety Authority) regulations for registering a feed additive [15].

At the beginning of the study, a veterinarian checked the health status of the animals through a general physical examination. All the recruited animals were healthy with no underlying conditions or drugs administered in the past 15 days. In the case of a change in health status, pregnancy, pharmacological treatments, diet modification, pathological symptoms, and/or death, the dogs would have to be excluded from the study. Dogs were housed in kennels with two or three animals per cage. The cage area was 6 ($\pm$2) square meters in size, with an open space of the same size, considering the principles of animal welfare and the Italian regulations, thus avoiding any stress. The breeders were informed of the design of the study and signed a written informed consent form. The experimental procedures used were approved by the Bioethics Committee of the University of Turin, Italy (approval 156895, 14 April 2020).

2.2. Data Collection

All the tests aimed at evaluating the nutritional status of the dogs were performed by the same veterinarian, following standard guidelines [16]. Daily water intake (WI) in liters was reported at each time point by the breeders. Body Weight (BW), Body Condition Score (BCS), Fecal Score (FS), FS measured with a penetrometer (FSp), Fecal Dry matter (DM), Fecal Humidity (UM), and Fecal IgA (IgA) were selected as the parameters for this study. The BW (kg) was recorded at T_0 (day 0), T_1 (day 7), T_2 (day 14), and T_3 (day 21), T_4 (day 28), and T_5 (day 35). The BCS [17] was recorded in scores ranging between 1 and 9, with the points being assigned after a visual examination and palpation of the animal at T_0 and T_5. A score of 4 or 5 represents the ideal [17] (WSAVA 2013). Four fecal parameters, FS (1-7), FSp (1-7), DM (%), UM (%), and IgA (mg/g), were evaluated at each time point.

FS was determined by direct examination of fresh feces using a 7-point scale, with an ideal score between 3 and 4 [18,19].

The measurement of fecal hardness in kg/cm^2, was performed on fresh stool with a Penetrometer 53220 FTA (GUSS Manufacturing, PTY Ltd., Cape Town, South Africa) using a technique already described by Davies and colleagues (Davies et al., 1986) [18]. The amount of stool used for each analysis was at least 40–50 g depending on the type and shape of the stool itself. The fecal hardness was defined as the mean of three measurements. Fecal hardness was converted using a validated scale to obtain the FSp.

For the assessment of the UM, a 5–10 g stool sample was weighed, dried in an oven at 105–110 °C for 20–24 h, and cooled down in a desiccator for 20–24 h. Then, the dried sample was weighed again. The final UM result was defined as the mean of two different measurements and expressed as percentage.

DM was calculated from the UM through the following method: DM = 100 − UM. The result was expressed as a percentage.

Lastly, the IgA level was measured on wet fecal samples using a specific ELISA kit (Dog IgA ELISA Quantitation Set, Bethyl Laboratories Inc., Montgomery, TX, USA).

2.3. Statistical Analysis

The statistical analysis conducted in the study varied depending on the type of parameter considered (categorical or numerical). In more detail, BW, FSp, IgA, DM, and UM were assessed using the analysis of variance (ANOVA) based on a repeated measures model, implemented through the mixed procedure (PROC MIXED MODEL SAS 9.4, 2013).

The statistical model was structured as follows:

$$y = \mu + Si + Gj + Tk + GTjk + ejkn \quad (1)$$

where y = dependent variable; μ = overall mean; Si = fixed effect of the sex (i = F; M); G = fixed effect of the treatment (j = 0.1); Tk = fixed effect of the kth time (k = 1.6). GTjk = fixed effect of the interaction between the jth treatment and kth time; ejkn = error.

Time was treated as a repeated measurement and replicated with groups as repeated subjects. The autoregressive covariance method was used for the covariance structure. Least square means were separated using Student's *t*-test.

The BCSs and FSs of the CTR and SACC groups were compared using the Kruskal–Wallis test (UCLA) for both the overall experimental period and the supplementation period, using PROC NPAR1WAY (SAS 9.4). If significant results were found, multiple comparison analyses based on pairwise two-sample Wilcoxon comparisons were conducted. Test results from the two-tailed tests with *p*-values < 0.10 were considered significant.

3. Results

All the dogs involved in the trial were healthy during the study and did not need any drugs or supplements at any time. Moreover, they did not receive any pharmacological treatment in the 15-day period before the start of the study, and no change in feed administration or in relevant consumption was highlighted.

In Table 1, the effects of the supplementation of *S. boulardii* on BW and BCS by breed are summarized. In more detail, at the beginning of the evaluation (T_0) the least square means (±SE) for the BW in the WT group were 8.2 ± 0.3 (CTR) and 7.50 ± 0.3 (SACC), while in the GS group they were 32.1 ± 1.6 (CTR) and 31.6 ±0.9 (SACC). The differences in BW between the SACC and CTR groups at T_0 were not significant over time in both dog breeds and treatment groups.

Table 1. Body weight (BW) and Body Condition Score (BCS) parameters measured in treated (SACC) versus untreated (CTR) West Highland White Terriers (WTs) and German Shepherds (GSs) at different times during the study.

		West Highland White Terrier (WT n = 28)			German Shepherd (GS n = 25)		
		CTR	SACC	*p*-Value	CTR	SACC	*p*-Value
BW (±SE) (kg)	T_0	8.2 ± 0.3	7.5 ± 0.3	0.1	32.1 ± 1.6	31.6 ± 0.9	0.6
	T_1	8.0 ± 0.3	7.6 ± 0.3	0.2	31.8 ± 1.6	32.0 ± 0.9	1.0
	T_2	8.0 ± 0.3	7.4 ± 0.3	0.2	31.7 ± 1.6	31.8 ± 1.0	0.9
	T_3	8.1 ± 0.3	7.5 ± 0.3	0.2	31.8 ± 1.6	31.9 ± 0.9	0.9
	T_4	8.3 ± 0.3	7.6 ± 0.3	0.1	31.7 ± 1.5	32.1 ± 0.9	0.9
	T_5	8.3 ± 0.3	7.7 ± 0.3	0.1	31.6 ± 1.5	32.1 ± 0.9	0.8
BCS (±SE) (1–9)	T_0	5.6 ± 0.1	5.5 ± 0.1	0.7	5.5 ± 0.2	5.6 ± 0.1	0.5
	T_1	5.5 ± 0.1	5.0 ± 0.1	<0.1	5.4 ± 0.1	5.2 ± 0.1	0.5
	T_2	5.5 ± 0.1	4.9 ± 0.1	<0.1	5.4 ± 0.1	4.6 ± 0.1	<0.1
	T_3	5.6 ± 0.1	4.9 ± 0.1	<0.1	5.3 ± 0.1	4.4 ± 0.1	<0.1
	T_4	5.7 ± 0.1	4.8 ± 0.1	<0.1	5.3 ± 0.2	4.4 ± 0.1	<0.1
	T_5	5.6 ± 0.1	4.6 ± 0.1	<0.1	5.2 ± 0.2	4.4 ± 0.1	<0.1

Focusing on the BCS, the mean values were 5.6 ± 0.1 (CTR) and 5.5 ± 0.1 (SACC) in the WT group, while they were 5.5 ± 0.2 (CTR) and 5.6 ± 0.2 (SACC) in the GS group at the beginning of the study with no significant differences reported (Table 1). Interestingly, for both dog breeds, the BCSs of SACC group reduced significantly over time (i.e., from T_0 to T_5), but such a decrease was not evident in the CTR group (Table 1). Overall, significant differences were observed in WT and GS between CTR and SACC from T_1 to T_5 (WT) and T_2 to T_5 (GS).

The FSp in both breeds in the SACC group showed a statistically significant decrease over time, with a mean value near to 3 at the end of the study. The CTR group did not show a significant reduction in the values measured. The FSp was significantly different between the two groups (T_1–T_5) in both breeds, and in the SACC groups it significantly decreased from T_1 to T_5 (Table 2).

Table 2. Fecal Score (FS), FS measured with a penetrometer (FSp), Fecal Dry matter (DM), Fecal Humidity (UM), and Fecal IgA (IgA) parameters measured in treated versus untreated West Highland White Terriers (WTs) and German Shepherds (GSs) at different times during the study.

		West Highland White Terrier (WT n = 28)			German Shepherd (GS n = 25)		
		CTR	**SACC**	***p*-Value**	**CTR**	**SACC**	***p*-Value**
FS (±SE) (1–7)	T_0	4.3 ± 0.2	4.3 ± 0.2	0.9	4.3 ± 0.4	4.2 ± 0.3	0.9
	T_1	4.3 ± 0.2	4.1 ± 0.2	0.4	4.4 ± 0.2	3.8 ± 0.2	<0.1
	T_2	4.3 ± 0.2	3.6 ± 0.2	<0.1	4.0 ± 0.2	3.4 ± 0.1	<0.1
	T_3	4.1 ± 0.3	3.8 ± 0.2	0.4	4.6 ± 0.4	3.3 ± 0.2	<0.1
	T_4	4.3 ± 0.3	3.6 ± 0.2	<0.1	4.8 ± 0.3	3.5 ± 0.1	<0.1
	T_5	4.1 ± 0.2	3.4 ± 0.1	<0.1	4.3 ± 0.2	3.2 ± 0.1	<0.1
FSp (±SE) (1–7)	T_0	4.5 ± 0.2	4.3 ± 0.3	0.7	4.4 ± 0.3	4.3 ± 0.3	0.7
	T_1	4.5 ± 0.1	3.9 ± 0.2	<0.1	4.6 ± 0.2	3.9 ± 0.2	<0.1
	T_2	4.6 ± 0.3	3.2 ± 0.2	<0.1	4.3 ± 0.2	3.3 ± 0.1	<0.1
	T_3	4.6 ± 0.3	3.5 ± 0.2	<0.1	4.6 ± 0.2	3.3 ± 0.1	<0.1
	T_4	4.6 ± 0.3	3.2 ± 0.1	<0.1	4.8 ± 0.3	3.3 ± 0.1	<0.1
	T_5	4.3 ± 0.3	3.0 ± 0.2	<0.1	4.3 ± 0.2	3.1 ± 0.1	<0.1
DM (±SE) (%)	T_0	30.4 ± 2.2	32.6 ± 2.8	0.5	29.2 ± 4.3	31.7 ± 2.8	0.6
	T_1	32.3 ± 2.0	34.2 ± 1.5	0.6	32.5 ± 1.9	37.3 ± 1.3	<0.1
	T_2	31.9 ± 2.1	42.6 ± 1.5	<0.1	36.1 ± 1.1	41.8 ± 1.4	<0.1
	T_3	33.3 ± 2.9	39.8 ± 2.4	<0.1	36.1 ± 3.1	41.4 ± 2.1	<0.1
	T_4	31.4 ± 2.7	42.5 ± 1.8	<0.1	34.6 ± 3.0	39.8 ± 1.7	<0.1
	T_5	33.1 ± 2.0	44.5 ± 1.2	<0.1	37.9 ± 1.7	44.0 ± 1.2	<0.1
UM (±SE) (%)	T_0	68.6 ± 1.1	65.9 ± 2.8	0.4	66.0 ± 4.3	67.1 ± 2.8	0.7
	T_1	66.8 ± 1.0	64.5 ± 1.5	0.5	68.1 ± 2.0	61.5 ± 1.2	<0.1
	T_2	67.1 ± 1.0	56.8 ± 1.5	<0.1	67.8 ± 1.2	57.5 ± 1.4	<0.1
	T_3	65.8 ± 1.4	59.6 ± 2.4	<0.1	70.1 ± 3.1	57.9 ± 2.1	<0.1
	T_4	67.8 ± 1.4	57.0 ± 1.8	<0.1	72.6 ± 2.9	59.3 ± 1.7	<0.1
	T_5	65.9 ± 1.0	55.0 ± 1.2	<0.1	66.5 ± 1.7	55.7 ± 1.2	<0.1
IgA (±SE) (mg/g)	T_0	0.6 ± 0.1	0.69 ± 0.1	0.3	0.57 ± 0.08	0.58 ± 0.04	0.8
	T_1	0.6 ± 0.1	0.68 ± 0.1	0.2	0.51 ± 0.07	0.62 ± 0.04	0.3
	T_2	0.6 ± 0.1	0.74 ± 0.1	0.2	0.51 ± 0.08	0.71 ± 0.06	<0.1
	T_3	0.6 ± 0.1	0.8 ± 0.1	<0.1	0.50 ± 0.08	0.78 ± 0.07	<0.1
	T_4	0.6 ± 0.1	0.9 ± 0.1	<0.1	0.52 ± 0.08	0.85 ± 0.07	<0.1
	T_5	0.5 ± 0.1	1.0 ± 0.1	<0.1	0.50 ± 0.07	0.92 ± 0.08	<0.1

In the same way, the FS observed by the veterinarian showed a statistically significant decrease in the SACC groups with the progression of the trial, with a value near to 3 for most of the animals at the end of the study period (T_5). In the WT group, the FS was different between the CTR and SACC groups at T_2, T_4, and T_5 (Table 2), with the SACC group showing a significant decrease.

The DM measured for both breeds in the SACC group showed a statistically significant increase overtime from T_1 (GS) or T_2 (WT) to T_5 and compared to the CTR group from T_1 (GS) or T_2 (WT) (Table 2).

The UM significantly decreased from T_1 (GS) or T_2 (WT) to T_5 in the SACC group in both breeds (Table 2).

The Fecal IgA level in both breeds showed a significant increase from T_1 to T_5 in the SACC groups, while all the CTR groups did not show any significant increase. The IgA measurements pointed out statistically significant differences between groups at T_3, T_4, and T_5 (Table 2).

Finally, the WI in the CTR and SACC groups within the same breed did not show any statistically significant variation during the whole study period (Table 3).

Table 3. Water intake (WI) expressed in liter (lt) measured in treated (SACC) versus untreated (CTR) dogs West Highland White Terriers (WTs) and German Shepherds (GSs) at different time points during the study.

		West Highland White Terrier (WT n = 28)		German Shepherd (GS n = 25)	
		CTR	SACC	CTR	SACC
WI (±SE) (Lt)	T_0	0.38 ± 0.02	0.37 ± 0.02	1.58 ± 0.06	1.58 ± 0.05
	T_1	0.38 ± 0.02	0.38 ± 0.02	1.60 ± 0.05	1.59 ± 0.08
	T_2	0.39 ± 0.01	0.38 ± 0.02	1.57 ± 0.06	1.59 ± 0.08
	T_3	0.37 ± 0.01	0.38 ± 0.02	1.57 ± 0.05	1.57 ± 0.04
	T_4	0.38 ± 0.02	0.39 ± 0.01	1.60 ± 0.06	1.57 ± 0.04
	T_5	0.38 ± 0.02	0.38 ± 0.02	1.58 ± 0.05	1.59 ± 0.05

4. Discussion

The most novel research on pets emphasizes the benefits of yeast products on the modulation of the intestinal microbiota (i.e., with potential increases in *Bifidobacterium* or *Lactobacillus*), with these compounds being able to enhance immune function, reduce potentially pathogenic microorganisms, and improve the animal's antioxidant status [7]. For example, *S. boulardii*, i.e., the probiotic yeast used in this study, can inhibit the colonization of pathogenic microorganisms, improve intestinal barrier function, and regulate immunity [13]. Very recently, its efficacy in promoting intestinal health and microbiome composition has also been demonstrated in kittens [2].

In the present study, *S. boulardii* was tested as a probiotic feed additive in healthy dogs for a period of 35 days. It did not cause any short-term adverse effects, as already reported by other authors [3,14]. At the beginning of the trial, all the dogs involved were healthy and no significant differences in the parameters selected as the outputs of the evaluation were highlighted.

In recent years, an increased use of probiotics in animal diets as supplements to maintain optimal gastro-intestinal health in both healthy pets and pets with disorders has been reported [1].

Moreover, yeast and yeast-based products have been shown to potentially promote gut health in both humans and animals [7]. However, there is still limited scientific literature discussing the impact of yeast products in dogs and cats, although the number of publications in this respect is slowly increasing [7].

Saccharomyces cerevisiae is one of the predominant yeast products used in various applications [7]. Various research works have demonstrated its beneficial effects on the intestinal health and microbiota of dogs, in both healthy animals [3] and in those with various gastrointestinal disorders [14]. Its administration improved dogs' fecal consistency (higher FS) resulting in a higher DM [20] when compared to the control groups, although the values remained within the ideal score range in some of the studies [7,14,21].

In this work, at the end of the treatment period, no significant differences in BW were reported in the SACC groups compared to the CTR groups nor over time, while the BCS showed a significant variation in the treated versus untreated dogs. At T_0 the two groups were similar in values (5.5–6), and then the BCS gradually decreased only in the treated groups, reaching a medium value between 4.5 and 5 at T_5 [17]. This finding highlights the positive effect of the supplement on BCS as it remains within the physiological range reflecting a good maintenance of the nutritional condition in the two dog breeds, which are known to have the tendency to become overweight, as also previously reported [7].

In addition, the data collected demonstrated the positive impact of *S. boulardii* on all fecal parameters under evaluation (FS, Fsp, DM, UM, IgA) for both dog breeds.

Indeed, FS as well as FSp presented significant differences in the two groups; specifically, they decreased in the SACC one, which was characterized by harder feces compared to the CTR groups, whose values were between 3 and 4 (Table 2).

A significant improvement in the other two fecal parameters (DM and UM) was reported, further strengthening the positive effect of *S. boulardii* supplementation on fecal consistency due to the minor water content in the samples.

In this trial, water intake (WI) should not have modified the fecal consistency parameters, as all animals, regardless of their treatment group, had similar values throughout the trial. Moreover, there was no statistically significant variation in WI between the CTR and SACC groups in both breeds during the study. This result supports the positive effect of *S. boulardii* on the previously discussed fecal parameters.

Interestingly, the administration of *S. boulardii* led to a significant increase in fecal IgA in the SACC group (Table 2). Fecal IgA is a biomarker for intestinal immunity or inflammation in dogs and puppies [22] and has been recently proposed as a non-invasive marker of canine intestinal health [3,7,22]. Indeed, secretory IgA is the most important humoral protective immune factor in the intestine. It inhibits the adhesion, colonization, and penetration of microorganisms, as well as the absorption of food antigens [23]. The data collected in this work, consistent with findings from other studies, confirmed that the use of supplementation with yeast products in dogs increases ileal and Fecal IgA, indicating enhanced mucosal immunity and immunomodulatory properties [7,20,24]. This study has some limitations; specifically, it would benefit from a longer trial period, the inclusion of different dog breeds to avoid potential breed biases, and the evaluation of other blood and fecal inflammatory parameters. Moreover, further analysis of changes in the gut microbiota would be useful to further support the role of *S. boulardii* as a gut flora stabilizer in healthy dogs.

5. Conclusions

The supplementation of *S. boulardii* for 35 days, at the recommended dietary dosage of 5.0×10^9 CFU/kg of feed in healthy dogs of different breeds (WH and GS), significantly improved the BCS, FS, FSp, DM, UM, and Fecal IgA parameters. Based on the results obtained, feed supplementation with the above-mentioned yeast product could also be recommended in clinical settings for an improvement in the fecal parameters of healthy dogs, also promoting an enhanced mucosal immunity. Further studies are recommended in dogs and cats suffering from gastrointestinal diseases.

Author Contributions: Conceptualization, N.B.; methodology, N.B. and A.C.; software, A.C. and F.P.; validation, F.P. and E.M.; formal analysis, A.C.; investigation, N.B., N.L., E.M., F.P. and M.B.; resources, N.B.; data curation, N.L., A.C., F.P., and E.M.; writing—original draft preparation, E.M. and F.P.; writing—review and editing, N.L., F.P., N.B., M.B., I.L., A.C., E.P., A.M., L.Z., G.M. and E.M.; visualization, A.C., E.M., and F.P.; supervision, N.B., project administration, N.B.; funding acquisition, N.B. All authors have read and agreed to the published version of the manuscript.

Funding: This research received no external funding. The APC was funded by ACEL pharma S.r.l.

Institutional Review Board Statement: This trial was reviewed and approved by the Bioethics Committee of the University of Turin (approval 156895, 14 April 2020).

Informed Consent Statement: Informed consent was obtained from the breeders involved in the study.

Data Availability Statement: The data that support the findings of this study are available on request from the corresponding author (EM).

Acknowledgments: We want to thank Selena Massa for her support in the data collection.

Conflicts of Interest: One of the authors is an employee of the Candioli Pharma S.r.l. Three of the authors are scientific consultants for the Candioli Pharma S.r.l. Also One of the authors is from Monge & C. S.p.A.

References

1. Bruni, N.; Martello, E.; Fusi, E.; Meineri, G.; Giardini, A. Study of faecal parameters and body condition in dogs with a diet supplemented with *Lactobacillus acidophilus* D2/CSL (CECT 4529). *Ital. J. Anim. Sci.* **2020**, *19*, 704–711. [CrossRef]
2. Zhang, M.; Mo, R.; Li, M.; Qu, Y.; Wang, H.; Liu, T.; Liu, P.; Wu, Y. Comparison of the Effects of Enzymolysis Seaweed Powder and *Saccharomyces boulardii* on Intestinal Health and Microbiota Composition in Kittens. *Metabolites* **2023**, *13*, 637. [CrossRef] [PubMed]
3. Meineri, G.; Martello, E.; Atuahene, D.; Miretti, S.; Stefanon, B.; Sandri, M.; Biasato, I.; Corvaglia, M.R.; Ferrocino, I.; Cocolin, L.S. Effects of *Saccharomyces boulardii* Supplementation on Nutritional Status, Fecal Parameters, Microbiota, and Mycobiota in Breeding Adult Dogs. *Vet. Sci.* **2022**, *9*, 389. [CrossRef] [PubMed]
4. Pais, P.; Almeida, V.; Yılmaz, M.; Teixeira, M.C. Saccharomyces boulardii: What Makes It Tick as Successful Probiotic? *J. Fungi* **2020**, *6*, 78. [CrossRef]
5. Redfern, A.; Suchodolski, J.; Jergens, A. Role of the gastrointestinal microbiota in small animal health and disease. *Vet. Rec.* **2017**, *181*, 370. [CrossRef]
6. Tomicic, M.Z.; Colovic, R.R.; Cabarkapa, S.I.; Vukmirovic, M.Đ.; Đuragic, M.O.; Tomicic, M.R. Beneficial properties of probiotic yeast *Saccharomyces boulardii*. *Food Feed. Res.* **2016**, *43*, 103–110. [CrossRef]
7. Maturana, M.; Castillejos, L.; Martin-Orue, S.M.; Minel, A.; Chetty, O.; Felix, A.P.; Adib Lesaux, A. Potential benefits of yeast Saccharomyces and their derivatives in dogs and cats: A review. *Front. Vet. Sci.* **2023**, *10*, 1279506. [CrossRef]
8. Rostoll Cangiano, L.; Villot, C.; Amorin-Hegedus, R.; Malmuthuge, N.; Gruninger, R.; Guan, L.L.; Steele, M. *Saccharomyces cerevisiae boulardii* accelerates intestinal microbiota maturation and is correlated with increased secretory IgA production in neonatal dairy calves. *Front. Microbiol.* **2023**, *14*, 1129250. [CrossRef]
9. Parada, J.; Magnoli, A.; Isgro, M.C.; Poloni, V.; Fochesato, A.; Martínez, M.P.; Carranza, A.; Cavaglieri, L. In-feed nutritional additive probiotic *Saccharomyces boulardii* RC009 can substitute for prophylactic antibiotics and improve the production and health of weaning pigs. *Vet. World* **2023**, *16*, 1035–1042. [CrossRef]
10. Hiltz, R.L.; Steelreath, M.R.; Degenshein-Woods, M.N.; Hung, H.C.; Aguilar, A.; Nielsen, H.; Rezamand, P.; Laarman, A.H. Effects of *Saccharomyces cerevisiae boulardii* (CNCM I-1079) on feed intake, blood parameters, and production during early lactation. *J. Dairy Sci.* **2023**, *106*, 187–201. [CrossRef]
11. Kelesidis, T.; Pothoulakis, C. Efficacy and safety of the probiotic *Saccharomyces boulardii* for the prevention and therapy of gastrointestinal disorders. *Therap. Adv. Gastroenterol.* **2012**, *5*, 111–125. [CrossRef]
12. Szajewska, H.; Kołodziej, M. Systematic review with meta-analysis: *Saccharomyces boulardii* in the prevention of antibiotic-associated diarrhoea. *Aliment. Pharmacol. Ther.* **2015**, *42*, 793–801. [CrossRef] [PubMed]
13. Santacroce, L.; Charitos, I.A.; Bottalico, L. A successful history: Probiotics and their potential as antimicrobials. *Expert. Rev. Anti-Infect. Ther.* **2019**, *17*, 635–645. [CrossRef] [PubMed]
14. D'Angelo, S.; Fracassi, F.; Bresciani, F.; Galuppi, R.; Diana, A.; Linta, N.; Bettini, G.; Morini, M.; Pietra, M. Effect of *Saccharomyces boulardii* in dog with chronic enteropathies: Double-blinded, placebo-controlled study. *Vet. Rec.* **2018**, *182*, 258. [CrossRef] [PubMed]
15. EFSA Panel on Additives and Products or Substances used in Animal Feed (FEEDAP); Bampidis, V.; Azimonti, G.; Bastos, M.D.L.; Christensen, H.; Durjava, M.; Dusemund, B.; Kouba, M.; López-Alonso, M.; López Puente, S.; et al. Safety and efficacy of a feed additive consisting of *Saccharomyces cerevisiae* DSM 34246 (Canobios-BL) for cats and dogs (ACEL pharma s.r.l.). *EFSA J.* **2024**, *22*, e8802. [CrossRef] [PubMed]
16. Baldwin, K.; Bartges, J.; Buffington, T.; Freeman, L.M.; Grabow, M.; Legred, J.; Ostwald, D., Jr. AAHA nutritional assessment guidelines for dogs and cats. *J. Am. Anim. Hosp. Assoc.* **2010**, *46*, 285–296. [CrossRef]
17. WSAVA World Small Animal Veterinary Association Global Nutrition Committee. (body Condition Score). 2013. Available online: https://www.edu-veterinar.ro/files/download/prezentari/ghiduri-practice/WSAVA-Animal-Welfare-Guidelines-(2018).pdf (accessed on 15 November 2024).

18. Davies, G.J.; Crowder, M.; Reid, B.; Dickerson, J.W. Bowel function measurements of individuals with different eating patterns. *Gut* **1986**, *27*, 164–169. [CrossRef]
19. Greco, D.; Diagnosis and Dietary Management of Gastro-Intestinal Disease. Purina Vet. *Diets.* 2011. Available online: https://www.purinaveterinarydiets.com/clinic-support/clinicresources/for-your-clinic/diagnose-gi-problems-with-thequick-guide-referencetool/ (accessed on 15 November 2024).
20. Middelbos, I.S.; Godoy, M.R.; Fastinger, N.D.; Fahey, G.C. A dose-response evaluation of spray-dried yeast cell wall supplementation of diets fed to adult dogs: Effects on nutrient digestibility, immune indices, and fecal microbial populations. *J. Anim. Sci.* **2007**, *85*, 3022–3032. [CrossRef]
21. Lin, C.Y.; Alexander, C.; Steelman, A.J.; Warzecha, C.M.; De Godoy, M.R.C.; Swanson, K.S. Effects of a *Saccharomyces cerevisiae* fermentation product on fecal characteristics, nutrient digestibility, fecal fermentative end-products, fecal microbial populations, immune function, and diet palatability in adult dogs. *J. Anim. Sci.* **2019**, *97*, 1586–1599. [CrossRef]
22. Grellet, A.; Mila, H.; Heilmann, R.M.; Feugier, A.; Gruetzner, N.; Suchodolski, J.S.; Steiner, J.M.; Chastant-Maillard, S. Effect of age, gestation and lactation on faecal IgA and calprotectin concentrations in dogs. *J. Nutr. Sci.* **2014**, *3*, 41. [CrossRef]
23. Benyacoub, J.; Czarnecki-Maulden, G.L.; Cavadini, C.; Sauthier, T.; Anderson, R.E.; Schiffrin, E.J.; von der Weid, T. Supplementation of food with *Enterococcus faecium* (SF68) stimulates immune functions in young dogs. *J. Nutr.* **2003**, *133*, 1158–1162. [CrossRef] [PubMed]
24. Buts, J.P.; Bernasconi, P.; Vaerman, J.P. Stimulation of secretory igA and secretory component of immunoglobulins in small intestine of rats treated with *Saccharomyces boulardii*. *Dig. Dis. Sci.* **1990**, *35*, 251–256. [CrossRef] [PubMed]

veterinary sciences

MDPI

Article

A New Graphical Method for Displaying Two-Dimensional Echocardiography Results in Dogs: Comprehensive Analysis of Results of Diagnostic Imaging Organized in a BOX (CARDIOBOX)

Federico J. Curra-Gagliano [1,*], Martín Ceballos [2], José I. Redondo [3] and Javier Engel-Manchado [4,5]

[1] Cátedra de Medicina 1, Facultad de Ciencias Veterinarias, Universidad de Buenos Aires, Buenos Aires C1427CWO, Argentina
[2] Cátedra Anestesiología y Algiología, Facultad de Ciencias Veterinarias, Universidad de Buenos Aires, Buenos Aires C1427CWO, Argentina
[3] Departamento de Medicina y Cirugía Animal, Facultad de Veterinaria, Universidad Cardenal Herrera-CEU, CEU Universities, 46115 Valencia, Spain; nacho@uchceu.es
[4] Internal Medicine, Veterinary Medicine and Therapeutic Research Group, Faculty of Veterinary Science, Research Institute of Biomedical and Health Sciences (IUIBS), University of Las Palmas de Gran Canaria, 35413 Las Palmas de Gran Canaria, Spain; jengelmanchado@gmail.com
[5] HeartBeatVet, 46025 Valencia, Spain
* Correspondence: fcg@fvet.uba.ar

Simple Summary: Rapid and accurate interpretation of echocardiographic results in dogs is crucial to good clinical decision-making. In this study, we developed and evaluated a graphical method called CARDIOBOX, designed to simplify the visual representation of echocardiographic values. CARDIOBOX is a system consisting of three overlapping boxes, which house nine boxes. Each box represents a specific range of echocardiographic values. Its design is based on the results of an analysis of 2967 dogs. In this process, percentiles and cut-off points were established to delimit the boxes corresponding to normal and abnormal ranges. To evaluate the efficacy of CARDIOBOX, 55 veterinarians participated in the interpretation of echocardiographs of dogs, presented in two different formats: using CARDIOBOX and exclusively using numerical reference tables. The time taken to interpret and the accuracy of responses in each format were recorded. The results indicated that CARDIOBOX enabled veterinarians to interpret echocardiographic values more quickly without decreasing interpretation efficiency. This study shows that CARDIOBOX allows veterinarians to interpret and apply echocardiographic results more quickly and effectively, significantly facilitating clinical decision-making.

Abstract: Introduction and objective: Rapid and efficient interpretation of echocardiographic findings is critical in clinical decision-making. This study aimed to design and validate a new graphical method, called CARDIOBOX, to represent echocardiographic findings in dogs. Methods: A prospective, observational, exploratory cohort study was conducted over three years. The design of CARDIOBOX was based on baseline values obtained from 802 healthy dogs and 2165 ill dogs. Using these data, a graph consisting of nine boxes was built to show the intervals of the different echocardiographic measurements. Validation of the method was performed by a survey of 55 veterinarians, who compared the use of CARDIOBOX with the use of numerical tables. Results: CARDIOBOX demonstrated significantly faster interpretability ($p < 0.05$) without reducing its effectiveness. In addition, the staff surveyed considered it easy to use and interpret. Conclusions: The introduction of CARDIOBOX emerges as a resource that facilitates rapid and efficient interpretation of echocardiographic findings in dogs. This new graphical method is presented

https://doi.org/10.3390/vetsci12010034

as a valuable tool for veterinary professionals in clinical decision-making in the field of veterinary cardiology.

Keywords: echocardiography; dogs; graphic method; graphic result interpretation; veterinary cardiology; clinical decision-making

1. Introduction

The exact prevalence of cardiac disease in dogs is unknown, although it may be around 10% of cases seen in daily clinical practice [1]. Echocardiography is the technique of choice in diagnosing and staging most cardiac diseases [2,3]. Echocardiographic data are usually interpreted employing tables, which use allometric formulas to predict average values based on a dog's weight [4].

Interpreting numerical values can be challenging for the human brain, designed to interpret graphs more effectively than numbers [5–7]. Human medicine has developed new strategies to present data innovatively and impactfully [8]. The efficient design of new graphical representation systems requires fast, simple, and effective implementation and interpretation, and they have been successfully employed in clinical pathology [9].

Thus, a new way of displaying these complex echocardiographic measurements could be through their graphic representation, normalising the values beforehand so that they can be easily interpreted independently of the animal's weight, as in the allometric tables used in veterinary cardiology [4]. Given the challenges faced by patients of varying sizes and weights in a clinical setting, this study sought to create, standardise, explain, and assess a visual system for interpreting echocardiographic findings, called CARDIOBOX (Comprehensive Analysis of Results of Diagnostic Imaging Organized in a BOX).

The objectives of this study were as follows: (a) to design a method for graphical representation of standardised echocardiographic results and (b) to validate the method's usefulness in a clinical setting. We hypothesise that CARDIOBOX allows the representation of echocardiographic numerical results in a simple graph that is easy and efficient to interpret and that its interpretation is faster than the traditional numerical method.

2. Material and Methods

This project received authorisation from the Institutional Committee for the Care and Use of Experimental Animals (CICUAL) at the Faculty of Veterinary Sciences, University of Buenos Aires, Argentina (Project 2019/49). All animal owners were informed about this study's objectives, methods, and purpose and gave their informed consent, which was also signed by the veterinarians involved in this clinical study. This study comprised two phases: (1) the design and standardisation of the CARDIOBOX method using echocardiographic values in dogs as a reference and (2) assessing the method's effectiveness in a clinical context.

2.1. Design and Standardisation of the CARDIOBOX Method

This descriptive, cross-sectional study was conducted to create and standardise the CARDIOBOX graphical method for assessing dog echocardiographic results. This study included 2967 dogs.

2.1.1. Patient Selection, Assessment, and Classification

All animals underwent a comprehensive evaluation, including anamnesis, detailed physical examination, and complementary diagnostic tests. These tests involved measuring systolic, diastolic, and mean arterial blood pressure with a SunTech Vet 30 oscillometric monitor (SunTech Medical Inc., Morrisville, USA), performing a six-channel electrocardiogram using a TEMIS model TM-300 digital electrocardiograph (Temis Tech., Cordoba, Argentina), right lateral and ventrodorsal radiographic views, and haematology including complete blood count (red blood cells, white blood cells, and platelets, as well as haemoglobin and haematocrit). Biochemistry, including glucose, urea, creatinine, alkaline phosphate (ALKP), cholesterol, alanine aminotransferase (ALT), total protein, globulin, albumin, and electrolytes (sodium, potassium, and chloride), was also measured.

Based on these assessments, animals were assigned to one of two groups according to health status. The HEALTHY GROUP (802 dogs) comprised dogs with no cardiac or systemic disease signs, excluding breeds with specific echocardiographic reference values (e.g., greyhounds, Labradors). Dogs diagnosed with cardiac disease formed the CARDIAC GROUP (2165 dogs). Together, the HEALTHY GROUP and the CARDIAC GROUP were called the TOTAL GROUP (2967 dogs).

2.1.2. Echocardiographic Evaluation

Echocardiography was performed following standard clinical recommendations [10] using a SonoScape A6 ultrasound machine (SonoScape Medical Corp., Guangdong, China) with 5–9 MHz microconvex probes and a Vinno 5 ultrasound machine (Vinno, Suzhou City, China) using 1–5 MHz adult phased array and 5–11 MHz microconvex probes. The left atrium and ascending aorta diameters were measured in B-mode in the right parasternal short-axis view at early diastole to calculate the left atrium/aorta (LA/Ao) index [11]. Additionally, in the right parasternal short-axis M-mode view at the level of the papillary muscles, the following parameters were measured:

- Interventricular septal thickness in diastole and systole (IVSd and IVSs)
- Left ventricular internal diameter in diastole and systole (LVIDd and LVIDs)
- Left ventricular free-wall thickness in diastole and systole (PLVWd and PLVWs)
- The principal investigator (FCG) performed all echocardiographic scans, and one additional investigator (JEM), each with over ten years of experience and specialised accreditation in veterinary cardiology, reviewed the images. Images that did not meet quality standards for accurate measurements were excluded from the analysis. Echocardiographic values for each patient were standardised to account for body weight variations using a formula derived from Cornell et al. (2004) [4]:

Standardised Value = Absolute Value (cm)/Weight (kg) ^{B}, where B is the constant *b* in the allometric equation defined by Cornell et al. (2004) [4].

2.1.3. Construction of the CARDIOBOX Graphical Method

The CARDIOBOX is a visual tool developed to evaluate graphically echocardiographic parameters in dogs. Each parameter is represented within a row of nine boxes, divided into three categories: values below the normal range (boxes 1, 2, and 3), values within the normal range (central grey boxes 4, 5, and 6), and values above the normal range (boxes 7, 8, and 9). Each box represents a specific percentile range, allowing an intuitive interpretation of a patient's data against reference values derived from a healthy population (see Figure 1).

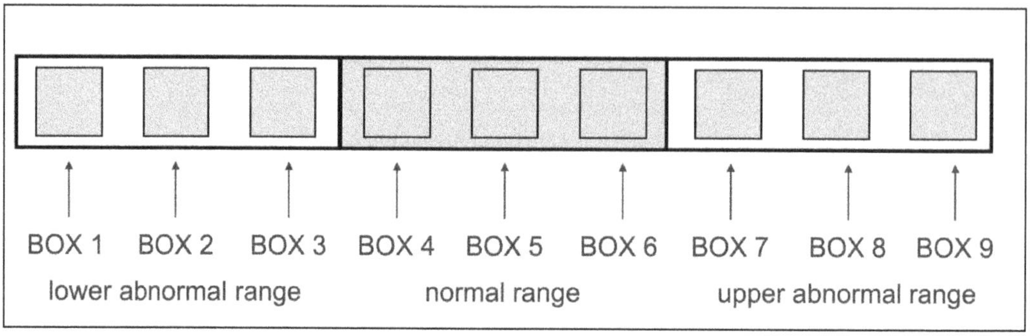

Figure 1. Basic structure of the CARDIOBOX. A marker indicates each parameter in one of the nine boxes: boxes 1, 2, and 3 for values below the normal range (white); boxes 4, 5, and 6 for values within the normal range (grey central rectangle); and boxes 7, 8, and 9 for values above the normal range (white).

The boundaries for each box in a specific parameter were determined as follows:

1. Define overall bounds (minimum and maximum): The minimum and maximum values were established using data from the TOTAL POPULATION, including HEALTHY and CARDIAC groups ($n = 2967$). The minimum value defines the lower bound of box 1, while the maximum value defines the upper bound of box 9.
2. Calculate boxes 4, 5, and 6 (normal range): The central normal range was established using data from the HEALTHY GROUP ($n = 802$). This standard range spans from the 1st to the 99th percentiles and consists of three central boxes. The intervals between boxes 4, 5, and 6 are equal, with box five centred precisely at the median:
    - Box 5 is centred on the median (50th percentile), representing the midpoint of the normal distribution.
    - Box 4 begins at the 1st percentile of the data of the HEALTHY GROUP and extends to a central boundary between the 1st percentile and the median, defining the transition from box 4 to box 5.
    - Box 6 begins at this equidistant boundary on the other side of the median and extends to the 99th percentile of the data of the HEALTHY GROUP.
3. Calculate boxes 1, 2, 3 and 7, 8, 9 (values outside the normal range): To represent values below and above the normal range, the following steps were applied:
    - Boxes 1, 2, and 3 capture values below the 1st percentile of the normal range (the lower bound of box 4). These boxes are set with equidistant boundaries that span the range from the minimum value (the lower bound of box 1) to the start of the normal range.
    - Boxes 7, 8, and 9 capture values above the 99th percentile of the normal range (the upper bound of box 6). These boxes are also set with equidistant boundaries, covering the range from the end of the normal range to the maximum value (the upper bound of box 9).

2.2. Validation of the CARDIOBOX

A survey was designed to assess the clinical utility of the CARDIOBOX graphical tool. This survey was created following the CHERRIES checklist [12]. The survey was developed and hosted on the JOTFORM platform (https://form.jotform.com/223278319852059, last accessed on 19 August 2023, also accessible on Supplementary Materials Figure S1), automatically recording each question's response time. The survey was active from 25 November to 6 December 2022. It was distributed in specialised veterinary groups, including cardiology, anaesthesiology, and general clinical practice. Participation was anonymous and voluntary.

The survey included four prominent clinical cases with echocardiographic findings: two patients without pathology and two with myxomatous mitral valve disease (MMVD) at different stages. Each case was presented using two formats: (a) a panel of numerical echocardiographic values accompanied by a reference table and (b) a CARDIOBOX chart displaying the values graphically.

The four cases were presented randomly, two using the CARDIOBOX method (one healthy and one pathological case) and two using the traditional numerical method with reference values (one healthy and one pathological case).

Key variables studied included response time (automatically measured by JOTFORM) and the percentage of correct answers for the seven variables under analysis. Additionally, participants were asked to rate the effectiveness and ease of use of the CARDIOBOX method compared to the traditional numerical format using a 5-point Likert scale. A final question gauged their willingness to adopt CARDIOBOX in clinical practice.

Statistical Analysis

The statistical analysis was done using the R language program (4.4.2). The first step is calculating the percentiles for the studied standardised echocardiographic variables to build the CARDIOBOX graph.

In the second step, a detailed statistical analysis was conducted to assess the effect of the presentation method (CARDIOBOX vs. numerical) and the presence of pathology on response time and accuracy. It was also used to analyse participants' perceptions using Likert-type questions.

The parameters analysed in the survey were response time (TIME) and percentage of correct answers (CORRECT), which were compared in two cases: CARDIOBOX and cases with numerical results and reference tables. Using the Shapiro–Wilk test, the null hypothesis of normality could be rejected. As the response data (time and correct answers) did not have a normal distribution, to validate the results of the form and detect statistical differences, the Wilcoxon test was used to compare the variables' percentage of correct answers and response time between groups. The significance level was set at 5% ($p < 0.05$).

Descriptive statistics, such as means and standard deviations, were calculated for response time and percentage of correct responses, grouped by method of presentation and presence of pathology. Shapiro–Wilk normality tests indicated that the data did not follow a normal distribution, justifying non-parametric tests in subsequent analyses.

The non-parametric Friedman test for repeated-measures data was applied to compare the median response time and percentage of correct responses among the different cases. After finding significant differences, post hoc comparisons were performed using the Wilcoxon paired signed-rank test with Bonferroni correction to adjust the significance level due to multiple comparisons. A linear mixed model was also used to analyse the effect of the method of presentation, presence of pathology, and order of presentation on response time and percentage of correct responses. These models allow for repeated measures of data and intra- and inter-subject variability. The lmer function of the lme4 R package was used, incorporating fixed effects for the method of assessment, the presence or absence of pathology, and the order in which cases were presented, as well as their interactions, in addition to a random effect for the participant's identifier (identifier variable). A similar model was fitted for the percentage of correct answers.

Cronbach's alpha coefficient was calculated to measure the reliability of the scales used in the questionnaire and assess the internal consistency of the Likert-type questions related to the perceived ease of use and effectiveness of CARDIOBOX.

3. Results

A total of 2967 dogs participated in this study. Of these, 802 were classified as healthy, forming the HEALTHY GROUP, while 2165 were diagnosed with cardiac diseases and were included in the CARDIAC GROUP. Table 1 shows the calculated percentiles for data parameters for the HEALTHY GROUP and TOTAL GROUP. The CARDIOBOX boxes' limits were determined and shown in Table 2 using this data.

Table 1. Calculated percentiles using the normalised data from 802 healthy dogs and the total population ($n = 2967$).

	Percentile										
HEALTHY GROUP	**Min**	**1**	**2.5**	**5**	**25**	**50**	**75**	**95**	**97.5**	**99**	**Max**
AIDn	0.49	0.56	0.59	0.62	0.70	0.75	0.82	0.92	0.96	1.00	1.23
IVSSDn	0.25	0.29	0.31	0.32	0.38	0.42	0.46	0.54	0.56	0.58	0.69
LVIDDn	0.84	1.12	1.19	1.22	1.38	1.47	1.55	1.66	1.68	1.69	1.74
PLVWDn	0.23	0.26	0.29	0.30	0.37	0.42	0.47	0.56	0.58	0.62	0.69
IVSSn	0.33	0.43	0.43	0.46	0.54	0.59	0.64	0.74	0.77	0.82	0.92
LVIDSn	0.28	0.47	0.54	0.58	0.73	0.82	0.90	1.01	1.04	1.06	1.09
PLVWSn	0.37	0.41	0.44	0.47	0.54	0.61	0.68	0.77	0.80	0.83	0.95
TOTAL POPULATION											
AIDn	0.35	0.58	0.61	0.64	0.73	0.80	0.91	1.20	1.37	1.52	2.22
IVSSDn	0.22	0.28	0.30	0.32	0.38	0.43	0.48	0.56	0.60	0.65	0.84
LVIDDn	0.68	1.07	1.16	1.22	1.41	1.54	1.69	2.12	2.29	2.46	3.04
PLVWDn	0.19	0.27	0.29	0.31	0.38	0.43	0.49	0.57	0.61	0.65	0.86
IVSSn	0.11	0.40	0.44	0.47	0.55	0.62	0.68	0.81	0.85	0.90	1.22
LVIDSn	0.24	0.45	0.52	0.59	0.75	0.86	0.97	1.18	1.34	1.58	2.55
PLVWSn	0.32	0.41	0.44	0.47	0.56	0.63	0.70	0.81	0.85	0.89	1.12

AIDn: normalised left atrial internal diameter in diastole. IVSSDn: normalised interventricular septal thickness in diastole. LVIDDn: normalised left ventricular internal diameter at diastole. PLVWDn: normalised left ventricular free-wall thickness in diastole. IVSSn: normalised interventricular septal thickness in systole. LVIDSn: normalised left ventricular internal diameter at systole. PLVWSn: normalised left ventricular free-wall thickness in systole.

Table 2. Limits for each box for constructing the CARDIOBOX for the studied echocardiographic variables.

	BOX 1		**BOX 2**		**BOX 3**		**BOX 4**		**BOX 5**		**BOX 6**		**BOX 7**		**BOX 8**		**BOX 9**	
AIDn	0.35	0.41	0.42	0.48	0.49	0.55	0.56	0.70	0.71	0.85	0.86	1.00	1.01	1.26	1.27	1.52	1.53	2.22
IVSSDn	0.22	0.24	0.25	0.26	0.27	0.28	0.29	0.39	0.40	0.48	0.49	0.58	0.59	0.61	0.62	0.65	0.66	0.84
LVIDDn	0.68	0.82	0.83	0.96	0.97	1.11	1.12	1.31	1.32	1.50	1.51	1.69	1.70	2.07	2.08	2.46	2.47	3.04
PLVWDn	0.19	0.21	0.22	0.23	0.24	0.25	0.26	0.38	0.39	0.50	0.51	0.62	0.63	0.63	0.64	0.65	0.66	0.86
IVSSn	0.11	0.21	0.22	0.32	0.33	0.42	0.43	0.56	0.57	0.69	0.70	0.82	0.83	0.85	0.86	0.90	0.91	1.22
LVIDSn	0.24	0.31	0.32	0.39	0.40	0.46	0.47	0.67	0.68	0.86	0.87	1.06	1.07	1.32	1.33	1.58	1.59	2.55
PLVWSn	0.32	0.35	0.36	0.38	0.39	0.40	0.41	0.55	0.56	0.69	0.70	0.83	0.84	0.86	0.87	0.89	0.90	1.12

AIDn: normalised left atrial internal diameter in diastole. IVSSDn: normalised interventricular septal thickness in diastole. LVIDDn: normalised left ventricular internal diameter at diastole. PLVWDn: normalised left ventricular free-wall thickness in diastole. IVSSn: normalised interventricular septal thickness in systole. LVIDSn: normalised left ventricular internal diameter at systole. PLVWSn: normalised left ventricular free-wall thickness in systole.

Validation of the CARDIOBOX

A total of 55 surveys were collected. Results indicated that evaluation times were significantly shorter with CARDIOBOX than with the traditional numerical method ($p < 0.0001$), with no significant time difference between cases with or without pathology (CARDIOBOX: $p = 0.28$; Numeric: $p = 0.80$). CARDIOBOX also yielded a higher percentage of correct answers compared to the numerical method ($p = 0.02793$), and there was no significant difference in accuracy between cases with and without pathology as shown in Table 3 (CARDIOBOX: $p = 1$; Numeric: $p = 0.224$).

Table 3. Mean time and standard deviation employed in the evaluation of echocardiographic data from the cases using the traditional numerical method and the proposed graphical method (CARDIOBOX) and percentage of correct answers in evaluating echocardiographic data of the cases using the conventional numeric method and the proposed graphical method (CARDIOBOX).

Method	Pathology	Mean Time	SD Time	Correct
CARDIOBOX	NO	65	35	97.7%
CARDIOBOX	YES	72	32	94.6%
Numeric	NO	109	52	92.7%
Numeric	YES	128	61	93.5%

Participants' perceptions of the CARDIOBOX graphical tool were also assessed using Likert-type questions, and the effect of the presentation method (traditional vs. CARDIOBOX) was analysed.

Mean scores and standard deviations were obtained for two critical aspects of the tool. Ease of use received a mean score of 4.49 (out of 5) with a standard deviation of 0.98, indicating that participants perceived the tool as easy to use. The tool's effectiveness received a mean score of 4.18 (out of 5) with a standard deviation of 1.19, suggesting that users consider CARDIOBOX effective for interpreting echocardiographic results.

The frequency analysis showed that most participants gave high scores in terms of ease of use: 40 people rated it a five, 7 a four, and only 8 gave scores of three or lower. For the tool's effectiveness, 32 participants rated it a five and 10 a four, while 13 scored three or less. Results are shown in Figure 2.

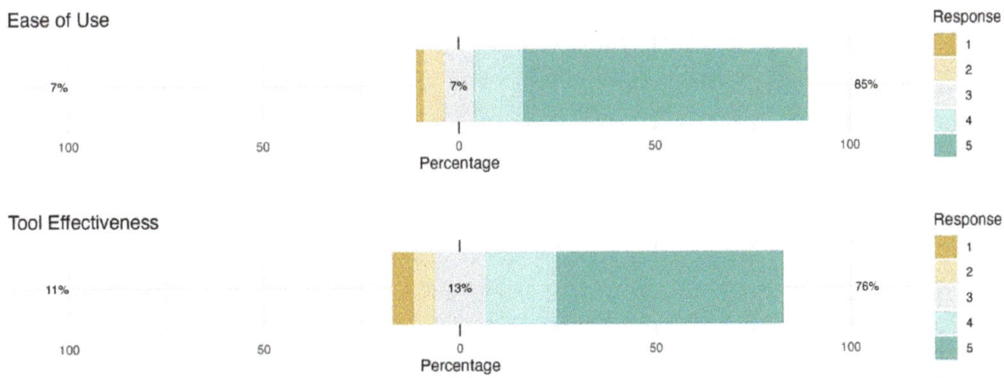

Figure 2. Likert graphs for perceived ease of use and tool effectiveness.

Interest in the future implementation of CARDIOBOX was high, with 44 respondents (80%) considering it beneficial for patient information. A smaller group of 10 respondents (18.18%) expressed uncertainty about its usefulness, while only 1 (1.82%) deemed it irrelevant.

Cronbach's alpha was calculated for the first two Likert-type questions (ease of use and effectiveness of the tool), obtaining a value of 0.76. This result indicates good internal consistency between items, supporting the measures' reliability.

For the analysis of turnaround time and accuracy, a linear mixed model was fitted to assess the effect of presentation method, presence of pathology, and order of presentation on turnaround time. The results showed that the traditional method was associated with a significant increase in turnaround time compared to CARDIOBOX. Specifically, using the conventional method increased response time by 49.49 s (standard error = 4.65, t-value = 10.64). Pathology had no significant effect on response time (t-value = −0.18), nor did pathology significantly affect the order of presentation (t-value = 1.21).

A linear mixed model was used to calculate the percentage of correct answers. The results indicated that the traditional method tended to slightly decrease the rate of correct answers compared to CARDIOBOX, with a reduction of 0.0299 (standard error = 0.0154, t-value = −1.94). However, this effect did not reach statistical significance. The presence of pathology showed no significant impact on the accuracy of responses (t-value = 1.47), and the order of presentation had no considerable effect either (t-value = 1.27). Analysis of the impact of the order of presentation revealed that this factor had no significant influence on either response time or the percentage of correct answers. This indicates that no learning or fatigue effects were observed that could bias this study's results.

The Likert-type questions indicated a positive trend in participants' perceptions of CARDIOBOX. Most participants gave high ratings for ease of use and effectiveness, and many showed interest in its future implementation.

4. Discussion

The CARDIOBOX graphical method was developed to provide a standardised, efficient, and intuitive means for interpreting dog echocardiographic parameters. Our findings support our hypothesis that CARDIOBOX could facilitate rapid and reliable clinical assessments. Clinicians found the tool user-friendly and effective in distinguishing between healthy and pathological cases, with significantly faster assessment times than traditional numerical methods, such as the Cornell allometric scaling charts [4].

Defining the range limits within the nine CARDIOBOX boxes was critical for clinical accuracy and utility. The method was constructed using a percentile-based approach, where the outer bounds (minimum and maximum) were defined from the TOTAL POPULATION, which included both healthy and cardiac groups. At the same time, the central boxes (4, 5, and 6) were determined exclusively from the HEALTHY GROUP. This combined approach aims to balance sensitivity for pathological cases with specificity for healthy cases, providing a reliable diagnostic aid for veterinarians.

Using the TOTAL POPULATION to define the minimum and maximum values (boxes 1 and 9) ensures that CARDIOBOX captures the full range of values encountered in practice, enhancing sensitivity by encompassing the entire data spectrum. This is essential, as a narrow outer range might exclude clinically relevant values. Including data from both healthy and cardiac groups at the extremes allows CARDIOBOX to reflect the broad echocardiographic variation seen in practice, making it suitable for diverse cases [4,13]. However, defining the central boxes (4, 5, and 6) from the HEALTHY GROUP ensures that standard range cutoffs remain strictly derived from non-pathological values, preserving high specificity. Previous research suggests that using only healthy populations for reference intervals is crucial for avoiding skewed ranges, which could misrepresent the normal distribution [3].

Another potential approach to constructing the CARDIOBOX would be to define the central and extreme boxes exclusively from the HEALTHY GROUP data. While this might yield a tighter distribution, such a configuration would likely result in pathological values clustering at the outermost edges (boxes 1 and 9). This clustering could reduce diagnostic efficacy, as it may limit the method's ability to differentiate degrees of abnormality, making distinguishing between mild and severe pathology harder. Practical diagnostic tools benefit from capturing a spectrum of abnormal values, enabling clinicians to detect deviations at varying stages [8]. By using healthy-only data to set the normal range while relying on the broader total population for the extremes, CARDIOBOX enhances both specificity and sensitivity across clinical cases.

CARDIOBOX was validated by 55 veterinarians, who responded positively to its ease of use and effectiveness. While the feedback suggests high user satisfaction, expanding this study to include veterinarians with varying experience in cardiology would provide more comprehensive validation. Additional insights from general practitioners and specialists would help confirm that CARDIOBOX is universally accessible. Given that veterinary technology can sometimes face resistance due to learning curves, the ease-of-use scores indicate that this tool could integrate smoothly into clinical practice, regardless of users' prior experience [13].

A further consideration is this study's focus on dogs under 20 kg, which restricts generalisability to larger breeds with differing cardiac morphologies. The variability in cardiac parameters across breeds underscores the importance of expanding CARDIOBOX to include larger dogs and diverse morphologies and validating its applicability for other canine populations. Further research in this direction would strengthen CARDIOBOX as a versatile tool and extend its clinical relevance.

Standardised echocardiographic values adjusted for weight through allometric scaling ensure consistency in measurement across dogs of different body sizes (see Supplementary Materials Tables S1 and S2). However, a more detailed explanation of how allometric scaling integrates with the graphical method could provide a more robust theoretical basis. This explanation would enhance readers' understanding of the method's applicability across different body sizes, further supporting CARDIOBOX's validity as a tool for diverse veterinary populations.

Although this study provides valuable cross-sectional data, longitudinal studies could offer additional insights into how well CARDIOBOX performs over time, especially in monitoring disease progression or assessing treatment efficacy. Applying CARDIOBOX in a longitudinal context could reveal its potential for tracking individual changes in echocardiographic parameters, broadening its utility in veterinary practice.

In human medicine, graphical data representation is widely recognised as beneficial for diagnostic accuracy, interpretive speed, and error reduction, mainly when it balances central tendencies with extremes [14]. CARDIOBOX follows these principles by adopting percentile-based limits, enhancing sensitivity and specificity for a reliable, clinically useful tool. By employing a clear, percentile-based visual differentiation, CARDIOBOX supports prompt and accurate decision-making in clinical settings [15]. Additionally, user feedback in this study highlights the positive reception of CARDIOBOX's layout, suggesting that graphical tools of this type could be beneficial across veterinary and human medicine for complex data interpretation [13].

The unique advantages of this new graphical method are summarized as follows: (A) It solves a critical problem. Traditional numerical representation of echocardiographic data often leads to delays and errors in interpretation, particularly in scenarios requiring quick decisions or for practitioners without immediate access to reference materials (see Supplementary Materials Figure S2). By providing a straightforward graphical representation, the method significantly reduces interpretation time while maintaining diagnostic accuracy. Furthermore, (B) it fills an existing gap. While echocardiography and its result-representation methods are well-established, no simple and easily adaptable method exists for future integration into ultrasound software. This approach bridges the gap by enabling automated, reliable, and simplified interpretations directly within ultrasound devices, empowering clinicians to make immediate decisions without external resources. Lastly, (C) it simplifies and modernizes an existing solution. Beyond its graphical design, it is built to seamlessly integrate with emerging technologies such as Deep Learning and Artificial Intelligence (AI). These advancements promise faster and more accurate interpretations, while the system's scalability allows practitioners to build personalized reference tables tailored to their patients and clinical approaches, making it a dynamic and evolving tool for modern veterinary practice. It has the potential to be integrated into ultrasound software, facilitating real-time graphical representation and automatic interpretation of echocardiographic data, which further enhances its utility in clinical settings. CARDIOBOX could integrate AI and machine learning to enhance diagnostic capabilities. Developing automated algorithms for segmentation and measurement could provide clinicians with immediate, standardized echocardiographic assessments. Tests in simulated environments show that using computer vision in ultrasound devices can streamline data acquisition and offer real-time feedback in CARDIOBOX outputs. These improvements would aid in monitoring disease progression and treatment efficacy, expanding CARDIOBOX's impact in veterinary cardiology.

This study has certain limitations. At present, this first study does not consider all the echocardiographic parameters. In this publication, we have focused on introducing the CARDIOBOX system using the main absolute parameters that vary with body weight. In addition, the CARDIOBOX system is validated on breeds with standard weight-adjusted reference tables, excluding Greyhounds, Labradors, and other giant breeds that are not represented in the demographic studied, as well as specific brachycephalic breeds that require customised reference tables. Future research should investigate the applicability of this system to these excluded breeds, also exploring variations between breed types (brachycephalic, mesocephalic, and dolichocephalic) and size (small vs. giant). In addition, extending the research to other species, such as cats and even humans, is essential to realise the full potential of this system. In conclusion, CARDIOBOX represents an effective tool for interpreting echocardiographic results in dogs, significantly improving interpretation speed without compromising diagnostic accuracy. The balance between sensitivity and specificity through the combined use of healthy and total population data enhances its applicability across clinical scenarios, from general practice to specialised cardiology. However, expanding the tool's validation to larger breeds, and conducting longitudinal studies would further strengthen CARDIOBOX's potential. These findings suggest that CARDIOBOX could become a widely applicable alternative to traditional methods, advancing clinical data interpretation and optimising patient care in veterinary medicine.

Supplementary Materials: The following supporting information can be downloaded at: https://www.mdpi.com/article/10.3390/vetsci12010034/s1; Figure S1: Form template used to validate the Cardiobox (jotform English.docx). Table S1: Table showing reference values of echocardiographic parameters using the CARDIOBOX system for dogs weighing up to 17 kg; Table S2: Table showing reference values of echocardiographic parameters using the CARDIOBOX system for dogs weighing 20 to 50 kg. Figure S2: (CARDIOBOX REPORT.doc) ecochardiographic report including the template of Cardiobox.

Author Contributions: F.J.C.-G. and J.E.-M. study design, interpretation of data and drafting of manuscript; F.J.C.-G. and J.I.R. data analysis; F.J.C.-G., J.E.-M., J.I.R. and M.C. revision of manuscript. All authors have read and agreed to the published version of the manuscript.

Funding: This research received no external funding.

Institutional Review Board Statement: This project received authorisation from the Institutional Committee for the Care and Use of Experimental Animals (CICUAL) at the Faculty of Veterinary Sciences, University of Buenos Aires, Argentina (Project 2019/49).

Informed Consent Statement: Informed consent was obtained from all subjects involved in this study.

Data Availability Statement: Data supporting the reported results can be sent to anyone interested by contacting the corresponding author.

Acknowledgments: The authors would like to thank the persons included in this study for their contribution to this work.

Conflicts of Interest: The authors declare no conflicts of interest.

References

1. Atkins, C.; Bonagura, J.; Ettinger, S.; Fox, P.; Gordon, S.; Haggstrom, J.; Hamlin, R.; Keene, B.; Luis-Fuentes, V.; Stepien, R. Guidelines for the Diagnosis and Treatment of Canine Chronic Valvular Heart Disease. *J. Vet. Intern. Med.* **2009**, *23*, 1142–1150. [CrossRef] [PubMed]
2. Fuentes, V.L.; Abbott, J.; Chetboul, V.; Côté, E.; Fox, P.R.; Häggström, J.; Kittleson, M.D.; Schober, K.; Stern, J.A. ACVIM Consensus Statement Guidelines for the Classification, Diagnosis, and Management of Cardiomyopathies in Cats. *J. Vet. Intern. Med.* **2020**, *34*, 1062–1077. [CrossRef] [PubMed]
3. Ghilardi, S.; Pradelli, D.; Rizzi, R.; Polli, M.; Bagardi, M.; Santilli, R.A.; Brambilla, P.G.; Bussadori, C.M. A Study of the Inter- and Intra-Operator Variability on Selected Echocardiographic Measurements in Dogs. *Vet. Res. Commun.* **2023**, *47*, 2323–2331. [CrossRef] [PubMed]
4. Cornell, C.C.; Kittleson, M.D.; Torre, P.D.; Häggström, J.; Lombard, C.W.; Pedersen, H.D.; Vollmar, A.; Wey, A. Allometric Scaling of M-Mode Cardiac Measurements in Normal Adult Dogs. *J. Vet. Intern. Med.* **2004**, *18*, 311–321. [CrossRef] [PubMed]
5. Kadosh, R.C.; Bahrami, B.; Walsh, V.; Butterworth, B.; Popescu, T.; Price, C.J. Specialization in the Human Brain: The Case of Numbers. *Front. Hum. Neurosci.* **2011**, *5*, 62. [CrossRef] [PubMed]
6. Damarla, S.R.; Just, M.A. Decoding the Representation of Numerical Values from Brain Activation Patterns. *Hum. Brain Mapp.* **2013**, *34*, 2624–2634. [CrossRef] [PubMed]
7. Nara, S.; Raza, H.; Carreiras, M.; Molinaro, N. Decoding Numeracy and Literacy in the Human Brain: Insights from MEG and MVPA. *Sci. Rep.* **2023**, *13*, 10979. [CrossRef] [PubMed]
8. Bauer, D.T.; Guerlain, S.; Brown, P.J. The Design and Evaluation of a Graphical Display for Laboratory Data. *J. Am. Med. Inform. Assoc.* **2010**, *17*, 416–424. [CrossRef] [PubMed]
9. Henry, J.B.; Kelly, K.C. Comprehensive Graphic-Based Display of Clinical Pathology Laboratory Data. *Am. J. Clin. Pathol.* **2003**, *119*, 330–336. [CrossRef] [PubMed]
10. Thomas, W.P.; Gaber, C.E.; Jacobs, G.J.; Kaplan, P.M.; Lombard, C.W.; Vet, M.; Moise, N.S.; Moses, B.L. Recommendations for Standards in Transthoracic Two-Dimensional Echocardiography in the Dog and Cat. *J. Vet. Intern. Med.* **1993**, *7*, 247–252. [CrossRef] [PubMed]
11. Hansson, K.; Häggström, J.; Kvart, C.; Lord, P. Left atrial to Aortic Root Indices Using Two-Dimensional and M-Mode Echocardiography in Cavalier King Charles Spaniels with and without Left Atrial Enlargement. *Vet. Radiol. Ultrasound* **2002**, *43*, 568–575. [CrossRef]
12. Eysenbach, G. Improving the Quality of Web Surveys: The Checklist for Reporting Results of Internet E-Surveys (Cherries). *J. Med. Internet Res.* **2004**, *6*, e34. [CrossRef] [PubMed]
13. Schrodt, J.; Dudchenko, A.; Knaup-Gregori, P.; Ganzinger, M. Graph-Representation of Patient Data: A Systematic Literature Review. *J. Med. Syst.* **2020**, *44*, 86. [CrossRef] [PubMed]
14. Jaeger, M. Learning and Reasoning with Graph Data. *Front. Artif. Intell.* **2023**, *6*, 1124718. [CrossRef] [PubMed]
15. Chen, J.C.; Cooper, R.J.; McMullen, M.E.; Schriger, D.L. Graph Quality in Top Medical Journals. *Ann. Emerg. Med.* **2017**, *69*, 453–461.e5. [CrossRef] [PubMed]

veterinary sciences

MDPI

Article

The Usefulness of the Kidney-to-Aorta Ratio in Dogs with Chronic Kidney Disease †

Dario Costanza [1,*], Erica Castiello [1], Pierpaolo Coluccia [1], Camilla Sangiuliano [1], Maria Pia Pasolini [2], Micheletino Matarazzo [1], Giacomo Gnudi [3], Adelaide Greco [1,‡] and Leonardo Meomartino [1,*,‡]

1 Interdepartmental Centre of Veterinary Radiology, University of Napoli "Federico II", Via Federico Delpino 1, 80137 Napoli, Italy

2 Department of Veterinary Medicine and Animal Production, University of Napoli "Federico II", Via Federico Delpino 1, 80137 Napoli, Italy

3 Department of Veterinary Science, University of Parma, Strada del Taglio 10, 43126 Parma, Italy

* Correspondence: dario.costanza@unina.it (D.C.); leonardo.meomartino@unina.it (L.M.)

† This article is a revised and expanded version of a paper entitled "The usefulness of Kidney-to-Aorta Ratio in Dogs with Renal Disease" presented as a poster at the IVRA-EVDI Joint Conference on 18–23 June 2023 in Dublin, Ireland.

‡ These authors contributed equally to this work.

Simple Summary: Renal dimensions are one of the parameters to evaluate during abdominal ultrasound examinations, as they can vary in the presence of diseases, such as chronic kidney disease (CKD), which is often associated with reduced renal size. In dogs, the significant variability in size among different breeds makes absolute measurements less meaningful. To address this limitation, a method where the ratio between the renal length and aortic diameter is measured has been proposed. This study assesses this method's effectiveness in distinguishing healthy dogs from those with CKD. The results highlight the method's high specificity: a value below the lower reference limit strongly indicates the presence of disease. However, the sensitivity is low, with many diseased dogs falling within the normal range. Therefore, this method proves particularly useful for confirming the diagnosis in dogs with clinical signs and laboratory findings consistent with CKD rather than for early screening.

Abstract: The kidney length (KL) to aortic diameter (AoD) ratio (KL/AoD) has been proposed as an ultrasonographic objective method to assess renal dimensions. However, its wide range of normal values limits sensitivity. Of note, its clinical utility in detecting renal disease has not been investigated. The main aim of this study was to assess the sensitivity and specificity of KL/AoD in dogs with chronic kidney disease. Of the 227 dogs in the final sample, 185 were healthy and 42 were diseased. The obtained cut-off values for healthy dogs were similar to those previously reported. No differences were found between the KL of healthy and diseased dogs; however, using the KL/AoD ratio, diseased dogs showed a significantly smaller ratio ($p = 0.0003$), although a partial overlap between the two groups' values was present. Considering the Youden index at KL/AoD = 6.3, the ROC curve displayed a specificity of 83.24% and a sensitivity of 41.67%, while for KL/AoD = 5.6, the specificity was 97.57% and the sensitivity was 13.10%. The results of this study provide clinical usefulness for the KL/AoD ratio method, revealing excellent specificity but poor sensitivity.

Keywords: aorta; canine; kidney injury; ratio; renal length; renal disease; ultrasonographic; ultrasound

1. Introduction

Determining renal dimensions is one of the parameters to be considered during the ultrasonographic examination of the urinary system, as it could provide useful clinical information [1–4]. Some pathologic conditions, such as acute kidney disease, compensatory hypertrophy, portosystemic shunts, pyelonephritis, amyloidosis, hydronephrosis, and neoplasia, can cause slight to severe nephromegaly [2,5]. Other conditions such as chronic kidney disease (CKD), congenital hypoplasia, and renal dysplasia can determine slight to severe reduction [2,5]. In dogs, absolute renal measurements have poor diagnostic clinical value due to the considerable variation in body weight and conformation [2,3,6]. To account for this variability, linear values of healthy kidneys, using body weight and conformations as variables, were proposed [4]. Methods where the kidney length (KL) is compared to another anatomical structure have also been proposed [7,8]. In the method by Mareschal et al., the KL is divided by the aortic luminal diameter (AoD) [7]. The proposed KL/AoD ratio method resulted in a wide range of normal cut-off values (5.5–9.1), resulting in possible overlap between healthy dogs and those with renal disease, consequently limiting the method's clinical usefulness [2,3,9]. Furthermore, the method's efficacy in detecting dogs with CKD has not been established.

Based on our clinical experience, which includes many dogs with clinical and laboratory findings consistent with CKD but KL/AoD values within normal limits, we hypothesized that the KL/AoD ratio is a method with high specificity but low sensitivity. The primary aim of this study was to investigate the clinical usefulness of the KL/AoD method in detecting dogs with clinical and laboratory results consistent with CKD. The secondary aims were to verify if the previously reported cut-off values [7] were respected in our sample and to test the effects of variables such as side, body weight, and body size on the KL, AoD, and KL/AoD ratio in the healthy group.

2. Materials and Methods

2.1. Study Design

This study was prospective, single-institutional, analytical cross-sectional, and reference interval in design [10]. It received approval from the Clinical Ethical Review Board of the University of Napoli "Federico II" (PG/2023/PG0002892).

2.2. Ultrasonographic Examination Protocol

All the ultrasonographic (US) exams were performed at the Interdepartmental Center of Veterinary Radiology of the University of Napoli "Federico II" with the patient in dorsal or lateral recumbency, using the same ultrasound device (MyLab Class C Vet, Esaote, Genova, Italy). A microconvex or linear, multifrequency electronic probe (model SC3123, 3.5–10 MHz and model LA533, 3–13 MHz, respectively, Esaote, Genova, Italy) was used depending on the ultrasonographer's preferences, which were mainly based on the best resolution achievable according to the patient's size. All the US exams were performed by the same operator (L.M.), a professor of veterinary radiology with more than 26 years of experience in ultrasonography. The operator was unblinded regarding the patient's clinical status and laboratory blood analysis results, if available during the exam. The KL/AoD ratio has been included in the institutional abdominal ultrasound examination protocol since May 2017 and has been assessed in all patients undergoing abdominal ultrasound using previously established methods [7]. Briefly, the KL was determined in the dorsal plane, from a subcostal or intercostal acoustic window, depending on patient conformation, and measured on a still frame acquired when the distance between the two poles was maximum and with the renal pelvis clearly delineated (Figure 1A). The aorta was assessed from the left flank, and images were acquired in a longitudinal scan just caudal to the root of the left renal artery. The measurements of the AoD were acquired at maximal luminal diameter, excluding the vessel walls, after reviewing the last frames using the cineloop function of the ultrasound device (Figure 1B).

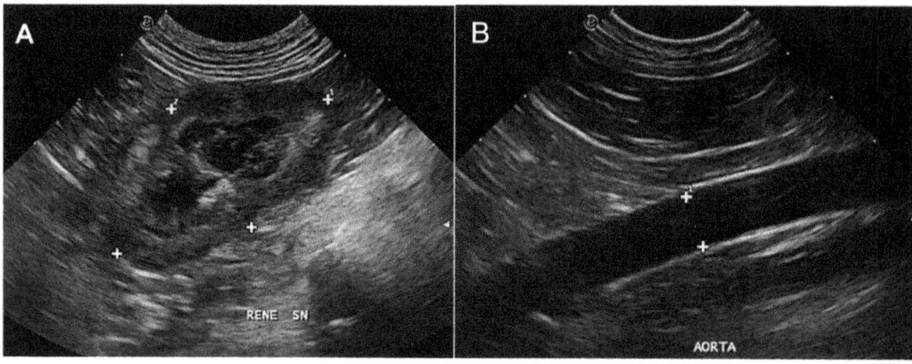

Figure 1. Dorsal ultrasonographic image of the left kidney (**A**) and of the aorta (**B**) acquired just caudal to the origin of the renal artery. In (**A**), the renal length is measured at the point of maximal cranio-caudal renal length (indicated by the measurement cursors with number 1). In (**B**), the maximal luminal diameter is obtained after reviewing cineloop frames to account for pulsation of the aorta; measurement cursors were placed at the margin of the lumen, excluding the vessel walls.

2.3. Exclusion Criteria

Exclusion criteria were as follows: (a) US exam performed by a different operator; (b) measurements of both KL and AoD not obtained at the time of the US examination; (c) images deemed to be of inadequate quality (e.g., borders poorly defined, blurred contours); (d) the patient was less than twelve months of age; (e) serum renal function test (creatinine and urea) were not collected within 48 h before or after the US examination; (f) the patient was under sedation or general anesthesia; (g) the patient had ultrasonographic renal alterations consistent with congenital or acquired disease different from CKD (e.g., renal tumors, hydronephrosis, polycystic kidney disease, pyelonephritis, renal hematoma, renal abscess).

2.4. Data Recording

The images, signalment, history, and clinical data (including serum creatinine and urea levels) of patients undergoing US examination of the abdomen between May 2017 and May 2023 were retrieved from the picture archiving and communication system (dcm4chee-arc-light version 5.11.1, http://www.dcm4che.org) [11] and institutional electronic medical record.

The KL and AoD dimensions (in centimeters), breed, sex, weight (in kilograms), age (in years), and results of serum creatinine and urea levels (both expressed in mg/dL) were recorded for each dog included in the preliminary sample in an electronic spreadsheet (Microsoft Excel version 16.52 2021, Microsoft Corp., Redmond, WA, USA). The decision on whether to include or exclude patients from the final sample was made by one of the authors (D.C.), a veterinarian with a Ph.D. in Veterinary Sciences and 5 years of expertise in ultrasonography.

After being included in the final sample, the same author divided dogs into two main groups: "healthy" and "diseased". The subjects included in the diseased group were required to have at least two consecutive tests with creatinine (threshold value of 1.4 mg/dL) and urea (threshold value of 60 mg/dL) levels above the reference values, with the last determination obtained within 48 h before or after the ultrasound examination [12].

Furthermore, dogs from the healthy group were subdivided into four sub-groups according to their body weight: toy (≤5.4 kg), small (5.5–10 kg), medium (10–25.9 kg), and large (≥26 kg).

2.5. Statistical Analysis

Statistical analyses were performed by one of the authors (D.C.), using commercial statistics software (Prism version 9.5.0 (525), GraphPad Software, San Diego, CA, USA and MedCalc version 19.2.6, MedCalc Software Ltd., Ostend, Belgium).

The D'Agostino–Pearson test was used to assess data for normality, and the data were further analyzed according to their distribution. Descriptive statistics, including the mean or median, range (minimum to maximum), standard deviation (SD), and 95% confidence interval (C.I.) of the mean, were calculated for age, body weight, right kidney length (RKL), left kidney length (LKL), AoD, RKL/AoD ratio, and LKL/AoD ratio.

In the healthy group, differences between the RKL and LKL were evaluated using the paired *t*-test and differences between the RKL/AoD and the LKL/AoD using the Wilcoxon signed-rank test. In the same group, differences in the KL and AoD among the toy, small, medium, and large dogs were assessed using one-way analysis of variance (ANOVA) with Welch's correction, as variance results were significantly different at the F-test and further analyzed using the Dunnett's multiple comparisons post hoc test. Differences in the KL/AoD ratio were assessed using the Kruskal–Wallis test, and the results were further analyzed using Dunn's multiple comparison post hoc test. Furthermore, in the healthy group, the correlation between the body weight and KL, AoD, and the KL/AoD ratio was investigated using Spearman's rank correlation coefficient (r_s) and simple linear regression.

Finally, in the healthy group, the reference range for the KL/AoD ratio was calculated according to the American Society for Veterinary Clinical Pathology guidelines for reference intervals [13]. Outliers were automatically identified according to Reed et al., and then, data distribution was tested automatically with the D'Agostino–Pearson test [14]. Reference lower and upper limits, corresponding to the 2.5th and 97.5th fractiles and the corresponding 90% C.I., were then calculated employing nonparametric methods following Clinical Laboratory and Standards Institute recommendations (CLSI C28-A3) [15,16].

Differences between healthy and diseased groups in the KL and KL/AoD ratio were assessed using the Mann–Whitney U test. Finally, the diagnostic performance of the KL/AoD ratio to discriminate between healthy and diseased dogs was evaluated by receiver operating characteristic (ROC) curves, following the method proposed by DeLong et al., to obtain the associated area under the curve (AUC) and Youden index (J = sensitivity + specificity − 1). A J value < 0.2 indicates poor diagnostic performance, values between 0.2 and 0.5 indicate moderate diagnostic performance, and values > 0.5 indicate good diagnostic performance [17–21]. In all analyses, $p < 0.05$ was considered statistically significant.

3. Results

Of the 1703 US exams performed in the considered period, 227 dogs met the inclusion criteria (87 males, 19 castrated males, 62 females, 59 neutered females). In total, 185 (81.5%) (72 males, 16 castrated males, 54 female, 43 spayed females) were rated as healthy, and 42 (18.5%) (15 males, 3 castrated males, 8 female, 16 spayed females) as diseased.

In the healthy group, the mean age was 8 (±3) years, and the median body weight was 14 kg (range 1.8–65 kg). In the diseased group, the mean age was 9 (±4) years, and the median body weight was 18 kg (range 3–35 kg). The absolute number of dogs and the number of diseased dogs for each breed included in the final sample are summarized in Table 1. Descriptive statistics for the RKL, LKL, AoD, RKL/AoD, and LKL/AoD for both the healthy and diseased groups are summarized in Table 1.

Table 1. Descriptive statistics for right and left kidney length, aortic luminal diameter, and kidney-to-aorta ratios in the healthy and diseased groups.

	Healthy Group (n = 185)					Diseased Group (n = 42)				
	RKL (cm)	LKL (cm)	AoD * (cm)	RKL/AoD *	LKL/AoD	RKL (cm)	LKL (cm)	AoD (cm)	RKL/AoD	LKL/AoD
Mean (±SD)	5.7 ± 1.4	5.6 ± 1.4	0.79 ± 0.21	7.5 ± 1.1	7.2 ± 1.1	5.5 ± 1.2	5.4 ± 1.2	0.82 ± 0.19	6.9 ± 1.2	6.8 ± 1.1
Median	5.6	5.6	0.8	7.3	7.1	5.7	5.4	0.85	6.8	6.9
95% C.I.	5.5–5.9	5.4–5.8	0.76–0.82	7.3–7.6	7.1–7.4	5.2–5.9	5.1–5.8	0.76–0.87	6.6–7.3	6.5–7.2
Range (min–max)	3–9.7	2.8–9.1	0.4–1.3	4.6–12	4.5–11	2.6–7.9	2.7–7.8	0.47–1.1	4.5–9.3	4.7–9.7

Values indicated with asterisk (*) are non-normally distributed. Abbreviations: AoD, aortic luminal diameter; C.I. = confidence interval; cm = centimeters; LKL = left kidney length; LKL/AoD = left kidney-to-aorta ratio; max = maximum; min = minimum; RKL = right kidney length; RKL/AoD = right kidney-to-aorta ratio; SD = standard deviation.

In the healthy group, the paired t-test revealed no statistical differences between the RKL and LKL (p = 0.29). Similarly, the Wilcoxon signed-rank test showed no statistical differences between the RKL/AoD and LKL/AoD ratios (p = 0.05). Therefore, in the subsequent statistical analyses, the KL and KL/AoD ratios of the left and right sides were pooled together.

In the healthy group, based on their body weight, 30 (16.22%) were classified as toy-sized, 46 (24.86%) as small-sized, 69 (37.3%) as medium-sized, and 40 (21.62%) as large-sized. Descriptive statistics for the KL, AoD, and KL/AoD for each sub-group are summarized in Table 2.

Table 2. Descriptive statistics for kidney length, aortic luminal diameter, and kidney-to-aorta ratio (pooled data) for the healthy group's four size sub-groups (toy, small, medium, and large).

Size Sub-Group		Mean (±SD)	Median	95% CI	Range (min–max)
Toy size (n = 30)	KL (cm) (n = 60)	3.8 ± 0.54	3.8	3.7–4	2.8–5
	AoD (cm) (n = 30)	0.51 ± 0.07	0.5	0.48–0.53	0.4–0.7
	KL/AoD (n = 60)	7.7 ± 1.1	7.8	7.4–8	5.7–10
Small size (n = 46)	KL (cm) (n = 92)	4.8 ± 0.62	4.8	4.7–4.9	3–6.6
	AoD (cm) (n = 46)	0.65 ± 0.08	0.7	0.63–0.68	0.5–0.8
	KL/AoD * (n = 92)	7.5 ± 1.2	7.4	7.3–7.8	5.3–11
Medium size (n = 69)	KL (cm) (n = 138)	6.1 ± 0.82	6	5.9–6.2	4.4–8.7
	AoD (cm) (n = 69)	0.87 ± 0.13	0.9	0.83–0.90	0.60–1.2
	KL/AoD (n = 138)	7.2 ± 0.95	7.2	7–7.3	4.5–9.6
Large size (n = 40)	KL (cm) (n = 80)	7.3 ± 0.85	7.2	7.1–7.5	5.7–9.7
	AoD (cm) (n = 40)	1 ± 0.13	1.0	0.98–1.1	0.80–1.3
	KL/AoD * (n = 80)	7.2 ± 1.2	6.9	7–7.5	5.2–12

Values indicated with asterisk (*) are non-normally distributed. Abbreviations: AoD, aortic luminal diameter; C.I. = confidence interval; cm = centimeters; KL = kidney length; KL/AoD = kidney-to-aorta ratio; max = maximum; min = minimum; SD = standard deviation.

Furthermore, in the healthy group, the one-way ANOVA and Dunnett's multiple comparison tests revealed statistically significant differences among all the four different sub-groups considered for KL (all $p < 0.0001$; Figure 2A) and AoD (all $p < 0.0001$; Figure 2B). Similarly, the Kruskal–Wallis and Dunn multiple comparison tests revealed a statistically significant difference in the KL/AoD ratio between the toy and medium groups and between the toy and large groups ($p = 0.01$ and $p = 0.009$, respectively; Figure 2C).

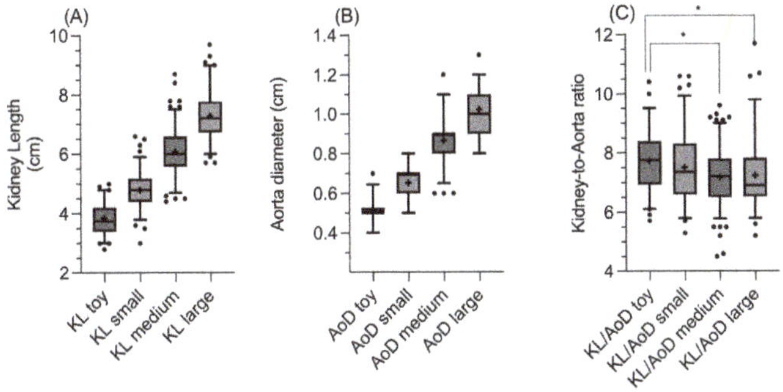

Figure 2. Box and whisker plot of one-way ANOVA of (**A**) KL (in cm), (**B**) AoD (in cm), and (**C**) Kruskal–Wallis test of KL/AoD ratio, for toy-, small-, medium-, and large-sized sub-groups. The line within the box represents the median, while the cross the mean, upper and lower sides of the box are the lower and upper quartiles, and the two extreme horizontal lines represent 5–95 percentiles. The asterisks (*) in Figure (**C**) indicate the statistically significant differences.

The Spearman's rank correlation coefficient revealed a positive correlation between the body weight and KL ($p < 0.0001$; $r_s = 0.89$; Figure 3A) and AoD ($p < 0.0001$; $r_s = 0.90$; Figure 3B), while a slight negative correlation was found between the body weight and KL/AoD ($p = 0.0002$; $r_s = -0.19$; Figure 3C).

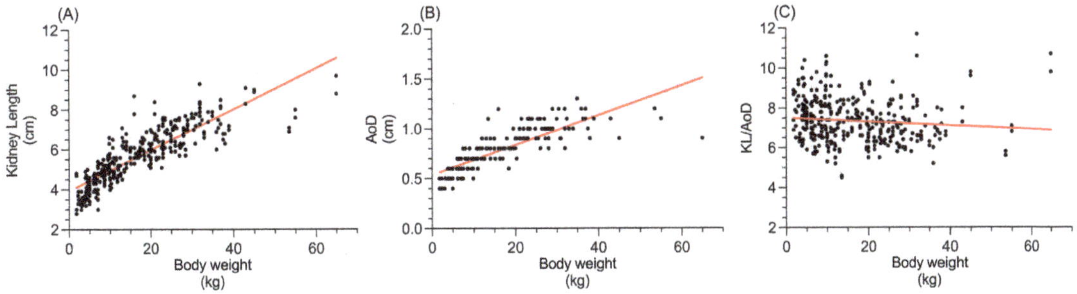

Figure 3. Correlation between body weight (in kg) and (**A**) KL (in cm), (**B**) AoD (in cm), and (**C**) KL/AoD ratio. The red solid line represents the simple linear regression.

The reference intervals, median, and range (minimum to maximum) with the relative 90% C.I. for the KL/AoD ratio in the healthy group are summarized in Table 3.

Table 3. Median, upper, and lower limits of the reference value (and corresponding 90% C.I.) of the KL/AoD ratio in the healthy group.

	Median	Lower Limit (90% C.I.)	Upper Limit (90% C.I.)
KL/AoD	7.2	5.6 (5.3–5.8)	9.94 (9.6–10.6)

Values are non-normally distributed. Abbreviations: C.I. = confidence interval; KL/AoD = kidney-to-aorta ratio.

The unpaired t-test did not reveal significant differences in the KL ($p = 0.53$; Figure 4A) and AoD ($p = 0.44$) between healthy and diseased dogs, while the Mann–Whitney U test found significant differences between the two groups for the KL/AoD ratio ($p = 0.0003$; Figure 4B), with diseased dogs having overall lower values.

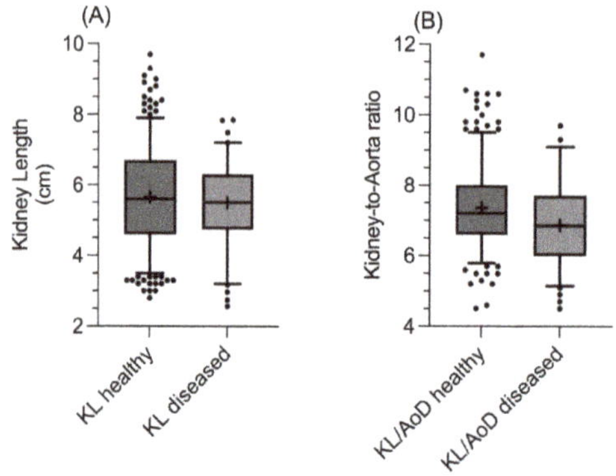

Figure 4. Box and whisker plot comparing (**A**) **the** kidney length (in cm) and (**B**) the KL/AoD ratio in healthy and diseased dogs. The line within the box represents the median, while the cross the mean, upper and lower sides of the box are the lower and upper quartiles, and the two extreme horizontal lines represent 5–95 percentiles.

The generated ROC curve for the KL/AoD ratio (Figure 5) displayed a fair AUC (AUC = 0.62; 95% C.I.: 0.55–0.69; $p = 0.0003$) and a moderate diagnostic performance (J = 0.25), considering the Youden index at KL/AoD = 6.3, where the ROC curve showed a specificity of 83.24% (95% C.I.: 79.10–86.70%) and a sensitivity of 41.67% (95% C.I.: 31.71–52.35%). For the lower value of the obtained reference range (KL/AoD = 5.6), the specificity increased to 97.57% (95% C.I.: 95.44–98.72%), while the sensitivity decreased to 13.10% (95% C.I.: 7.47–21.95%).

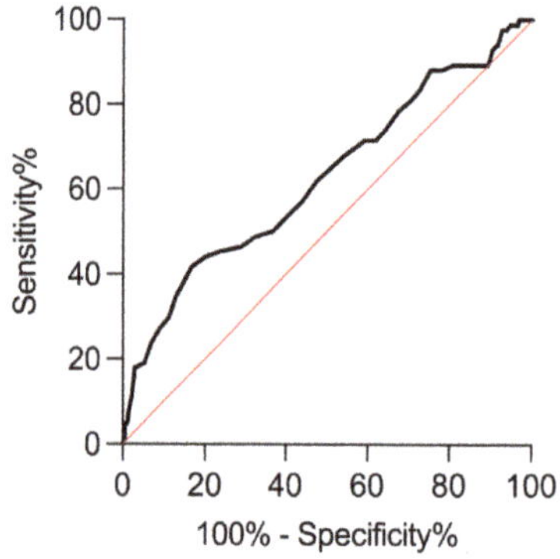

Figure 5. Receiver operating characteristic (ROC) curve of KL/AoD ratio.

4. Discussion

This study evaluated the clinical usefulness of the KL/AoD ratio. It assessed the method's sensitivity and specificity by comparing the results obtained from healthy dogs with those from clinical and laboratory results consistent with CKD. The study's results supported our hypothesis.

The ROC curve demonstrated that for a KL/AoD ratio = 5.6 (i.e., the obtained lower reference limit), the specificity is very high (97.57%), while on the other hand, the sensitivity is very low (13.10%). The diagnostic performance of the method was barely moderate considering the Youden index (which combines sensitivity and specificity into a single measure of diagnostic test effectiveness) at a KL/AoD ratio = 6.3, where specificity remained excellent (83.24%), while the sensitivity increased considerably (41.67%).

As expected, the comparison of the KL between healthy and diseased dogs, without considering other variables, showed no significant differences, with overlapping values between the two groups. On the other hand, for the KL/AoD ratio, a statistically significant difference was found between healthy and diseased dogs, with the latter showing a lower median and 95% C.I. These results support using this method to normalize the renal dimensions among dogs of different sizes while also suggesting that dogs with CKD generally tend to exhibit a lower KL/AoD ratio. However, the KL/AoD ratio values obtained from the healthy and diseased groups overlap to a certain extent, limiting the clinical usefulness of the method in a clinical scenario. Indeed, dogs with initial CKD may not exhibit significant alterations in renal linear dimensions, causing partial overlap in the KL/AoD ratio values between the two groups and limiting the method sensitivity [22]. The obtained sensitivity and specificity values result in low false positives, making it highly likely that subjects with values below the reference range are genuinely affected by CKD. Therefore, the authors retain that the KL/AoD ratio is a valuable tool for confirming the diagnosis of CKD when clinical, laboratory, and additional ultrasonographic findings are consistent with the condition, allowing for greater diagnostic confidence while excluding other causes of elevated serum urea and creatinine levels, since these increased values are not exclusively indicative of CKD [23].

From our results, and as expected, the KL/AoD ratio is not suitable as a standalone screening method for CKD due to the partial overlap of values between healthy and diseased subjects, which leads to low sensitivity, and many diseased subjects may present values within the range of healthy individuals (false negatives). The reference ranges in our sample for the KL/AoD ratio (i.e., 5.6–9.9; Table 3) are broad and similar to those previously described (i.e., 5.5–9.2) [7]. This wide range is probably related to the randomness of the sample, as it is proved that dogs of the same breed tend to have narrower reference intervals [9].

To ensure the most accurate measurements, only images in which the cranial and caudal profiles of each kidney were clearly visible were included in the sample, and all measurements were taken during the examination rather than retrospectively. Although, in deep-narrowed chest dogs, the cranial profiles of the RK can be more difficult to delineate due to its subcostal position [2], in the current study, similar to previous studies [7–9] and differently from others [4,24,25], there were no differences between the RKL and LKL. Consequently, a single reference interval can be used for both kidneys.

In the original study by Mareschal et al., the AoD measurements acquired on longitudinal scans were more accurate than those acquired on transversal scans, showing a higher degree of inter-observer agreement [7]. On the contrary, in a later study on growing puppies, the authors suggested acquiring the AoD on transverse scans since, according to them, those are less affected by motion artefacts [25]. In a study that established the KL/AoD ratio reference interval in clinically healthy whippets, the authors did not find a significant difference between the two scanning planes [9]. In the present study, the AoD was acquired on longitudinal scans since, according to the literature, transversal scans can be affected by refraction artefacts that can alter image quality [26,27]. Although the method proved to have a good inter-operator agreement in the original study [7], we decided to include only the measurements performed by a single observer to limit the influence of this variable. Furthermore, we also decided to exclude from the final sample dogs that underwent US examination under sedation or general anesthesia, as the different sedation/anesthesia adopted protocols could affect blood pressure and, possibly, the AoD. Additionally, only adult dogs were included in the final sample because, as previously demonstrated, growing dogs, particularly those younger than six months, tend to have a higher KL/AoD ratio [25], and consequently alter the reference values.

The further subdivision of dogs by body size (toy, small, medium, large) revealed significant differences between the groups in KL and AoD. As expected, large dogs exhibited the largest dimensions for both KL and AoD. Interestingly, toy breeds showed statistically higher mean KL/AoD ratio values than medium and large breeds. This finding aligns with previous studies suggesting that renal dimensions tend to increase progressively with body size, though the rate of increase diminishes at the extremes of body weight, especially compared to aortic dimensions [4,28]. Consistently, in the present study, a positive correlation between body weight and KL was observed, though it became less pronounced at the higher extremes of body weight. These observations are further supported by the negative correlation between body weight and the KL/AoD ratio, confirming that at higher body weights, the AoD increases proportionally more than the KL.

The main limitations of our study include the lack of histopathological confirmation of the underlying cause of CKD affecting the dogs in the diseased group since this procedure is rarely performed at our institution. Although unlikely, it is possible that some dogs with altered serum urea and creatinine levels for causes unrelated to CKD were erroneously included in the diseased group.

A second limitation is the reduced number of dogs in the diseased group, resulting from our exclusion criteria where only dogs with a US study performed by the same operator and with laboratory analysis performed within 48 h from the US study were included. This low number of subjects included in the final sample and the lack of additional data necessary for the staging of the CKD, such as symmetric dimethylarginine levels, urine protein to creatinine ratio, and blood pressure, also prevented the staging of the dogs with CKD according to the IRIS guidelines [12] and the subdivision of the diseased dogs in sub-groups as for the healthy ones. Indeed, it would be interesting to assess whether the KL/AoD ratio tends to decrease progressively in subjects with advanced stages of CKD and if the subdivision among sub-groups could help increase the sensitivity of the method. Another limitation was the lack of information regarding the dogs' hydration status and the presence of cardio-circulatory diseases that may alter the aorta distensibility and, thus, the KL/AoD ratio.

5. Conclusions

This study showed reference ranges for the KL/AoD ratio similar to those previously described for a sample of mixed subjects. This range is broad and limits the clinical usefulness of the method, suggesting the determination and use of breed-specific ratios that may be narrower. In the general canine population, a KL/AoD ratio = 6.3, although higher than the lower cut-off value, has good specificity and acceptable sensitivity in confirming CKD in dogs with clinical, laboratory, and additional US findings consistent with the disease. It must be kept in mind that the assessment of the renal length and renal linear dimensions, in general, is only one of the parameters to be considered during the US exam of the kidneys, and other alterations affecting the renal echotexture, echogenicity, and perfusion must also be taken into account [1–3,22]. Further studies involving a larger cohort of subjects with confirmed CKD are needed to evaluate variations in the KL/AoD ratio. These studies should consider additional factors such as breed, systemic blood pressure, inter-observer variability, different stages of CKD, operator experience, and the effects of systemic blood pressure and hydration on the measurements.

Author Contributions: Conceptualization, D.C. and L.M.; methodology, D.C. and L.M.; software, D.C.; validation, D.C., E.C., P.C., C.S., M.P.P., M.M., G.G., A.G. and L.M.; formal analysis, D.C. and E.C.; investigation, L.M., A.G., D.C., P.C., E.C. and C.S.; resources, D.C., L.M., A.G., E.C., P.C., M.M. and C.S.; data curation, D.C., E.C., P.C. and L.M.; writing—original draft preparation, D.C.; writing—review and editing, E.C., P.C., C.S., M.P.P., M.M., G.G., A.G. and L.M.; visualization, D.C.; supervision, L.M., A.G., M.P.P. and G.G.; project administration, D.C., L.M. and A.G.; funding acquisition, L.M. and A.G. All authors have read and agreed to the published version of the manuscript.

Funding: This research received no external funding.

Institutional Review Board Statement: The animal study protocol was approved by the Institutional Ethics Committee of the University of Napoli "Federico II" (PG/2023/PG0002892).

Informed Consent Statement: Informed consent was obtained from the owners of all subjects involved in the study.

Data Availability Statement: Anonymized data supporting reported results are available from the corresponding author upon reasonable request.

Conflicts of Interest: The authors declare no conflicts of interest.

Abbreviations

ANOVA, analysis of variance; AoD, aortic luminal diameter; AUC area under the curve, C.I., confidence interval; CKD, chronic kidney disease; KL, kidney length; LKL, left kidney length; RKL, right kidney length; ROC receiver operating characteristic; SD, standard deviation; US, ultrasonographic.

References

1. Seiler, G.S.; Cohen, E.B.; d'Anjou, M.-A.; French, J.; Gaschen, L.; Knapp, S.; Salwei, R.M.; Saunders, H.M. ACVR and ECVDI consensus statement for the standardization of the abdominal ultrasound examination. *Vet. Radiol. Ultrasound* **2022**, *63*, 661–674. [CrossRef] [PubMed]
2. Widmer, W.R.; Mattoon, J.S.; Vaden, S.L. 16—Urinary tract. In *Small Animal Diagnostic Ultrasound*, 4th ed.; Mattoon, J.S., Sellon, R.K., Berry, C.R., Eds.; W.B. Saunders: St. Louis, MO, USA, 2021; pp. 583–634. [CrossRef]
3. d'Anjou, M.-A.; Penninck, D. Kidneys and ureters. In *Atlas of Small Animal Ultrasonography*, 2nd ed.; Wiley Blackwell: Hoboken, NJ, USA, 2015; pp. 331–361.
4. Barr, F.J.; Holt, P.E.; Gibbs, C. Ultrasonographic measurement of normal renal parameters. *J. Small Anim. Pract.* **1990**, *31*, 180–184. [CrossRef]
5. Seiler, G.S. Chapter 41—Kidneys and Ureters. In *Textbook of Veterinary Diagnostic Radiology*, 7th ed.; Thrall, D.E., Ed.; W.B. Saunders: Philadelphia, PA, USA, 2018; pp. 823–845. [CrossRef]
6. Hecht, S.; Henry, G.A. Ultrasonography of the Urinary Tract. In *Nephrology and Urology of Small Animals*; John Wiley & Sons: Hoboken, NJ, USA, 2011; pp. 128–145. [CrossRef]
7. Mareschal, A.; d'Anjou, M.A.; Moreau, M.; Alexander, K.; Beauregard, G. Ultrasonographic measurement of kidney-to-aorta ratio as a method of estimating renal size in dogs. *Vet. Radiol. Ultrasound* **2007**, *48*, 434–438. [CrossRef]
8. Barella, G.; Lodi, M.; Sabbadin, L.A.; Faverzani, S. A new method for ultrasonographic measurement of kidney size in healthy dogs. *J. Ultrasound* **2012**, *15*, 186–191. [CrossRef]
9. Costanza, D.; Pasolini, M.P.; Greco, A.; Mennonna, G.; Auletta, L.; Lamagna, F.; Meomartino, L. Ultrasonographic measurement of kidney-to-aorta parameters in Whippets. *Vet. Radiol. Ultrasound* **2021**, *62*, 476–482. [CrossRef]
10. Scrivani, P.V.; Erb, H.N. Invited review: Study design considerations for clinical research in veterinary radiology and radiation oncology. *Vet. Radiol. Ultrasound* **2013**, *54*, 317–325. [CrossRef]
11. Costanza, D.; Coluccia, P.; Castiello, E.; Greco, A.; Meomartino, L. Description of a low-cost picture archiving and communication system based on network-attached storage. *Vet. Radiol. Ultrasound* **2022**, *63*, 249–253. [CrossRef]
12. International Renal Interest Society (IRIS) Ltd. IRIS Staging of CKD Staging of CKD—Modified 2023. Available online: http://www.iris-kidney.com (accessed on 20 November 2024).

13. Friedrichs, K.R.; Harr, K.E.; Freeman, K.P.; Szladovits, B.; Walton, R.M.; Barnhart, K.F.; Blanco-Chavez, J. ASVCP reference interval guidelines: Determination of de novo reference intervals in veterinary species and other related topics. *Vet. Clin. Pathol.* **2012**, *41*, 441–453. [CrossRef]
14. Reed, A.H.; Henry, R.J.; Mason, W.B. Influence of statistical method used on the resulting estimate of normal range. *Clin. Chem.* **1971**, *17*, 275–284. [CrossRef]
15. CLSI. Defining, Establishing and Verifying Reference Intervals in the Clinical Laboratory. In *Approved Guideline*, 3rd ed.; CLSI: Wayne, PA, USA, 2008.
16. Horn, P.S.; Pesce, A.J. Reference intervals: An update. *Clin. Chim. Acta* **2003**, *334*, 5–23. [CrossRef]
17. Metz, C.E. Basic principles of ROC analysis. *Semin. Nucl. Med.* **1978**, *8*, 283–298. [CrossRef] [PubMed]
18. Metz, C.E. ROC methodology in radiologic imaging. *Investig. Radiol.* **1986**, *21*, 720–733. [CrossRef]
19. Zweig, M.H.; Campbell, G. Receiver-operating characteristic (ROC) plots: A fundamental evaluation tool in clinical medicine. *Clin. Chem.* **1993**, *39*, 561–577. [CrossRef] [PubMed]
20. DeLong, E.R.; DeLong, D.M.; Clarke-Pearson, D.L. Comparing the areas under two or more correlated receiver operating characteristic curves: A nonparametric approach. *Biometrics* **1988**, *44*, 837–845. [CrossRef] [PubMed]
21. Youden, W.J. Index for rating diagnostic tests. *Cancer* **1950**, *3*, 32–35. [CrossRef] [PubMed]
22. Bragato, N.; Borges, N.C.; Fioravanti, M.C.S. B-mode and Doppler ultrasound of chronic kidney disease in dogs and cats. *Vet. Res. Commun.* **2017**, *41*, 307–315. [CrossRef]
23. Lefebvre, H.P. Renal Function Testing. In *Nephrology and Urology of Small Animals*; John Wiley & Sons: Hoboken, NJ, USA, 2011; pp. 91–96. [CrossRef]
24. Sohn, J.; Yun, S.; Lee, J.; Chang, D.; Choi, M.; Yoon, J. Reestablishment of radiographic kidney size in Miniature Schnauzer dogs. *J. Vet. Med. Sci.* **2017**, *78*, 1805–1810. [CrossRef]
25. Kawalilak, L.T.; Pease, A.P.; Nelson, N.C. Evaluation of ultrasonographically determined ratios of kidney length to aorta diameter for assessment of kidney size in healthy young dogs. *Am. J. Vet. Res.* **2019**, *80*, 764–770. [CrossRef]
26. Casella, I.B.; Presti, C.; Porta, R.M.; Sabbag, C.R.; Bosch, M.A.; Yamazaki, Y. A practical protocol to measure common carotid artery intima-media thickness. *Clinics* **2008**, *63*, 515–520. [CrossRef]
27. Pasolini, M.P.; Spinella, G.; Del Prete, C.; Valentini, S.; Coluccia, P.; Auletta, L.; Greco, M.; Meomartino, L. Ultrasonographic assessment of normal jugular veins in Standardbred horses. *BMC Vet. Res.* **2019**, *15*, 343. [CrossRef]
28. Olesen, S.; Genster, H.G. Estimation of renal volume and function in dogs from the radiological appearance. *Invest. Urol.* **1970**, *7*, 363–370.

veterinary sciences

MDPI

Article

Changes in Biomarkers of Inflammation and Oxidative Status in Dogs Subjected to Celiotomy or Video-Assisted Ovariohysterectomy

Fabíola Dalmolin [1], Camila Peres Rubio [2,*], Carla Sordi Furlanetto [1], Rafael Steffens [1], Najla Ibrahim Isa Abdel Hadi [1], Adriellen de Lima da Silva [3], Paloma Tomazi [1], Bernardo Nascimento Antunes [4], Fabiana Elias [3], Elizabeth Moreira dos Santos Schmidt [5] and Maurício Veloso Brun [4]

1 Programa de Pós-Graduação em Saúde, Bem-estar e Produção Animal Sustentável na Fronteira Sul (PPG-SBPAS), Universidade Federal da Fronteira Sul (UFFS), Realeza 85770-000, Brazil; fabiola.dalmolin@uffs.edu.br (F.D.); carla.sordi@hotmail.com (C.S.F.); rafaelsteffens@msn.com (R.S.); najlahadi@hotmail.com (N.I.I.A.H.); palomatomazi.medvet@gmail.com (P.T.)

2 Department of Animal Surgery and Medicine, University of Murcia, 30100 Murcia, Spain

3 Curso de Medicina Veterinária, Universidade Federal da Fronteira Sul (UFFS), Realeza 85770-000, Brazil; adriellenlima97@gmail.com (A.d.L.d.S.); fabiana.elias@uffs.edu.br (F.E.)

4 Programa de Pós-Graduação em Medicina Veterinária (PPGMV), Universidade Federal de Santa Maria (UFSM), Santa Maria 97105-900, Brazil; bernardonascimentoantunes@gmail.com (B.N.A.); mauriciovelosobrun@hotmail.com (M.V.B.)

5 School of Veterinary Medicine and Animal Science, São Paulo State University (FMVZ-UNESP), Campus Botucatu, Botucatu 18618-681, Brazil; elizabeth.schmidt@unesp.br

* Correspondence: camila.peres@um.es

Simple Summary: There are contrasts in the recovery and response to surgical stress in canine after open and video surgery. The main difference includes incision size, intra- and postoperative pain, recovery time regarding eating, urination, defecation, and movement, as well as changes in acute phase proteins, oxidative metabolism, and leukocytosis, among others, with advantages to laparoscopy. The aim of this study was to clarify the effect of the access and hemostatic techniques regarding inflammation and oxidative stress. There is some evidence that video surgery, due to the carbonic gas, provokes higher oxidative stress regardless of lower inflammation compared to open surgery. This study was carried out including one of the most used techniques in video surgery, the video-assisted two portals technique, a conventional technique involving manual rupture of the ovarian suspensory ligament and ligation with surgical threads and an open access without ligament rupture and bipolar coagulation. This study demonstrates similar oxidative stress between techniques and advantages to the video-assisted technique, even when using a pneumoperitoneum.

Abstract: We evaluated the surgical stress response of dogs undergoing three ovariohysterectomy (OVH) techniques. Twenty-nine healthy females were allocated into groups: celiotomy and ligature (CelioSut), celiotomy and bipolar energy (CelioBip), and a video-assisted technique using two portals and bipolar energy (VidBip). Clinical evaluation was performed, and the following blood analyses were determined: acute phase proteins (C-reactive protein and haptoglobin), white blood cell counting (WBC), and biomarkers of oxidative status. The VidBip required more time despite a smaller incision, lower heart rate, and earlier feeding and urination. All groups had high white blood cells counts; the C-reactive protein (CRP) levels peaked at 6 and 12 h in all groups and was higher in the CelioBip and VidBip groups compared to CelioSut; haptoglobin concentrations peaked at 48 h in all groups and increased in dogs of the CelioSut group. Trolox equivalent antioxidant capacity, ferric reducing ability of plasma, cupric reducing antioxidant capacity, and advanced oxidation protein were not significantly different among the groups or time points. Total thiol concentrations were lower in CelioBip and CelioSut groups. All surgical techniques induced an inflammatory and oxidative stress response, but the video-assisted technique produced early clinical recovery. The bipolar device produces fewer disturbances than suspensory ligament rupture and ligature.

Keywords: homeostasis; inflammation; laparoscopy; neutering; oxidative stress

1. Introduction

The ovariohysterectomy (OVH) is one of the most common surgeries performed in dogs [1,2]. It can be accessed in several ways, such as open midline abdominal incision and laparoscopy [1,3,4], and can leverage various hemostatic methods such as a suture ligature, vascular clips, and bipolar energy, among others [1,3].

Previous studies in dogs have evaluated the response to surgical stress to understand the effects of different techniques and to identify, manage, or prevent complications [4–6]. In general, patients experience trauma and anesthesia-induced immune activation in the postoperative period, which exacerbates the inflammatory response [4]. Inflammation is a fundamental defense mechanism against infection and the initiator of important tissue repair. In contrast, prolonged and uncontrolled inflammation can have negative effects such as excessive pain, immunosuppression, organ dysfunction, and death [7]. The response to surgery consists of hormonal, metabolic, and inflammatory responses that occur after surgery and allow the body to adapt to trauma and repair damaged tissues [8].

During the acute phase response, there is an increase in the synthesis and release of acute phase proteins, driven by proinflammatory cytokines produced primarily by the liver, such as C-reactive protein (CRP) and haptoglobin (Hp) [9]. In addition to inflammation, oxidative stress due to surgical procedures has also been shown in dogs [5,6]. Oxidative stress is an imbalance between oxidants and antioxidants in favor of oxidants, leading to disruption of redox signaling and control and/or molecular damage [10]. A change in the oxidative balance is recognized by cell damage and the cause of several diseases. In major surgeries, it occurs acutely mainly during ischemia followed by reperfusion, as shown after laparoscopy [11]. The choice of intraoperative approaches, techniques, and materials often depends on efforts to reduce trauma [7] and shorten operative time [2]. Minimally invasive procedures, such as video-assisted procedures, had advantages over open surgery, resulting in less tissue injury and reduced release of inflammatory mediators, reduced pain, and thus early recovery [12], because it reduces the inflammatory response [3,5]. The hemostatic harmonic scalpel is an effective and efficient tool for hemostasis, although its cost is high. Bipolar electrocoagulation has been used instead because it is considered safe, less expensive [6], and promotes a shorter operative time and rare complications compared to conventional suspensory ligament and ligament rupture [2]. A bipolar device has been described in open [2] and video-assisted canine OVH [3,6].

The effect of traditional and bipolar hemostatic methods and the celiotomy and endoscopic access on inflammation and the oxidative response after canine OVH is unknown. Thus, the objectives of this study were the following: (i) to evaluate the dog's response to surgical stress after OVH with mid-abdominal celiotomy and hemostatic ligature (CelioLig); ventral median celiotomy and bipolar coagulation (CelioBip); and video-assisted dual port and bipolar coagulation (VidBip); and (ii) to investigate the effect of different approaches (celiotomy and endoscopy) and hemostatic techniques (ligature and bipolar coagulation) on the surgical stress response of OVH in dogs by evaluating blood analytes such as acute phase protein concentrations (CRP and Hp), WBC, and the oxidative stress biomarkers (total antioxidant capacity (TAC), Trolox equivalent antioxidant capacity (TEAC), copper-reducing antioxidant capacity (CUPRAC), ferric reducing ability of plasma (FRAP), serum total thiol, and the advanced oxidation protein products biomarker (AOPP).

2. Materials and Methods

2.1. Animals

All the animal handling and procedures were approved by the Faculty's Animal Experimentation Ethics Committee of the Universidade Federal da Fronteira Sul, Brazil (protocol number 23205.003951-2018-77 CEUA). The study population consisted of twenty-nine healthy female dogs of different breeds. The animals included were considered healthy based on physical and laboratory examinations (total white blood count (WBC), serum activity of alkaline phosphatase, alanine aminotransferase, serum concentrations of creatinine, and total plasma protein measurements). Those with a history of previous illness, skin abnormalities, ectoparasites, endoparasites, or chronic conditions (e.g., dermatological, respiratory, allergic, gastrointestinal) were excluded. All selected animals were dewormed, vaccinated, and limited to 4 years of age or younger to avoid issues related to degenerative or age-associated diseases. Abdominal ultrasound (Sonosite M Turbo, Bothell, United State), electrocardiogram (InCardio X InPulse, Florianópolis, Brazil), absence of significant macroscopic changes in the uterus, ovaries, and uterine tubes after surgery were also considered healthy. Patients were randomly divided into three groups according to the surgical procedure: i. CelioLig (n = 9, 2.05 ± 0.80 years; 15.40 ± 3.35 kg); ii. CelioBip (n = 10, 2.44 ± 1.50 years; 14.16 ± 3.73 kg); and iii. VidBip (n = 10, 1.80 ± 0.63 years; 13.26 ± 3.33 kg).

2.2. Surgical Preparation and Anesthesia

Forty-eight hours before surgery, patients were placed in boxes in the same location where postoperative assessments were performed. Animals were given water and food ad libitum and had access to an external restricted area for defecation and urination. Patients were exposed to researchers and other laboratory animals only 96 h after surgery. The same team performed all procedures, avoiding pattern changes. After a 12 h food and 8 h water fast, the dogs received acepromazine (0.05 mg/kg i.m.), (Acepran 0.2%, Univet, São Paulo, Brazil), and 15 min later, a broad trichotomy of the abdomen and venous access with Ringer's Lactate solution (5 mL/kg/h i.v.), (Equiplex, Aparecida do Goiás, Brazil) were performed. The anesthesia was induced via propofol (6 mg/kg) administered intravenously (IV), (Fresofol, Frenesius-Kabi, São Paulo, Brazil) followed by tracheal intubation and 100% oxygen vaporized isoflurane in a calibrated vaporizer. Fentanyl (2.5 µg/kg IV), (Janssen-Cilag Farmacêutica LTDA, São Paulo, Brazil) analgesic was used before the right ovary manipulation. Morphine was predicted if an elevation of 30% or more of the initial cardiac rate or systolic blood pressure happened. During anesthesia, a 16 G catheter was introduced into the jugular vein, fixed via a suture and instantaneous adhesive and protected by a sterile bandage to obtain blood samples at the postoperative time points.

2.3. Surgical Procedures

All procedures were started with patients in dorsal recumbency. The CelioSut group underwent a retro-umbilical celiotomy involving one-third of the distance between the umbilicus and the pubis. After manual rupture of the ovarian suspensory ligament, polyglactin 910 (Shalon, Goiás, Brazil) double ligatures (circular and fixed) were applied to the ovarian pedicles and uterine body using a modified three-clamp technique [12]. Celiorrhaphy was performed with the same thread and cross mattress pattern. Subcutaneous reduction was performed with continuous suture and the same material, and dermorraphy was performed with an isolated horizontal mattress template and nylon suture. CelioBip was performed via celiotomy. Additionally, 5 mm laparoscopic bipolar forceps (Gyrus ACMI®, Southborough, MA, USA) were used for hemostasis of the ovary and uterine body. No ovarian suspensory ligament rupture was performed in this group. Abdominal sutures were performed as described above.

For VidBip, we used two ports (11 mm; length 10.5 cm), one through the umbilicus and the other in the pre-pubic region. The abdomen was inflated with CO_2 (1/min) to 10 mmHg and a 10 mm 0° rigid endoscope was used. Patients were placed in the lateral recumbent position, and the ovary was temporarily fixed to the abdominal wall with a transparietal suture using a traumatic curved needle (2 cm) and 2-0 nylon. Hemostasis was performed via bipolar coagulation as in CelioBip. Vascular hemostasis of the uterine body was performed extracorporeally. Residual gas was removed, and abdominal sutures were applied as described above [3]. The dimensions of the sutures were adjusted according to the size of the patient. The size of the incisions was measured with a pachymeter; in the VidBip group, the size of two incisions was recorded.

2.4. Postoperative and Blood Sampling

Immediately after surgery, patients received tramadol hydrochloride (5 mg/kg subcutaneously, every 8 h/48 h) (Medley Indústria Farmacêutica Ltd.a, Campinas, São Paulo), chlorhexidine for wounds (every 24 h), and protective clothing. The patients were evaluated at 2, 6, 12, 24, and 48 h after surgery with a pain scale [13] and analgesic rescue was predicted with morphine (0.3 mg/kg intramuscularly) if the patients reached a score of six or more points.

Rectal temperature, respiratory rate (RR), heart rate (HR), and blood samples were taken immediately before surgery and at 2, 6, 12, 24, 48 h and seven days later. For WBC count, 0.5 mL of blood was collected via jugular venipuncture and placed in EDTA tubes (Labor 0.5 mL) in the same time points and processed immediately. Another aliquot (5 mL) was obtained, placed in a serum separator tube, and centrifuged at 3000× g for 10 min. Each serum sample was aliquoted (400 μL/Eppendorf type tube) and stored at −80 °C until analysis. Frozen aliquots (−80 °C) were sent to INTERLAB—University of Murcia (UMU) for analysis of biomarkers of inflammation and oxidative stress.

2.5. Laboratory Analysis

White blood cells were automatically counted (Bio–2900 Vet—Bioeasy Diagnostical, São Paulo, Brazil). Differential leukocyte counts were performed via slide preparation stained with the Diff Quick method and an optical microscope reading which included segmented neutrophils, lymphocytes, monocytes, eosinophils, and basophils [14].

Serum C-reactive protein (CRP) concentration was determined via immunoturbidimetric analysis (CRP OSR6147, Olympus Life and Material Science, Lismeehan, Ireland). Serum haptoglobin (Hp) concentration was measured using a commercial kit (Tridelta Development, Maynooth, Ireland). The methods were previously validated for use in dogs [14,15]. Biomarkers of oxidative status previously validated in canine serum samples were determined: total antioxidant capacity (TAC) [16,17], Trolox equivalent antioxidant capacity (TEAC) [16,18], copper-reducing antioxidant capacity (CUPRAC) [18,19], ferric reducing ability of plasma (FRAP) [20,21], and serum total thiol [22,23]. Products of oxidation of proteins were evaluated via the advanced oxidation protein products biomarker (AOPP) [24]. The analyses were performed using an automated biochemistry analyzer (Olympus AU400, Olympus Diagnostica GmbH, Hamburg, Germany).

2.6. Statistical Analysis

Data were assessed for their distribution pattern with the Shapiro–Wilk test and for homogeneity of variances with the Levene test. Comparisons between groups were made using two-way ANOVA and Tukey's multiple comparison test. Comparison between timepoints was performed using two-way ANOVA and Dunnett's multiple comparison test. Normally distributed data were presented as means ± standard deviation. Nonparametric data were presented as median and interquartile range. Differences were considered significant at 5% ($p < 0.05$).

3. Results

3.1. Surgery Data

All animals had no surgical complications, and all recovered after eight days. During the CelioSut surgery, a spontaneous pressure opening was observed in the uterine body (n = 1). Sharp paramedian incision (n = 2), bipolar device damage to the peritoneum (estimated as small as 0.3 cm) (n = 1), and difficulty with ovarian exposure (n = 2) were noted during CelioBip. In addition, events included a VidBip-shaped paramedian portal (n = 3), portal displacement (n = 3), transparietal suture difficulty (n = 5), superficial splenic injury (n = 2), bipolar peritoneal injury (n = 4) (estimate as small as 0.2 cm), and difficulties with exteriorization of the uterus and the ovaries ($n = 2$); these events occurred mainly in two obese and large dogs, which increased the difficulty of the procedure and the surgical time. In the endoscopic surgery, a longer operative time ($p < 0.01$) and smaller incision ($p < 0.01$) were observed.

3.2. Animal Recovery and Clinical Examination

Analgesic rescue was not necessary for any animal during the evaluation time. Earlier time to first ingestion of solids ($p < 0.01$) and urination ($p = 0.03$) were observed in the VidBip group compared to others (Table 1). The first urination was earlier in the CelioBip group than in the CelioSut group ($p = 0.03$). No difference was found between groups in time to first bowel movement ($p = 0.28$) (Table 1). Regarding clinical parameters, rectal temperature was significantly different between groups between 6 h ($p = 0.04$) and 12 h ($p = 0.04$); hypothermia was observed in CelioSut and CelioBip and was more pronounced in the latter. Over time, CelioSut (37.4 and 38 °C, $p < 0.01$) and CelioBip animals (37.7 and 37.9 °C, $p = 0.03$) showed hypothermia (range 38–39 °C) at two and six hours. VidBip showed a normal temperature (~38 °C, $p = 0.01$). Regarding RR, VidBip animals had a lower effect at 24 h compared to other groups ($p = 0.01$). No difference was observed between CelioSut and CelioBip ($p > 0.05$). Regarding HR, a statistically significant difference was observed between groups at 6 h ($p = 0.01$), 12 h ($p < 0.01$), 24 h ($p < 0.01$), and 48 h ($p = 0.03$); VidBip animals showed lower values, and no differences were observed between the other groups. Over time, only VidBip animals showed a difference; after baseline, HR decreased at 6, 12, and 24 h and increased at 48 h ($p < 0.01$) (Table 2).

Table 1. Surgery time, incision size, and time to first solid intake, urination, and defecation after canine OVH via ventral median celiotomy and ligature (CelioSut), ventral median celiotomy and bipolar coagulation (CelioBip), and video-assisted technique with two ports and bipolar coagulation (VidBip).

	CelioSut	CelioBip	VidBip	*p* Value
	M ± SD	M ± SD	M ± SD	
Surgery time (min) [a]	21.44 ± 5.47 A	20.20± 5.02 A	58.40 ± 16.17 B	$p < 0.01$
Incision size (cm) [a]	5.96 ± 0.66 A	5.95 ± 1.26 A	2.91 ± 0.62 B	$p < 0.01$
	Md (Interquartile Range)	**Md (Interquartile Range)**	**Md (Interquartile Range)**	
First solid intake (h) [b]	11.82 (5.91–13.38) A	6.55 (6.23–10.38) A	1.88 (1.21–3.77) B	$p < 0.01$
First urination (h) [b]	7.38 (5.72–8.97) A	6.59 (3.5–7.06) AB	4.00 (2.83–5.46) B	$p = 0.03$
First defecation (h) [b]	18.77 (11.95–25.55) A	9.19 (6.8–22.15) A	11.08 (10.19–12.25) A	$p = 0.28$

The normality of data distribution was verified with the Shapiro–Wilk test. [a] Parametric data (means [M] ± standard deviation [SD]). [b] No parametric data (median [Md] and interquartile range [Q1–Q3]). Capital letters indicate statistical differences between groups.

3.3. Laboratorial Analysis

White blood cell count is shown in Figure 1. Over time, WBC counts peaked at 6 h and 12 h for CelioSut ($p < 0.05$) and returned to the base line at 24 h; the CelioSut WBC counts increased at 6 h ($p < 0.01$), peaked at 12 h ($p < 0.001$), and decreased at 24 h ($p < 0.05$), returning to the base line at 48 h; the VidBip group had no alteration. The segmented neutrophil counts in CelioSut peaked at 6 h ($p < 0.001$), decreased at 12 h ($p < 0.01$) and at 24 h ($p < 0.05$), and returned to basal levels at 48 h; the CelioBip group presented a difference at 2 h ($p < 0.05$), peaked at 6 h ($p < 0.001$) and 12 h ($p < 0.001$), reduced at 24 h ($p < 0.01$), and returned to the basal level at 48 h. The VidBip group increased just at 12 h ($p < 0.05$) and returned to normal at 24 h. Lymphopenia and eosinopenia was observed just at 6 h in CelioSut animals ($p < 0.05$). The lymphocytes for the CelioSut and CelioBip groups decreased at 2 h, 6 h, and 12 h, followed by increases at 48 h and seven days ($p < 0.01$; $p < 0.01$). No difference was observed for VidBip ($p = 0.08$). There were no differences in eosinophils between groups ($p > 0.06$). Over time, CelioSut animals had eosinopenia at 6 and 12 h ($p < 0.01$), followed by an increase at 24 and 48 h and a return to baseline at 48 h. There were no differences between groups or times for monocytes ($p > 0.10$).

Table 2. Rectal temperature, respiratory rate, and heart rate after canine OVH via ventral median celiotomy and ligature (CelioLig), ventral median celiotomy and bipolar coagulation (CelioBip), and video-assisted technique with two ports and bipolar coagulation (VidBip).

Times	CelioLig	CelioBip	VideoBip	*p* Value **
		Rectal temperature (°C)		
Basal	38.40 a	38.55 a	38.75 a	0.29
2	37.40 b	37.70 b	38.05 b	0.06
6	38.00 bAB*	37.90 bB	38.20 abA	0.04
12	38.20 aAB	38.05 bB	38.65 aA	0.04
24	38.20 a	38.15 ab	38.65 a	0.10
48	38.50 a	38.25 ab	38.65 a	0.06
72	38.45 a	38.30 a	38.50 a	0.30
p value **	$p < 0.01$	$p = 0.03$	$p = 0.01$	
		Respiratory rate (mpm)		
Basal	38	34	34	0.42
2	28	28	24	0.79
6	31	32	28	0.33
12	36	34	29	0.07
24	36 A*	38 A	28 B	0.01
48	34	32	28	0.26
72	32	34	32	0.48
p value **	0.06	0.33	0.06	
		Heart rate (bpm)		
Basal	116	108	120 a*	0.60
2	110	100	110 a	0.46
6	118 A**	122 A	96 bB	0.01
12	128 A	114 A	98 bB	0.01
24	124 A	124 A	96 bB	0.01
48	112 A	112 A	102 abB	0.03
72	116	104	102 ab	0.18
p value **	0.27	0.38	0.01	

The normality of data distribution was verified with the Shapiro–Wilk test. Later, the data were investigated via the * Dunn–Bonferroni and ** Kruskal–Wallis Tests. Capital letters and small letters indicate statistical difference between groups.

The results of inflammatory biomarkers are presented in Figure 2. Over time, CelioSut animals increased haptoglobin (Hp) concentration at 2 h and peaked at 48 h ($p < 0.05$; $p < 0.01$, respectively). The CelioBip increased at 2 h, peaked at 24 h and 48 h ($p < 0.05$, $p < 0.01$, $p < 0.001$), and returned to basal levels at 7 days. VidBip increased and peaked at 24 h and 48 h ($p < 0.01$). The differences between groups were at 48 h, with the CelioSut group being higher than CelioBip ($p < 0.01$) and with CelioSut higher than VidBip ($p < 0.5$).

In relation to the CRP levels, the CelioBip group showed increased levels at 12 h and 24 h after surgery ($p < 0.01$). Regarding the VidBip group, differences were identified at 12 h and 24 h ($p < 0.05$), and then CRP returned to baseline values. No significant differences were observed between groups for CRP ($p > 0.05$).

Considering the oxidative stress biomarkers (Figure 3), the TEAC increased at 96 h in CeliSut ($p < 0.05$) and at 7 days in CelioBip ($p < 0.05$); while in the VidBip group, TEAC presented no alteration. Regarding CUPRAC, in the CelioBip group the level was higher at 2 h ($p < 0.05$) and in the CelioSut at 6 h and 12 h ($p < 0.05$); the VidBip group showed no alteration in TEAC. FRAP showed no differences between the different time-points of the surgical procedures. The thiol was higher in the CelioSut at 2 h, 12 h, and 24 h ($p < 0.001$, $p < 0.001$ and $p < 0.01$, respectively) but showed no changes in the CelioBip and VidBip groups. The AOPP was higher at 6 h and 12 h in the CelioSut ($p < 0.05$) group, and at 24 h in VidBip ($p < 0.05$); however, in the CelioBip group, there were no differences.

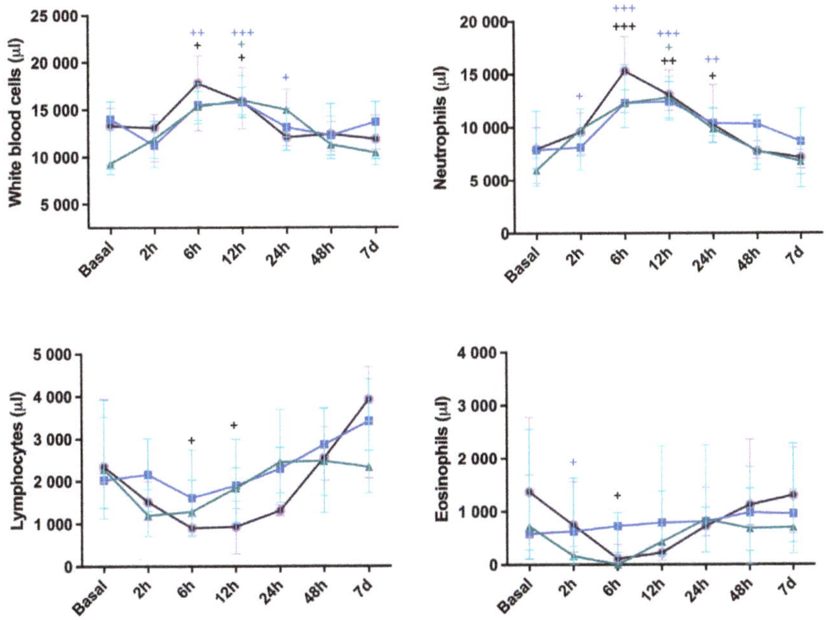

Figure 1. Total leukocytes, neutrophils, and lymphocytes median score and eosinophils median score in canines submitted to ovariohysterectomy via celiotomy and ligature (CelioSut; purple lines), celiotomy and bipolar coagulation (CelioBip; green lines), or the video-assisted technique with two ports and bipolar coagulation (VidBip; blue lines) along the evaluation time. Plus (+) means difference between the time-point and the basal time (+, $p < 0.05$; ++, $p < 0.01$; +++, $p < 0.001$).

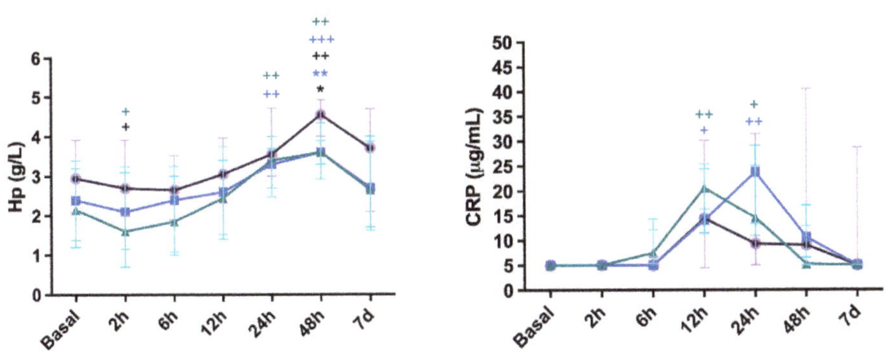

Figure 2. Haptoglobin (Hp) and C-reactive protein (CRP) concentrations in canines submitted to ovariohysterectomy via celiotomy and ligature (CelioSut; purple lines), celiotomy and bipolar coagulation (CelioBip; green lines), or the video-assisted technique with two ports and bipolar coagulation (VidBip; blue lines) along the evaluation time. Black asterisk (*) indicates difference between CelioSut and VidBip ($p < 0.05$). Blue asterisks (**), indicate differences between CelioSut and CelioBip ($p \leq 0{,}01$). Plus (+) means a difference between the time-point and basal time (+, $p < 0.05$; ++, $p < 0.01$; +++, $p < 0.001$).

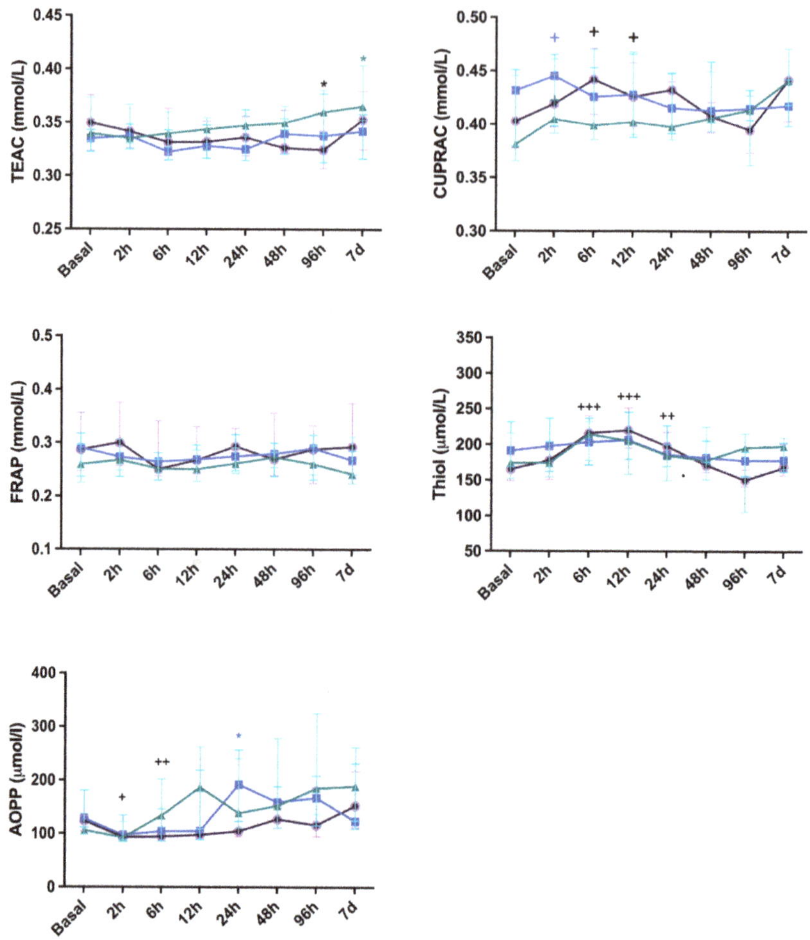

Figure 3. Trolox equivalent antioxidant capacity (TEAC), cupric reducing antioxidant capacity (CUPRAC), ferric reducing ability of plasma (FRAP), and thiol and the advanced oxidation protein products biomarker (AOPP) concentration in canine submitted to ovariohysterectomy via celiotomy and ligature (CelioSut; purple lines), celiotomy and bipolar coagulation (CelioBip; green lines), or the video-assisted technique with two ports and bipolar coagulation (VidBip; blue lines) immediately before surgery until seven days after. Black asterisks mean difference between CelioSut and VidBip; green asterisks mean difference between CelioBip and VidBip; blue asterisks mean difference between CelioSut and CelioBip ($p < 0.05$). Plus (+) means difference between the time-point and basal time (+, $p < 0.05$; ++, $p < 0.01$; +++, $p < 0.001$).

4. Discussion

This study proposed to compare surgical stress after three OVH techniques, with different access and hemostatic methods, considering the importance of rapid recovery and fewer postoperative complications after a routine surgery in canines. Different OVH techniques have advantages and disadvantages, and they are described for female dogs [1,2,12]. In previous studies, the incidence of minor intraoperative complications after conventional canine OVH ranged from 7.5% to 19%. They were mainly associated with incision site inflammation and gastrointestinal upset [12]. Major complications included granulomas in remnants of the uterine or ovarian pedicle (28%). The complications which required intervention or treatment included inflammation or infection (8 of 142 dogs; 5.6%) and bleeding (4 of 142 dogs; 2.8%) [12]. Although studies with a large number of dogs are not available in the literature, the complication rate after laparoscopy of OVH was approximately 2% [4].

In our results, all techniques had small intercurrences during the procedures, which were more frequent in the VidBip group, probably because the surgeon had more practice in celiotomy than in celioscopy [6]. For the same reason, a longer operating time was required for the VidBip technique; probably due to the presence of two obese dogs (86 min and 74 min), a portal introduced to the vesical median ligament (86 min), and ovarian vascular coagulation difficulties in one patient (76 min) as previously reported [3]. Additionally, just a small needle size (2 cm) was available for the transparietal suture [3] and the lack of a specific surgical table for endoscopic surgery, which allows aseptic care and faster repositioning [6], may have lengthened the video-assisted procedure in this study.

When comparing the CelioSut and CelioBip groups, the advantage of the second one is that bipolar energy avoids foreign bodies entering the abdomen compared to suture material [2] and suture complications [1]. Bipolar electrosurgical devices can safely block vessels up to seven millimeters in diameter and uterine bodies below nine millimeters [12]. In this study the use of bipolar coagulation and ligature was equally difficult because the same instrument (video surgical bipolar) was required. Although, it has been associated with reduced surgical time when using a hand instrument [1,2]. Thus, the careful use of the laparoscopic device in open surgery was necessary due to the length of the instrument; the same device was used for the open and the video surgery to standardize the coagulation trauma. Ovarian exposure was difficult in CelioBip animals, and the surgical procedure took longer to perform, mainly in younger dogs, presumably due to the no rupture of the ovarian suspensory ligament [6].

Lower HR and RR were observed after laparoscopy compared to open procedures, which may be related to lower stress-induced cortisol and adrenaline release [25]. Mild hypothermia was observed after celiotomy, in contrast to normothermia after endoscopy. In humans, perioperative hypothermia is associated with higher intraoperative hemorrhage, adverse events such as surgical infection, longer postoperative chills, postanesthetic recovery, and hospitalization, although mortality is significantly increased [26]. Although hypothermia was more severe in this study when the bipolar device was used, no complications were observed probably due to the healthy conditions of the dogs [26]. In addition, the duration of anesthesia was associated with hypothermia [26], which was not observed in this study after laparoscopy, probably because the abdominal organs were not exposed, the low blood loss [3], and the position of the animals in lateral recumbency which minimizes body temperature changes [26]. The first solid intake and urination occurred earlier after video-assisted OVH than celiotomies, as previously described [4–6]. The video-assisted OVH is associated with less manipulation and injury, resulting in less cortisol and catecholamine release [26] and less pain. Thus, it promotes early recovery compared to open procedures [3,5].

In this study, leukocytosis was observed at 6, 12, and 24 h after surgery, being one of the changes observed for the acute inflammatory response in dogs [27]. The WBC peaked at 6 h to 24 h and returned to baseline values in all groups seven days after surgery, and both OVH techniques induced prolonged neutrophilia. The number of neutrophils increased and decreased similarly to WBC values due to a systemic inflammatory response [6,25,27]. In this study, lymphopenia and eosinopenia were observed just for the CelioSut group and were thus associated with the suspensory ligament rupture and ligature application. This result reinforces previous data, that the stress hemogram is present in canines subjected to open OVH [5].

The acute phase response is considered a part of the innate host defense system. The systemic inflammatory response is induced by the release of pro-inflammatory cytokines (interleukin-1, interleukin-6, and alpha tumor necrosis factor), which stimulate the synthesis and release of CRP and Hp by the liver [9]. In this study, haptoglobin concentrations increased 48 h after CelioSut compared to other groups, which could be related to suspensory ligament trauma [6]. The CRP concentrations which peaked at 12 h in CelioSut and at 24 h in VidBip, being higher for the CelioSut group, could be associated with surgical trauma [4], suggesting an initial although reduced inflammatory response for video surgery. Our results are similar to those previously described, where lower and higher CRP concentrations were associated with laparoscopy and open surgery, respectively [5,25]. Significant differences in CRP concentrations were observed over time when bipolar coagulation was applied via celiotomy and the video-assisted technique. A peak of CRP concentrations was observed at 6 and 12 h after surgery, suggesting that this protein was useful for monitoring patients after bipolar coagulation.

An imbalance between oxidants and antioxidants has been associated in humans with postoperative complications such as myocardial injury, sepsis, pulmonary edema, renal and hepatic failure, and increased mortality [28]. In this study, oxidative status was assessed using different integrated markers according to previous recommendations [21,28]. The antioxidant response can be monitored via the TAC assay, which is used to assess antioxidant levels in biological samples and to assess the antioxidant response against free radicals [17]. More specifically, it has been described that the thiol group is mainly responsible for the antioxidant effect of plasma proteins [22]. To the author's knowledge, the combination of oxidative and antioxidant biomarkers used here has not been evaluated in dogs or humans in order to compare open and video-assisted surgery with the bipolar coagulation and ligature method.

No significant differences were found for FRAP in this study. The CUPRAC increased at 2 h in the CelioBip group and at 6 h and 12 h in the CelioSut group and the thiol concentrations increased at 6 h, 12 h, and 24 h for the CelioSut group, similar to the changes in WBC and neutrophils in the animals that had open surgery, confirming that oxidative stress and inflammation are part of the surgical stress response [7,10]. The AOPP increased at 6 h and 12 h for the CelioSut group, and it could be associated with the higher trauma of abdominal access and disruption of the suspensory ligament, as observed with the thiol, TEAC, and hemogram data in this study. The AOPP was significantly different at 24 h after laparoscopy; it could be related to the pneumoperitoneum, which causes a decrease in splanchnic blood flow through both systemic and local effects in a pressure- and time-dependent manner [28]. There is evidence in animal studies that pneumoperitoneum causes end-organ ischemia and damage. Abdominal insufflation and increased intra-abdominal pressure can cause significant organ ischemia followed by reperfusion injury upon gastric emptying, resulting in an imbalance between oxidants and antioxidants [28]. Also, it has been shown in canines that heating CO_2 can generate a greater inflammatory response and formation of reactive oxygen species at the plasma and peritoneal levels [29]. In the same way, the thiol concentrations did not contribute to differentiating the groups, supporting the assumption that minimizing operative time and iatrogenic lesions are necessary to reduce the oxidative protein damage [10,28].

Considering the oxidant and antioxidant biomarkers used in this study, it could be assumed that no oxidative stress was present in the patients that underwent the different OVH techniques. Similarly, no changes in total serum oxidation state and total antioxidant capacity were observed in the previous study [30]. A similar protocol has been used to evaluate pyometra in queens, and it was considered efficient [31].

In general, the duration of the surgical stress response is proportional to the injury and complications [25]. As observed in this study, compared to the celiotomy, the video-assisted procedure improves the cosmetic effect, reduces the risk of complications due to a smaller incision, which significantly reduces recovery time [5,6], and promotes a less intense response to surgical stress [4,5], as related before.

It should be noted that cortisol concentrations were not measured in this study. Further research should incorporate cortisol levels alongside the analytes assessed here to provide a more comprehensive understanding on the utility of these markers in characterizing surgical stress.

5. Conclusions

The OVH techniques produced a similar inflammatory and oxidative stress response, but the video-assisted technique promoted earlier clinical recovery. The bipolar device has been presented as a safe device, less injurious than the traditional method of suspensory ligament rupture and ligature.

Author Contributions: Conceptualization, F.D.; methodology, F.D., C.P.R. and M.V.B.; formal analysis, C.P.R.; investigation, F.D., C.S.F., R.S., N.I.I.A.H., A.d.L.d.S., P.T., B.N.A., F.E., E.M.d.S.S. and M.V.B.; writing—original draft preparation F.D., C.P.R. and E.M.d.S.S.; supervision, F.D.; project administration, F.D.; funding acquisition, C.P.R. and F.D. All authors have read and agreed to the published version of the manuscript.

Funding: This research was supported by the "Fundação Araucária de Apoio ao Desenvolvimento Científico e Tecnológico do Estado do Paraná (FA)", Paraná, Brazil (Ed. Nº 1010/GR/UFFS/2018). C.P.R. has a postdoctoral fellowship "Ramon y Cajal" (RYC2021-034764-I) supported by the "Ministerio de Ciencia e Innovación, Agencia Estatal de Investigación" (MCIN/AEI/10.13039/501100011033), Spain, and The European Next Generation, EU/PRTR.

Institutional Review Board Statement: The procedures were approved by the ethics committees on Animal Experimentation of Universidade Federal da Fronteira Sul (CEUA CR 005/CEUA/UFFS/2019).

Informed Consent Statement: Informed consent was obtained from all dog owners.

Data Availability Statement: The raw data supporting the conclusions of this article will be made available by the authors on request.

Acknowledgments: The C.P.R. was granted with a post-doctoral fellowship "Capes-Print-UNESP Program (03-2019)" within the project "Establishing international, transdisciplinary, collaborative networks of the One health leaders of tomorrow: educations at the interface of human, animal, and environmental sustainability" (FMVZ, UNESP, Campus of Botucatu, Brazil).

Conflicts of Interest: The authors declare no conflicts of interest.

References

1. Schwarzkopf, I.; Van Goethem, B.; Vandekerckhove, P.M.; De Rooster, H. Vessel Sealing versus Suture Ligation for Canine Ovarian Pedicle Haemostasis: A Randomised Clinical Trial. *Vet. Rec.* **2015**, *176*, 125. [CrossRef]
2. Watts, J. The Use of Bipolar Electrosurgical Forceps for Haemostasis in Open Surgical Ovariectomy of Bitches and Queens and Castration of Dogs. *J. Small Anim. Pract.* **2018**, *59*, 465–473. [CrossRef] [PubMed]
3. Brun, M.V. Cirurgias no Aparelho Reprodutor Feminino de Caninos. In *Videocirurgia em Pequenos Animais*; Editora Guanabara Koogan Ltd.a: Rio de Janeiro, Brazil, 2014; pp. 186–213.
4. Freeman, L.J.; Rahmani, E.Y.; Al-Haddad, M.; Sherman, S.; Chiorean, M.V.; Selzer, D.J.; Snyder, P.W.; Constable, P.D. Comparison of Pain and Postoperative Stress in Dogs Undergoing Natural Orifice Transluminal Endoscopic Surgery, Laparoscopic, and Open Oophorectomy. *Gastrointest. Endosc.* **2010**, *72*, 373–380. [CrossRef] [PubMed]
5. Dalmolin, F.; Lhamas, C.L.; Pinto Filho, S.T.L.; Feranti, J.P.S.; Poerschke, A.; Beck, R.C.; Abdalla, F.H.; Andrade, C.M.; Brun, M.V. Biomarcadores Inflamatórios e de Estresse Oxidativo Em Cadelas Submetidas à Ovário-Histerectomia Videoassistida Ou Convencional. *Arq. Bras. Med. Veterinária Zootec.* **2016**, *68*, 687–694. [CrossRef]
6. Devitt, C.M.; Cox, R.E.; Hailey, J.J. Duration, Complications, Stress, and Pain of Open Ovariohysterectomy versus a Simple Method of Laparoscopic-Assisted Ovariohysterectomy in Dogs. *J. Am. Vet. Med. Assoc.* **2005**, *227*, 921–927. [CrossRef] [PubMed]
7. Villano, J.; Collins, D.E.; Nemzek, J.A. Inflammatory Response. In *Veterinary Surgery: Small Animal*; Johnston, S.A., Tobias, K.M., Eds.; Elsevier: St. Louis, MO, USA, 2018; Volume 2, pp. 77–117.
8. Schmidt, E.M.S.; Rubio, C.P.; Thomas, F.; Ferreira, J.C.P.; Eckersall, D.P. Acute Phase Proteins in Bitches Subjected to Conventional and Minimally Invasive Ovariohysterectomy. *Pesqui. Veterinária Bras.* **2018**, *38*, 2124–2128. [CrossRef]
9. Cerón, J.J.; Eckersall, P.D.; Martínez-Subiela, S. Acute Phase Proteins in Dogs and Cats: Current Knowledge and Future Perspectives. *Vet. Clin. Pathol.* **2005**, *34*, 85–99. [CrossRef]
10. Sies, H. Oxidative Stress: A Concept in Redox Biology and Medicine. *Redox Biol.* **2015**, *4*, 180–183. [CrossRef]
11. Rosenfeldt, F.; Wilson, M.; Lee, G.; Kure, C.; Ou, R.; Braun, L.; De Haan, J. Oxidative Stress in Surgery. In *Systems Biology of Free Radicals and Antioxidants*; Laher, I., Ed.; Springer: Berlin/Heidelberg, Germany, 2014; pp. 3929–3946.
12. Fransson, B.A. Ovaries and Uterus. In *Veterinary Surgery: Small Animal*; Johnston, S.A., Tobias, K.M., Eds.; Elsevier: St. Louis, MO, USA, 2018; Volume 2, pp. 4884–5025.
13. Murrell, J.C.; Psatha, E.P.; Scott, E.M.; Reid, J.; Hellebrekers, L.J. Application of a Modified Form of the Glasgow Pain Scale in a Veterinary Teaching Centre in the Netherlands. *Vet. Rec.* **2008**, *162*, 403–408. [CrossRef]

14. Martínez-Subiela, S.; Cerón, J.J. Validación Analítica de Técnicas Comerciales Para La Determinación de Haptoglobina, Proteína C Reactiva y Amiloide A Sérico En Caninos Analytical. *Arch. Med. Vet.* **2005**, *37*, 61–66. [CrossRef]
15. Muñoz-Prieto, A.; Tvarijonaviciute, A.; Escribano, D.; Martínez-Subiela, S.; Cerón, J.J. Use of Heterologous Immunoassays for Quantification of Serum Proteins: The Case of Canine C-Reactive Protein. *PLoS ONE* **2017**, *12*, e0172188. [CrossRef] [PubMed]
16. Arnao, M.B.; Cano, A.; Hernández-Ruiz, J.; García-Cánovas, F.; Acosta, M. Inhibition Byl-Ascorbic Acid and Other Antioxidants of the 2,2′-Azino-Bis (3-Ethylbenzthiazoline-6-Sulfonic Acid) Oxidation Catalyzed by Peroxidase: A New Approach for Determining Total Antioxidant Status of Foods. *Anal. Biochem.* **1996**, *236*, 255–261. [CrossRef] [PubMed]
17. Rubio, C.P.; Hernández-Ruiz, J.; Martinez-Subiela, S.; Tvarijonaviciute, A.; Ceron, J.J. Spectrophotometric Assays for Total Antioxidant Capacity (TAC) in Dog Serum: An Update. *BMC Vet. Res.* **2016**, *12*, 166. [CrossRef] [PubMed]
18. Rubio, C.P.; Hernández-Ruiz, J.; Martinez-Subiela, S.; Tvarijonaviciute, A.; Arnao, M.B.; Ceron, J.J. Validation of Three Automated Assays for Total Antioxidant Capacity Determination in Canine Serum Samples. *J. Vet. Diagn. Investig.* **2016**, *28*, 693–698. [CrossRef] [PubMed]
19. Campos, C.; Guzmán, R.; López-Fernández, E.; Casado, Á. Evaluation of the Copper (II) Reduction Assay Using Bathocuproinedisulfonic Acid Disodium Salt for the Total Antioxidant Capacity Assessment: The CUPRAC–BCS Assay. *Anal. Biochem.* **2009**, *392*, 37–44. [CrossRef]
20. Benzie, I.F.F.; Strain, J.J. The Ferric Reducing Ability of Plasma (FRAP) as a Measure of "Antioxidant Power": The FRAP Assay. *Anal. Biochem.* **1996**, *239*, 70–76. [CrossRef]
21. Rubio, C.P.; Martínez-Subiela, S.; Hernández-Ruiz, J.; Tvarijonaviciute, A.; Cerón, J.J.; Allenspach, K. Serum Biomarkers of Oxidative Stress in Dogs with Idiopathic Inflammatory Bowel Disease. *Vet. J.* **2017**, *221*, 56–61. [CrossRef]
22. Costa, C.M.D.; Santos, R.C.C.D.; Lima, E.S. A Simple Automated Procedure for Thiol Measurement in Human Serum Samples. *J. Bras. Patol. E Med. Lab.* **2006**, *42*, 345–350. [CrossRef]
23. Jocelyn, P.C. Spectrophotometric Assay of Thiols. In *Methods in Enzymology*; Elsevier: Amsterdam, The Netherlands, 1987; Volume 143, pp. 44–67, ISBN 978-0-12-182043-5.
24. Witko-Sarsat, V.; Friedlander, M.; Capeillère-Blandin, C.; Nguyen-Khoa, T.; Nguyen, A.T.; Zingraff, J.; Jungers, P.; Descamps-Latscha, B. Advanced Oxidation Protein Products as a Novel Marker of Oxidative Stress in Uremia. *Kidney Int.* **1996**, *49*, 1304–1313. [CrossRef]
25. Desborough, J.P. The Stress Response to Trauma and Surgery. *Br. J. Anaesth.* **2000**, *85*, 109–117. [CrossRef]
26. Redondo, J.I.; Suesta, P.; Serra, I.; Soler, C.; Soler, G.; Gil, L.; Gómez-Villamandos, R.J. Retrospective Study of the Prevalence of Postanaesthetic Hypothermia in Dogs. *Vet. Rec.* **2012**, *171*, 374. [CrossRef] [PubMed]
27. Rizzi, T.E.; Meinkoth, J.H.; Clinkenbeared, K.D. Normal Hematology of the Dog. In *Schalm's Veterinary Hematology*; Weiss, D.J., Wardrop, K.J., Schalm, O.W., Eds.; Wiley-Blackwell: Ames, IA, USA, 2010; pp. 799–810. ISBN 978-0-8138-1798-9.
28. Sammour, T.; Mittal, A.; Loveday, B.P.T.; Kahokehr, A.; Phillips, A.R.J.; Windsor, J.A.; Hill, A.G. Systematic Review of Oxidative Stress Associated with Pneumoperitoneum. *Br. J. Surg.* **2009**, *96*, 836–850. [CrossRef] [PubMed]
29. Milech, V.; De Oliveira, J.S.; De Ataide, M.A.W.; Coradini, G.P.; Mann, T.R.; Sarturi, V.Z.; Hartmann, H.F.; Linhares, M.T.; Antunes, B.N.; De Andrade, C.M.; et al. Effects of Heated Pneumoperitoneum on Inflammation, Oxidative Stress, and Peritoneal Histology in Female Dogs That Underwent Video-Assisted Ovariohysterectomy. *Vet. Med. Int.* **2021**, *2021*, 5515559. [CrossRef] [PubMed]
30. Gautier, A.; Graff, E.C.; Bacek, L.; Fish, E.J.; White, A.; Palmer, L.; Kuo, K. Effects of Ovariohysterectomy and Hyperbaric Oxygen Therapy on Systemic Inflammation and Oxidation in Dogs. *Front. Vet. Sci.* **2020**, *6*, 506. [CrossRef]
31. Vilhena, H.; Figueiredo, M.; Cerón, J.J.; Pastor, J.; Miranda, S.; Craveiro, H.; Pires, M.A.; Tecles, F.; Rubio, C.P.; Dabrowski, R.; et al. Acute Phase Proteins and Antioxidant Responses in Queens with Pyometra. *Theriogenology* **2018**, *115*, 30–37. [CrossRef]

veterinary sciences

MDPI

Article

Evaluation of Haematological Ratios at: Different Stages of Canine Periodontal Disease

Carolina Silva [1,2], Ana Carolina Abrantes [1,2], Ana Carolina Fontes [1,2], Isabel Dias [1,2,3], Rosário Domingues [4,5], Francisco Peixoto [6,7] and Carlos Viegas [1,2,8,*]

1 Department of Veterinary Sciences, School of Agricultural and Veterinary Sciences (ECAV), University of Trás-os-Montes e Alto Douro (UTAD), Quinta de Prados, 5000-801 Vila Real, Portugal; al57891@alunos.utad.pt (C.S.); carolina.psca@gmail.com (A.C.A.); al61909@alunos.utad.pt (A.C.F.); idias@utad.pt (I.D.)
2 Animal and Veterinary Research Center (CECAV)—AL4AnimalS, University of Trás-os-Montes e Alto Douro (UTAD), 5000-801 Vila Real, Portugal
3 CITAB-Centre for the Research and Technology of Agro-Environmental and Biological Sciences—Inov4Agro, University of Trás-os-Montes e Alto Douro (UTAD), 5000-801 Vila Real, Portugal
4 Mass Spectrometry Centre, LAQV-REQUIMTE, Department of Chemistry, University of Aveiro, Campus Universitário de Santiago, 3810-193 Aveiro, Portugal; mrd@ua.pt
5 CESAM, Departament of Chemistry, University of Aveiro, Campus Universitário de Santiago, 3810-193 Aveiro, Portugal
6 Chemistry Center of Vila Real, University of Trás-os-Montes e Alto Douro (UTAD), Quinta de Prados, 5000-801 Vila Real, Portugal; fpeixoto@utad.pt
7 RISE-Health: Health Research Network, Faculty of Medicine, University of Porto, 4099-002 Porto, Portugal
8 CIVG—Vasco da Gama Research Center, University School Vasco da Gama (EUVG), Campus Universitário, Avenida José Rodrigues Sousa Fernandes, Lordemão, 3020-210 Coimbra, Portugal
* Correspondence: cviegas@utad.pt

Simple Summary: Periodontal disease is a highly prevalent disease in the canine population and has a considerable systemic impact. Blood count indices have been investigated in many different diseases but, so far, no studies have been carried out on their value in different degrees of canine periodontal disease. The aim of the present study was to investigate and analyse five different haematological ratios in healthy clinical dogs, dogs with gingivitis, and dogs with periodontitis and, secondarily, evaluate their potential as diagnostic biomarkers. This retrospective study included 80 dogs of a specific breed and with different stages of canine periodontal disease. We found statistically significant differences between the groups in three of the haematological ratios analysed. These results suggest the systemic impact that canine periodontal disease has and reinforces the need of further research in this area.

Abstract: This is a retrospective study about haematological ratios in different stages of canine periodontal disease in one single breed, more specifically the Portuguese Podengo. The aim of the study was to assess the clinical significance of five haematological ratios, namely neutrophil-to-lymphocyte ratio (NLR), platelet-to-lymphocyte ratio (PLR), mean platelet volume-to-platelet count ratio (MPV/PLT), monocyte-to-lymphocyte ratio (MLR), and platelet-to-neutrophil ratio (PNR), easily obtained through the parameters provided by the complete blood count, in the stages of gingivitis and periodontitis, compared to clinical healthy individuals. Receiver operating characteristic curves were generated to identify ideal sensitivity and specificity cut-offs for cases of gingivitis or periodontitis. Our study included 80 dogs, which comprised 24 healthy dogs, 26 diagnosed with gingivitis, and 30 with periodontitis. The median NLR was significantly lower in periodontitis than in clinical healthy dogs ($p = 0.040$) and in dogs with gingivitis ($p = 0.037$). The median PLR was significantly lower in cases of gingivitis ($p = 0.020$) and periodontitis ($p = 0.024$) than in healthy dogs. MPV/PLT and MLR did not demonstrate significant differences between any of the three groups. The median PNR was significantly lower in gingivitis than in control dogs ($p = 0.019$). PNR had the highest accurate results, in distinguishing between healthy individuals and those with gingivitis, with an area under the curve of 0.692 (95% CI [0.539–0.845], $p = 0.020$). However, even this index had weak discriminatory power between the groups in the analysis. The impact of systemic inflammation generated by canine periodontal disease can thus be seen through some of these indices. However, the results obtained here demonstrate the need for complementarity with other methods to diagnose canine periodontal disease and reinforce the need for additional studies with more markers of systemic inflammatory response.

Keywords: haematological ratios; dog; gingivitis; periodontal disease; periodontitis; Podengo Portuguese Bbreed; ratios; systemic inflammatory response

1. Introduction

Canine periodontal disease (PD) is one of the most prevalent conditions diagnosed in small animal clinic [1,2]. Its prevalence is so high that some studies show that 80% of dogs around two years old already have PD [3]. However, the disease is more common in older dogs [4] and the most frequently affected breeds are the small and toy breeds [5,6].

The susceptibility and clinical expression of PD are conditioned by its multifactorial aetiology, which includes behavioural, environmental, genetic, microbiological, and systemic factors [7].

PD is a progressive infectious-inflammatory disease characterised by two different stages: gingivitis and periodontitis, where destruction of the periodontium tissue is induced by the immune response generated by the individual, due to the bacterial biofilm on the tooth surface, and by the direct action of this same bacterial biofilm [2,5,8–10]. Gingivitis is considered the first stage of PD, in which inflammation can be seen along the gingival margin and haemorrhage can occur. However, there is no loss of connective tissue or alveolar bone attachment. In some cases, gingivitis can develop into periodontitis if it is not treated early and properly. Periodontitis, on the other hand, is an irreversible and painful condition. It is characterised by a chronic inflammation and destruction of the supporting tooth tissues, namely the gingiva, the alveolar bone, the periodontal ligament, and the cementum [3,11,12]. As the disease progresses, it can lead to tooth mobility, tooth loss, periodontal abscesses, oronasal fistulas, or even osteomyelitis of the maxilla and mandible [3,11,12].

A dog's general health may be significantly impacted by PD, even in its early stages [13,14]. An article by Whyte et al. [13] demonstrated this. The degree of dental plaque in dogs diagnosed with PD at an early stage, specifically stage 1, was correlated with alterations at platelet level and at liver level, through alanine aminotransferase [13]. This shows that even at an early stage, the impact of PD is already significant [13]. It is known that there is a systemic inflammatory response that is generated in PD patients, whether they are humans or dogs [15,16]. In the case of canine PD, it has been linked to endocardiosis and endocarditis, hepatitis, interstitial nephritis, chronic bronchitis, and pulmonary fibrosis [17,18]. These comorbidities are boosted by the dissemination of bacterial metabolic products from periodontopathogens into the systemic circulation, which triggers the host's responses [15,16]. In turn, this systemic inflammatory response has been associated with an increase in systemic inflammatory mediators in dogs and humans [19–22]. Consequently, PD must be seen as a disease that can seriously jeopardise the individual's life. Veterinarians should express to the owners the real impact of this disease, which is certainly not a localised dental pathology [8].

Complete blood count (CBC) tests are routinely used in clinical small animal practice. Using the CBC, it is possible to easily calculate various ratios, including the neutrophil-to-lymphocyte ratio (NLR), platelet-to-lymphocyte ratio (PLR), mean platelet volume-to-platelet count ratio (MPV/PLT), monocyte-to-lymphocyte ratio (MLR), and platelet-to-neutrophil ratio (PNR), which can be used as important systemic inflammatory indicators. When compared to individual inflammatory cells, these haematological ratios (HRs) provide a more comprehensive evaluation of the inflammatory response. This is achieved because the interactions between various types of cells are taken into account and leads to greater precision in determining the immunological and inflammatory state [23,24]. In light of this, these indices are good indicators of systemic inflammatory responses and have the potential to be used as biomarkers for diagnosis and prognosis in cardiovascular diseases and inflammatory and neoplastic conditions [25,26].

Neutrophils are fundamental elements of the innate immune system and participate in the body's defence against pathogens, as they are the first circulating phagocytic cells [27]. Their process of phagocytising potential pathogens takes place due to the intervention of inflammatory mediators, which stimulate the migration of neutrophils into the blood [27]. In turn, lymphocytes also deserve to be highlighted for their role in specific anti-inflammatory and antitumour immunity [16,28]. Monocytes, the largest group of white blood cells, are immature cells that are either freely moving through the bloodstream or on their way to a specific tissue [29]. Mature macrophages are created when a young monocyte enters tissue and leaves the bloodstream [29]. Due to the complex interaction between monocytes and macrophages with blood and tissues, they are referred to as a single system, or the mononuclear phagocyte system [29].

Leukocyte ratios, particularly the PLR and the NLR, have been the subject of much research in human medicine as biomarkers of inflammation and inflammation-related diseases [30]. These ratios have shown potential as prognostic indicators in patients with COVID-19, cancer, cardiovascular disease, pneumonia, and sepsis [30–34]. However, in contrast to human medicine, the NLR is not a reliable indicator of the prognosis for dogs with pneumonia and has no relation to the duration of hospitalisation or mortality in dogs with septic peritonitis [35,36]. NLR in veterinary medicine has already been studied in pathologies such as lymphomas [37], mast cell tumours [38], and sarcomas [39]. In dogs and cats with acute pancreatitis and acute diarrhoea, NLR is a helpful indicator of the prognosis and severity of the condition [40]. Moreover, the use of NLR makes it possible to distinguish dogs with meningoencephalitis of unknown origin from other forebrain disorders [41].

Platelets are another blood element worth mentioning. They are fundamental for homeostasis but also play a significant role in the inflammatory process [29,42,43]. They actively participate in the induction of inflammation and tissue repair, as well as coregulate the host's defence against microbes [29,42,43]. Thrombocytosis is associated with various inflammatory, metabolic, and neoplastic processes [44]. In human medicine, thrombocytosis is even used as a prognostic marker in many neoplastic diseases [45]. PLR is calculated as a ratio between platelet and lymphocyte counts. These biomarkers reflect the balance between immunomodulatory and inflammatory processes (as reflected by the platelet count) and immunity response (lymphocyte count) [30]. In addition, PLR has been the subject of research into various canine pathologies [40,46,47], namely in canine parvovirosis, where it may even have prognostic value [30]. It has been shown that in dogs with serious haemorrhage, the PLR correlates with the period of hospitalisation in the intensive care unit. Moreover, the PLR and NLR correlate with the severity of the disease [30]. PLR and NLR are also inflammatory markers in canine inflammatory bowel disease and may be helpful in estimating the result of treatment [48–50]. Regarding platelets, another parameter is the mean platelet volume (MPV), which has also been the extensively studied, particularly in the field of oncology [51,52]. However, its role has yet to be clarified, so combining it with the total platelet count is recommended [16].

In the specific case of MLR, there are studies that prove that its increase, together with NLR, is associated with human cardiovascular diseases such as adverse cardiac events, mitral valve disease, coronary artery disease, and stroke [53,54]. However, in veterinary medicine, too, these HRs have been shown to have potential as diagnostic and prognostic biomarkers in myxomatous mitral valve disease in dogs [26].

Finally, the PNR resulting from the ratio obtained between platelets and neutrophils still needs to be investigated. However, there is already a study on this in the field of oncology, more specifically in canine diffuse large B-cell lymphoma [55].

Although HRs are being used more and more routinely in human medicine and in a wide variety of diseases, there is still a long way to go in veterinary medicine to demonstrate the potential value of these indices. So far, we know that blood count indices such as NLR, MLR, and PLR have shown potential as biomarkers of the systemic inflammatory response, such as in canine pancreatitis [40], neoplastic diseases [16,38], or sepsis [46]. However, no study has yet been carried out comparing certain blood count indices in different stages of canine PD, and even less so in a single breed. The aim of this study was to determine the clinical significance and compare the NLR, PLR, MPV/PLT, MLR, and PNR in dogs diagnosed with different degrees of canine PD. The authors of this study hypothesised that these HRs could be affected by the stage of canine PD and consequently be considered as potential biomarkers for distinguishing the different stages.

2. Materials and Methods

2.1. Data Collection and Analysis

The study is a retrospective case–control study. We chose to select just one breed for this study, the Portuguese Podengo, more specifically the medium-sized breed. The clinical records of 80 Portuguese Podengo dogs belonging to different owners were reviewed and analysed retrospectively from May 2020 to June 2024. The owners of the animals included in the study sample were asked to consent to the use of their animals' data for research purposes.

Blood samples were previously obtained by puncturing the jugular veins, collected in tubes with EDTA anticoagulant, and processed immediately. It should be mentioned that all the blood samples were taken during the daytime in order to avoid any possible influence of the circadian rhythm on the parameters being analysed. The analyses were carried out on an automatic analyser (BC-5000 Vet, Mindray, Fujifilm Portugal, S.A., Vila Nova de Gaia, Portugal) and a complete study of the leukocyte, erythrocyte, and platelet series was obtained. Blood analyses were carried out in cases of medical prophylaxis based on an appointment with the assistant veterinarian.

2.2. Inclusion Criteria

The animals were divided into three different groups according to their PD condition, i.e., healthy, gingivitis, or periodontitis. All the dogs in the three groups had to present an unaltered general clinical examination, be properly vaccinated and dewormed, and not present any obvious analytical alterations compatible with any infectious, neoplastic, endocrine, renal, or hepatic pathology.

2.3. Exclusion Criteria

The exclusion factors were as follows: age under 1 year, presence of systemic pathology, being under medical treatment and females in heat, pregnant, or lactating. In addition to these factors, we also considered that all animals that had undergone surgery or periodontal treatment in the last 12 months would not be considered for the study sample.

2.4. The Characterisation of Each of the Three Groups Under Study

The group of healthy animals, known as the control group (C group), included 24 dogs. None of them showed even the slight presence of dental plaque or dental calculus. Nor could they show any evidence of gingival inflammation or alterations to the height or architecture of the alveolar margin.

The second group of dogs consisted of 26 dogs with clinical signs of gingivitis and was therefore called the gingivitis group (G group). The animals included in this group had superficial inflammation of the gingiva, with marginal erythema and oedema. They may also show bleeding on touch or slight spontaneous bleeding. It should be emphasised that the size and architecture of the alveolar margin are normal and there is no loss of insertion, although there may be a slight accumulation of dental calculus.

The third and final group in this study corresponded to dogs diagnosed with periodontitis (P group). This group included 30 dogs. These animals had to show signs of periodontitis, such as gingival recession, furcation exposure, tooth mobility, or even tooth loss. In addition to these signs, those mentioned above for G group also had to be present.

2.5. Calculation of Haematological Ratios

The HRs calculated were NLR, PLR, MPV/PLT, MLR, and PNR. The NLR was calculated as the ratio between the absolute value of neutrophils to lymphocytes. PLR and MLR were calculated as the ratio of the absolute values of platelets and monocytes to lymphocytes, respectively. The MPV/PLT was calculated as the ratio between the mean platelet volume and the absolute value of platelets. Finally, PNR was calculated as the ratio between the absolute value of platelets and neutrophils. In addition to these data, information was also collected on the age, gender, and weight of each animal. In particular, with regard to age and gender, statistical analyses were carried out on the groups in the study in order to ascertain whether age and gender influences the values of the HR under study.

2.6. Statistical Analysis

Statistical analyses were performed using Jamovi statistical software (version 2.3.28) and IBM SPSS statistics (version 29.0.0.0). The continuous variables were assessed for normality using the Shapiro–Wilk test. Normally distributed variables were reported as mean ± standard deviation. On the other hand, not normally distributed data were presented by the median and 25th and 75th percentiles. Laboratory variables were compared between the different groups using the Student's *t*-test and the Mann–Whitney non-parametric test, depending on whether the variable was normally or not normally distributed, respectively. The influence of gender and age on the HRs studied in the three groups was analysed using the Mann–Whitney non-parametric test and simple linear regression, respectively. A receiver operating characteristic (ROC) curve analysis was used to identify cut-off levels of sensitivity and specificity for the HR that showed a statistically significant difference between the C group and one of the other two different groups. The area under the curve (AUC) was calculated and the cut-off values that optimised sensitivity and specificity were found using the Youden index. A p-value < 0.05 was considered statistically significant.

3. Results

3.1. Study Population

Table 1 summarises the main characteristics of the study sample (n = 80), comprising a total of 24 healthy dogs, 26 dogs with gingivitis, and 30 dogs with periodontitis.

Table 1. General characterisation of dog group sample (n = 80).

Group	Median (Range) Age of Dogs (Years)	Median (Range) Weight of Dogs (kg)	Number of Female Dogs (%)	Number of Male Dogs (%)
C	2 (1–8)	13.0 (10.0–16.0)	14 (58.3%)	10 (41.7%)
G	2.5 (1–9)	13.0 (9.0–17.0)	14 (53.8%)	12 (46.2%)
P	3 (1–11)	13.0 (9.0–17.0)	17 (56.7%)	13 (43.3%)

Abbreviations: C: control group; G: gingivitis group; P: periodontitis group.

All the parameters provided by the CBC and statistical analysis between the three groups are described in Table 2, where different letters represent statistically significant differences.

Table 2. CBC parameters in C, G, and P groups and respective significance.

Variable	Unit and Reference Range	Control Group	Gingivitis Group	Periodontitis Group
WBC	6.000–17.000 × 10^9/L	11.300 ± 3.410 a	13.300 ± 3.150 b	12.300 ± 3.920 b
NEU	3.620–12.300 × 10^9/L	7.560 ± 2.500 b	8.880 ± 2.690 b	7.770 ± 2.500 b
LYM	0.830–4.910 × 10^9/L	2.150 (1.690–2.720) a	2.600 (2.120–3.290) b	2.830 (2.170–3.630) b
MON	0.140–1.970 × 10^9/L	0.650 (0.520–0.778) b	0.735 (0.542–0.895) b	0.675 (0.505–0.925) b
EOS	0.040–1.620 × 10^9/L	0.595 (0.347–0.945) b	0.770 (0.578–0.938) b	0.560 (0.330–0.795) b
RBC	5.100–8.500 × 10^{12}/L	6.290 ± 1.180 b	6.340 ± 0.822 b	7.010 ± 0.808 a
HGB	110.000–190.000 g/L	151.000 ± 26.500 b	153.000 ± 19.000 b	168.000 ± 17.900 a
HCT	33.000–56.000%	42.600 ± 7.140 b	43.300 ± 5.080 b	48.000 ± 4.820 a
MCV	60.000–76.000 fL	68.100 ± 4.780 b	68.400 ± 3.870 b	68.700 ± 3.480 b
MCH	20.000–27.000 pg	24.100 ± 1.720 b	24.100 ± 1.470 b	24.100 ± 1.450 b
MCHC	300–380 g/L	351 (349–356) b	354 (347–358) b	352 (344–357) b
RDW-CV	0.125–0.172	0.138 (0.132–0.143) b	0.139 (0.133–0.146) b	0.139 (0.134–0.145) b
RDW-SD	33.200–46.300 fL	36.900 (35.300–38.300) b	37.200 (35.900–38.700) b	38.000 (36.400–39.100) b
PLT	117.000–490.000 × 10^9/L	326.000 (241.000–376.000) b	237.000 (156.000–274.000) b	260.000 (230.000–260.000) b
MPV	8.000–14.100 fL	10.100 (9.470–10.300) b	10.100 (9.430–11.300) b	10.000 (8.930–10.800) b
PDW	12.000–17.500	15.800 (15.500–16.300) b	15.700 (15.400–16.600) b	16.000 (15.700–16.800) b

Abbreviations: WBC: white blood cells; NEU: neutrophil count; LYM: lymphocyte count; MON: monocytes count; EOS: eosinophil count; RBC: red blood cells; HGB: haemoglobin; HCT: hematocrit; MCV: mean corpuscular volume; MCH: mean corpuscular haemoglobin; MCHC: mean corpuscular haemoglobin concentration; RDW-CV: red cell distribution width expressed as coefficient of variation; RDW-SD: red cell distribution width standard deviation; PLT: platelet count; MPV: mean platelet volume; PDW: platelet distributed width. Different letters mean statistically significant difference.

3.2. Haematological Ratios

Table 3 summarises the statistical results obtained by analysing gender and age in relation to the HRs in the C, G, and P groups.

Table 3. The statistical results obtained in the HR of the C, G, and P groups in relation to gender and age.

	Gender			Age		
	C Group	G Group	P Group	C Group	G Group	P Group
Variable	*p* Value					
NLR	0.666	0.560	0.113	0.227	0.124	0.444
PLR	0.403	0.527	0.133	0.696	0.073	0.111
MPV/PLT	0.464	0.131	0.477	0.139	0.384	0.453
MLR	0.931	0.494	0.213	0.826	0.335	0.009
PNR	0.796	0.899	0.711	0.170	0.415	0.100

Abbreviations: NLR: neutrophil-to-lymphocyte ratio; PLR: platelet-to-lymphocyte ratio; MPV/PLT: mean platelet volume-to-platelet count ratio; MLR: monocyte-to-lymphocyte ratio; PNR: platelet-to-neutrophil ratio.

The median of the five different HRs is shown in Table 4 for each of the three groups under study, covering a total of 80 individuals.

Table 4. Haematological ratios in C, G, and P groups.

	C Group (Median, 25th and 75th Percentiles)	G Group (Median, 25th and 75th Percentiles)	P Group (Median, 25th and 75th Percentiles)
NLR	3.460 (2.540–4.490)	3.320 (2.420–4.460)	2.430 (1.980–3.220)
PLR	119.000 (85.700–210.000)	76.900 (63.800–117.000)	78.300 (59.700–123.000)
MPV/PLT	0.031 (0.026–0.041)	0.046 (0.034–0.067)	0.040 (0.033–0.051)
MLR	0.331 (0.239–0.379)	0.272 (0.212–0.316)	0.228 (0.186–0.288)
PNR	40.400 (32.000–52.500)	26.800 (20.800–35.300)	31.800 (25.600–45.000)

Abbreviations: C: control group; G: gingivitis group; P: periodontitis group; NLR: neutrophil-to-lymphocyte ratio; PLR: platelet-to-lymphocyte ratio; MPV/PLT: mean platelet volume-to-platelet count ratio; MLR: monocyte-to-lymphocyte ratio; PNR: platelet-to neutrophil-ratio.

3.2.1. Neutrophil-to-Lymphocyte Ratio (NLR)

The NLR showed no statistically significant difference between the individuals in the C group and those in the G group ($p = 0.751$). However, this was not the case for the C group versus P group ($p = 0.040$) (Figure 1) and the G group versus P group ($p = 0.037$) (Figure 2). In these cases, there were statistically significant differences between the groups for the NLR, with the median of this haematological ratio being lower in the P group.

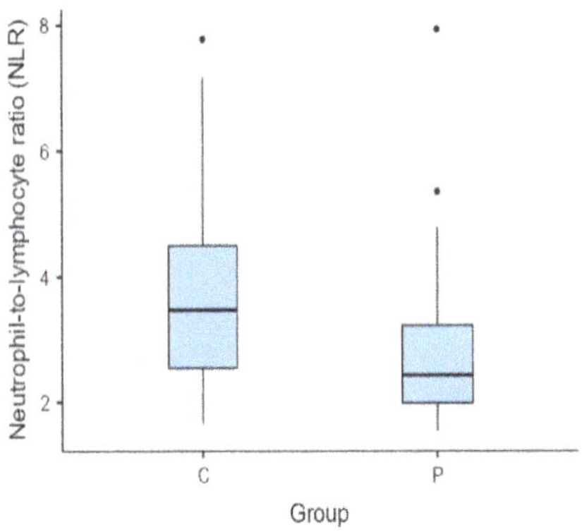

Figure 1. Box and whisker plot comparing the NLR for healthy individuals (C group) versus individuals diagnosed with periodontitis (P group). The central horizontal lines indicate the median value; the upper and lower lines of the box plot represent the first and the third quartiles of the values, respectively. The outliers are represented with circles.

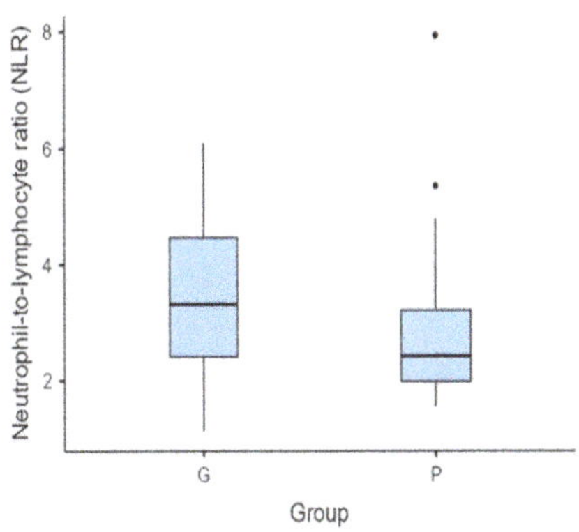

Figure 2. Box and whisker plot comparing the NLR for subjects diagnosed with gingivitis (G group) versus those with periodontitis (P group). The central horizontal lines indicate the median value; the upper and lower lines of the box plot represent the first and the third quartiles of the values, respectively. The outliers are represented with circles.

3.2.2. Platelet-to-Lymphocyte Ratio (PLR)

The median PLR in the C group was the highest (119 (85.7–210)). The PLR was significantly higher between the C group and the G ($p = 0.020$) (Figure 3) and P groups ($p = 0.024$) (Figure 4). However, there were no statistically significant differences between the G and P groups in terms of the PLR ($p = 0.864$).

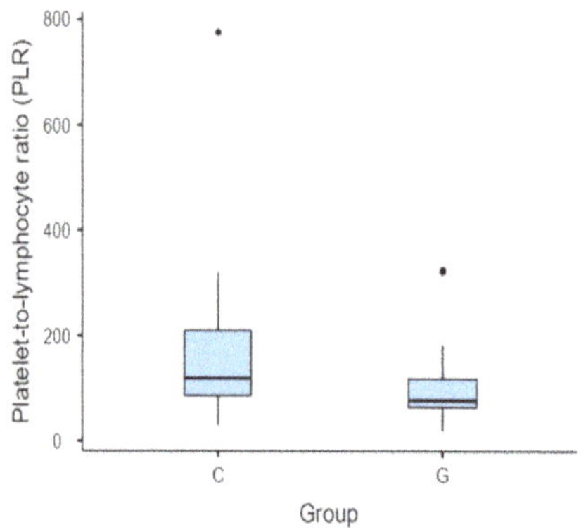

Figure 3. Box and whisker plot comparing the PLR between the C group and G group. The central horizontal lines indicate the median value; the upper and lower lines of the box plot represent the first and the third quartiles of the values, respectively. The outliers are represented with circles.

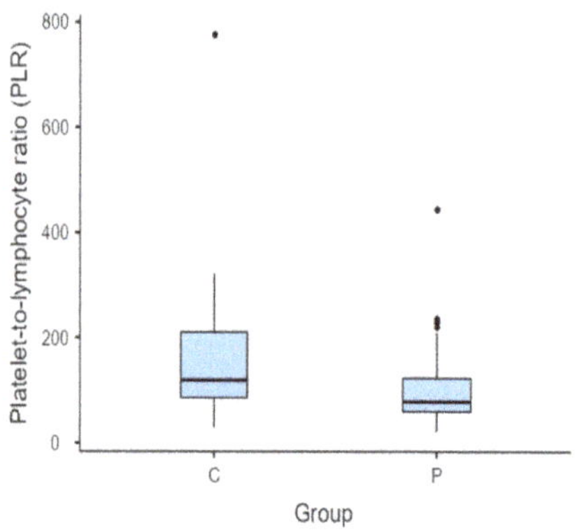

Figure 4. Box and whisker plot comparing the PLR between dogs considered clinically healthy (Group C) and dogs diagnosed with periodontitis (Group P). The central horizontal lines indicate the median value; the upper and lower lines of the box plot represent the first and the third quartiles of the values, respectively. The outliers are represented with circles.

3.2.3. Mean Platelet Volume-to-Platelet Count Ratio (MPV/PLT)

No statistically significant differences were found in any of the three MPV/PLT analyses carried out between the various groups.

3.2.4. Monocyte-to-Lymphocyte Ratio (MLR)

The MLR did not differ significantly in the comparisons made between the different groups, i.e., between group C and group G (p = 0.293), group C and group P (p = 0.054), and between group G and group P (p = 0.217).

3.2.5. Platelet-to-Neutrophil Ratio (PNR)

The PNR was significantly higher in C group in comparison with G group (p = 0.019) (Figure 5). On the other hand, there were no statistically significant differences in the comparisons made between C group versus P group (p = 0.185) and G group versus P group (p = 0.166).

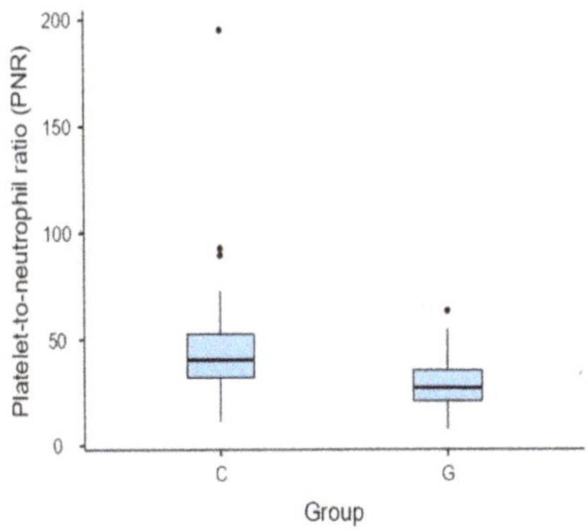

Figure 5. Box and whisker plot comparing the PNR between the C group versus P group. The central horizontal lines indicate the median value; the upper and lower lines of the box plot represent the first and the third quartiles of the values, respectively. The outliers are represented with circles.

3.3. ROC Curves

ROC curves were generated for the HR that demonstrated a statistically significant difference between healthy individuals and cases with gingivitis or periodontitis. It should be noted that the PNR showed the most accurate results, particularly in distinguishing between healthy individuals and those with gingivitis, with an AUC of 0.692 (95% CI [0.539–0.845], $p = 0.020$) (Figure 6). In this case, a cut-off of <34.490 demonstrated a 75.0% sensitivity and 73.1% specificity for individuals with gingivitis. On the other hand, the PLR was the only haematological ratio with statistically significant differences between the C group and the other two groups under study, which consequently led to the analysis of two different ROC curves. However, given the AUC values obtained for the HR analysed, the discriminating power of the models is considered weak, since all the AUCs are in the [0.5; 0.7] range [56].

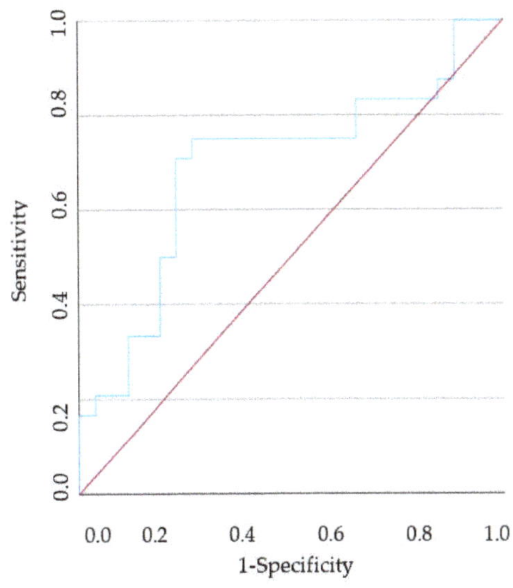

Figure 6. The ROC curve analysis of predicting the PNR between healthy dogs and dogs with gingivitis. The red line illustrates what a random classification would look like, and the blue line represents the model under study.

The AUC using the NLR was 0.0664 (95% CI [0.513–0.815], p = 0.040) (Figure 7). A cut-off of <2.577 showed a 75% sensitivity and 63.6% specificity for periodontitis cases.

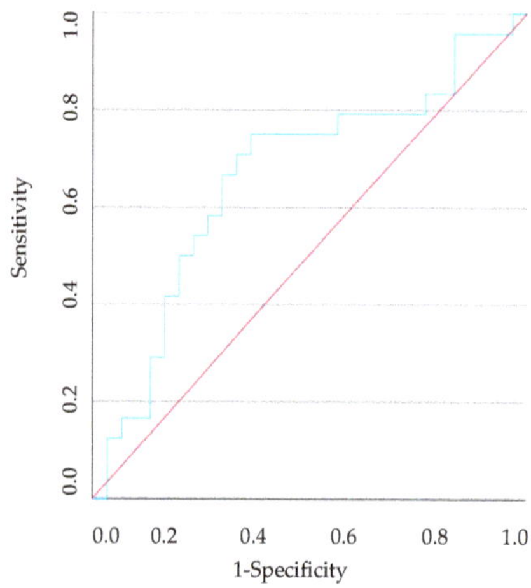

Figure 7. The ROC curve analysis of predicting the NLR between healthy dogs and dogs with periodontitis. The red line illustrates what a random classification would look like, and the blue line represents the model under study.

The ROC curve indicated that PLR could be used to predict the presence of an individual with gingivitis and another with periodontitis with a cut-off point of <81.492 (AUC = 0.691, 95% CI [0.540–0.842], p = 0.021, sensitivity 79.2%, specificity 61.5%) (Figure 8) and <82.118 (AUC = 0.679, 95% CI [0.533–0.825], p = 0.025, sensitivity 79.2%, specificity 60.0%) (Figure 9), respectively.

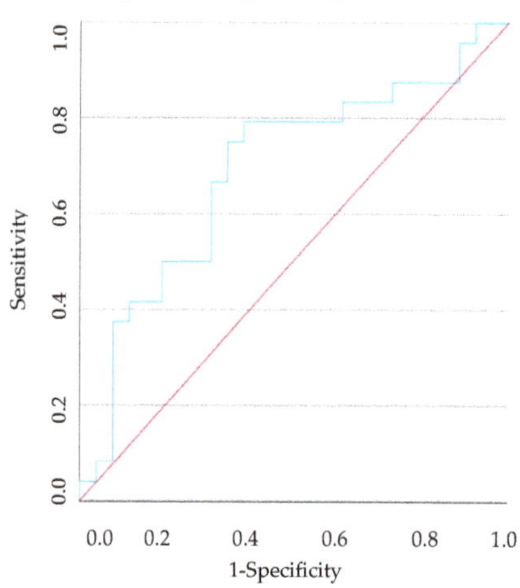

Figure 8. The ROC curve analysis of predicting the PLR between healthy dogs and dogs with gingivitis. The red line illustrates what a random classification would look like, and the blue line represents the model under study.

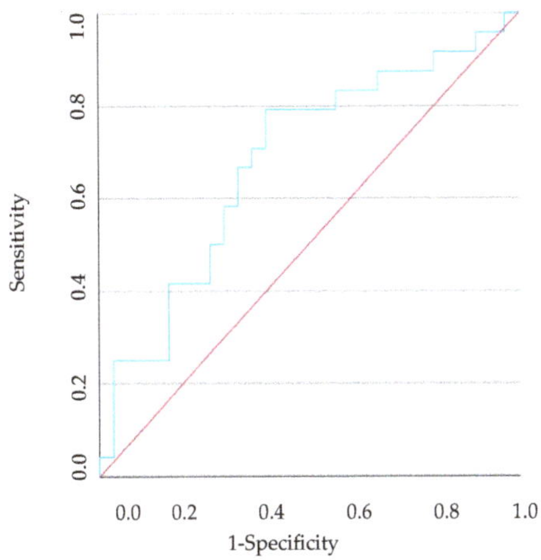

Figure 9. The ROC curve analysis of predicting the PLR between healthy dogs and dogs with periodontitis. The red line illustrates what a random classification would look like, and the blue line represents the model under study.

4. Discussion

The present study evaluated certain HRs in different stages of canine PD. A previous study examined the NLR, PLR, and MPV/PLT in dogs with periodontitis in comparison to healthy dogs and dogs with oropharyngeal tumours [16]. However, to date, no other study has been carried out on HRs in different stages of canine PD, much less in a specific canine breed. It should be emphasised that in this study, none of the dogs in the sample showed clinical signs or laboratory alterations that could be compatible with any other concomitant inflammatory, immune-mediated, or neoplastic pathology, which could influence the HR being analysed.

The HR refers to a ratio that shows the relationship between different types of cells. In our study, we aimed to demonstrate how these indices can vary depending on the degree of canine PD. Only two of the HRs assessed in this study showed no statistically significant differences between the three groups in the analysis.

One factor to consider when analysing our results, which likely influenced all the indices to varying degrees, is the possibility of periodic bacteraemia in dogs with PD [14,22]. Therefore, the more exacerbated periods will influence the HRs analysed here, since there is an activation of both the innate immune response and the acquired immune response, as a result of the absorption of toxins and other bacterial products into the bloodstream [14].

The initial purpose was to determine whether gender and age could influence the HRs analysed. It was concluded that there were no statistically significant differences between males and females in terms of the elements analysed in any of the three groups. In turn, age and HR did not show a statistically significant correlation in any of the study groups, with the exception of MLR in the P group. To date, no studies have analysed in detail the possible influence of gender and age on these HRs. However, in our specific case and in general, there seems to be no influence of gender and age on these parameters. The fact that the MLR in group P showed a statistically significant correlation with age, although this is considered moderate (Spearman's rho = −0.466), should be investigated in future studies. It has been described that the population of lymphocytes decreases as dogs age [57]. Furthermore, periodontitis, as the last stage of PD, is more prevalent in older dogs [4]. These facts would lead us to think that MLR would increase with age, which was not the case here. Therefore, more studies are needed to understand the possible influence of age on HR.

The first haematological ratio analysed, the NLR, reflects the ratio between neutrophils and lymphocytes, and is considered a biomarker of systematic inflammation, even showing correlations with inflammatory markers such as C-reactive protein [58,59]. Moreover, it reflects two complementing immune pathways [59,60]: lymphocytes carry out adaptive immunological responses, while neutrophils produce non-specific inflammation [59,61] and are cellular elements involved in the body's first line of defence against infections [62]. A high number of neutrophils would be expected in dogs with periodontitis, as has been observed in humans [59,63,64]. However, the NLR in P group was the lowest, and there were even statistically significant differences between C group versus P group and G group versus P group. Furthermore, it is important to note the decrease in this haematological ratio between clinically healthy dogs and dogs with periodontitis, is, the natural progression of the disease. Given the similarities between the dog, as animal model for PD [12], and humans, it is logical to extrapolate to our case. Thus, this decrease in NLR may be due to an increase in lymphocytes. During periodontitis, there is a chronic inflammatory phase in which the adaptive immune response also becomes relevant [65–67], mainly due to lymphocytes [68]. Inflammation induced by persistent microbial harm is what induces T lymphocytes to differentiate [69,70]. Through cytokine release and osteoclast formation regulation, activating distinct T lymphocyte subtypes regulates chronic inflammation. It is a major factor in deciding whether an inflammatory lesion will result in tissue-destructive periodontitis [70,71]. This specific type of lymphocyte has a major impact on the resorption of alveolar bone and is more prevalent [65,70,72]. Moreover, T lymphocytes are key elements to keep the gingival environment in equilibrium [70,71]. In addition to T lymphocytes, B lymphocytes must also be taken into account. B lymphocytes are essential for gingival homeostasis, due to the production of antibodies against different periodontal pathogens, what contributes to the host protection [71]. This leads us to conclude that as canine PD progresses, the role of lymphocytes is more prominent than that of neutrophils, given the chronic inflammatory stage of periodontitis compared to individuals with gingivitis or clinically healthy individuals. Furthermore, the NLR is a more reliable indicator than analysing neutrophils and lymphocytes individually [59,73], which was also demonstrated in this study, since the only statistically significant difference that existed in terms of cellular elements was between the lymphocytes in the C group and the P group.

In the present study, we identified statistically significant differences in the PLR between healthy individuals and individuals with gingivitis, as well as between healthy individuals and individuals with periodontitis. It can therefore be seen that the PLR value is significantly lower in G group and P group compared to C group. Once again, this may be due to the increase in lymphocytes as a result of the chronic stimulation of the immune system due to the inflammatory process that exists in cases of PD [65,66], which could play a key role in the development of other systemic diseases [22,74]. Note that locally produced chemokines and cytokines in inflammatory periodontal tissues contribute to the recruitment of B and T lymphocytes [14], which can stimulate, inhibit, or even kill microorganisms or infected host cells [66]. Inflammation and periodontal tissue damage may result from this action [66,75,76]. The trend towards a higher number of lymphocytes in cases of periodontitis had already been reported in human patients by Iqbal et al. [77], which is in line with our study. Platelet count and also MPV are markers of platelet activation [16,78], with platelet count elevated in chronic inflammatory states [14,42,79]. There are several studies associating platelet activation with PD [14,80,81], and our results lead us to believe that platelet activation and function are in fact more prominent in canine PD than the platelet count itself, as was described by Lu et al. [59]. In addition, the increase in the number of lymphocytes will be more evident, which consequently means that PLR is lower in the groups with gingivitis and periodontitis.

We did not find significant differences in MPV/PLT between any of the three different groups. For this reason, this haematological ratio cannot be considered as a biomarker of canine PD, since there are no statistically significant differences between the various stages. The absence of statistically significant differences in the three comparisons made between the different groups in relation to MPV/PLT can be explained by the fact that there were no significant haematological changes in the parameters used to calculate these ratios. This is contrary to information published in the literature, in which periodontitis patients tended to have an increased platelet count [14], size, and shape [80]. However, in our study, no distinction was made between the existing grades within the periodontitis stage (grade 2, 3 or 4), in contrast to what Mutthineni et al. determined [80], since a distinction was made between moderate periodontitis and severe periodontitis. This may have influenced the failure to obtain statistically significant differences using this ratio.

The other haematological ratio in which there were no statistically significant differences between the groups was MLR. In other words, it seems that canine PD has no impact on MLR. However, it should be remembered that the MLR was the only haematological ratio that showed a statistically significant correlation with age in the P group. Therefore, age may have had an effect on the MLR results. This ratio includes the values of monocytes and lymphocytes, both of which are considered relevant cells in the pathogenesis of PD [82]. In the specific case of monocytes, they are an important element of the innate immune system [46]. When there is an infection with a consequent inflammatory response, monocytes migrate to the inflamed tissues [46]. They then differentiate into macrophages, which kill pathogens by phagocytosis while at the same time producing inflammatory cytokines and reactive oxygen species [46]. In addition, the production of a wide range of pro-inflammatory cytokines also occurs when T lymphocytes are activated [82]. These cytokines cause a succession of events that eventually lead to the progressive loss of the alveolar bone, which supports teeth [71,82]. In the specific case of interleukin-1, this cytokine leads to an exacerbation of the inflammatory response [83–85]. It also induces enzymes that degrade the extracellular matrix and, consequently, promote osteoclastic bone resorption [83–85]. Although monocyte values were higher in the G group and P group than in the C group, they did not show statistically significant differences compared to the C group, which may have influenced the conclusions of statistical interest in the MLR. This increase in monocytes has been described previously by Buhlin et al. [86] and, in this case, in a statistically significant way. The lack of differences between the groups in terms of MLR suggests the presence of a certain balance between the innate immune response and the adaptive immune response, mediated, among other cells, by monocytes and lymphocytes, respectively. Unfortunately, there are few studies analysing MLR and its importance in canine pathologies [55,87], and our study is the only one to analyse this haematological ratio in canine PD. It would therefore be important to carry out more research into this ratio and it may be interesting, as in the case of MPV/PLT, to substage periodontitis in order to better understand the potential changes in this ratio with the disease.

The lower and statistically significant PNR in the G group compared to the C group may be a direct consequence of the increase in the number of neutrophils. It is widely recognised that the first stage of canine PD is gingivitis [2,5,9]. Later, in the absence of adequate treatment, periodontitis appears, which is considered a chronic inflammatory phase [3]. In cellular terms, neutrophils act as the first line of the innate immune system in the presence of acute infections [14,63], which aligns with this study. As gingivitis is the first stage of canine PD, it is only natural that there should be an increase in the number of neutrophils in group G compared to group C. However, it should be remembered that acute inflammation also leads to the activation of other cells involved in the immunity response, such as platelets [63]. Even so, in this specific case, the increase in neutrophils is still more prevalent. The fact that there were no statistically significant differences in the PNR between clinically healthy individuals and those diagnosed with periodontitis, as well as between the two stages of the disease, once again suggests an adaptive immune response [75], in which T helper lymphocytes contribute to the progression of the disease [67,68]. Thus, acquired immunity becomes more prominent as the disease progresses, causing neutrophils to lose prominence, as cellular elements belonging to the innate immune response.

This study also aimed to analyse the diagnostic accuracy of the HR in predicting the presence of gingivitis or periodontitis, with data obtained through the ROC curve. The NLR in the P group was found to be significantly lower than in the C group. Furthermore, an NLR < 2.577 could differentiate a dog with periodontitis from a clinically healthy dog. On the other hand, the PLR in G and P group was significantly lower than in the C group and a PLR < 81.492 could differentiate a dog with gingivitis from a clinically healthy dog, just as a PLR < 82.118 may identify a dog with periodontitis. Finally, the PNR in the G group was significantly lower than in the C group and could have the ability to discriminate between a dog with gingivitis and a healthy dog when the PNR < 34.490. The only drawback is the sensitivity and specificity obtained, as all from this analysis have poor discriminatory power. This leads us to believe that the exclusive use of these HRs will not be viable. A careful and detailed stomatological-dental assessment, including intraoral radiographs, will lead to a more accurate diagnosis of canine PD. However, it is also important to emphasise that in a veterinary appointment in which a blood count is performed, more specifically in dogs in which it is difficult to assess the oral cavity without sedation and in which there are no other diseases detectable, the use of these HRs could be an added value. A routine blood test will allow us to demonstrate the systemic impact caused by canine PD. Importantly, this fact will facilitate communication between the veterinarian and the owner, empowering them with knowledge of the importance of early diagnosis and intervention in canine PD, and their role in their pet's health.

Although it was not the main objective of this study, we also have to mention the CBC's red line. Although there are some studies that associate chronic stages of PD with a reduction in red blood cell parameters [14,88–90], in our case, exactly the opposite happened, since there is a significant and increasing trend in RBC, HGB, and HCT as the disease progresses. However, all parameters of the three groups under study were within the reference range, which is in line with what had already been reported by Aljohani et al. [91]. Therefore, additional studies will be necessary to understand the reason for this increase in RBC, HGB, and HCT in cases of periodontitis.

5. Limitations

It is essential to mention the limitations that the authors of this study encountered while preparing it. The main limitation, related to the fact that this is a retrospective study, was the lack of blood smears to evaluate potential platelet aggregates and to confirm differential leukocyte counts. However, the haematology analyser used in this study is suitable for obtaining the analyses in question and is widely used in small animal haematology [92]. In addition, the fact that this was a retrospective study meant that only clinical data that was properly discriminated in terms of the examination of the oral cavity was analysed to be able to fit it into one of the three groups under study. This meant that the number of dogs in the study sample was smaller than desirable.

6. Conclusions

To our knowledge, this is the first study to analyse five different HRs in canine PD in a single dog breed, in this case the Portuguese Podengo.

In veterinary medicine, the CBC is performed frequently and is considered a minimally invasive and relatively inexpensive complementary means of diagnosis. The HRs that are generated can be simply calculated and without additional cost. Consequently, these HRs could be useful as biomarkers for diagnosis and, eventually, for monitoring a given disease. Our study was able to identify statistically significant differences in three HRs in different stages of canine PD. The NLR was significantly lower in dogs with periodontitis compared to healthy dogs and dogs with gingivitis. Compared to dogs with gingivitis or periodontitis, clinically healthy canines had a much greater PLR. Finally, the value of the PNR showed noticeable differences between dogs who were healthy and dogs who had gingivitis.

Although there are significant differences between the groups in three of the HRs analysed, the results obtained in the present study demonstrate that the exclusive use of NLR, PLR, and PNR is not discriminatory enough to achieve an accurate diagnosis of gingivitis or periodontitis and should always be complemented with a detailed stomatological-dental examination. Nevertheless, given the results obtained, we can say that NLR, PLR, and PNR could be useful as inflammatory biomarkers in canine PD.

Due to the scarcity of studies on HRs in canine PD, we believe that this is an area that will be increasingly explored. Choosing a specific breed is an advantage, and studies carried out on other breeds should be implemented since canine PD has a strong genetic component. As has become a trend, there is an increasing use of specific breeds to determine specific reference intervals, which leads us to believe that this study could be the beginning of this. Therefore, by focusing on developing specific reference intervals for each breed, it will be possible to adopt a more precise and detailed approach to each animal. This trend can be seen in other areas of veterinary medicine, such as cardiology and small animal imaging, and there are already established parameters for certain breeds. In the future, it would be interesting to carry out one or more multi-centre studies with a much higher number of Portuguese Podengo dogs to determine the reference intervals for HRs in healthy dogs, which will facilitate comparative analysis with dogs with a certain pathology. Furthermore, it will also be interesting to determine these HRs in the substages of periodontitis and understand their evolution in dogs undergoing periodontal treatment.

Author Contributions: Conceptualization, C.V.; Methodology, C.S., A.C.A. and C.V.; Validation, C.V., R.D. and F.P.; Formal Analysis, C.S. and C.V.; Writing—Original Draft Preparation, C.S., A.C.F. and C.V.; Writing—Review and Editing, C.V., I.D., A.C.A., R.D. and F.P.; Supervision, C.V., R.D. and F.P.; Funding Acquisition, C.V., R.D. and F.P. All authors have read and agreed to the published version of the manuscript.

Funding: The authors acknowledge the University of Aveiro, Fundação para a Ciência e a Tecnologia, I. P. (FCT, Portugal) and Ministério da Ciência e Tecnologia (MCT), and CESAM (UIDP/50017/2020+ UIDB/50017/2020+LA/P/0094/2020) for the financial support through Portuguese funds and, where applicable, cofinanced by the European Regional Development Fund (FEDER), within the Portugal 2020 Partnership Agreement. FCT—Foundation for Science and Technology, by the projects UIDB/04033/2020, LA/P/0059/2020 and UIDB/CVT/00772/2020.

Institutional Review Board Statement: This is a retrospective study, so there was no need for ethical review and approval.

Informed Consent Statement: The owner of each animal was asked to sign a written informed consent form to authorise the use of their animal's clinical laboratory data for this research.

Data Availability Statement: The datasets generated for this study are available under request.

Conflicts of Interest: The authors declare no conflicts of interest.

References

1. Harvey, C.E. Periodontal Disease in Dogs: Etiopathogenesis, Prevalence, and Significance. *Vet. Clin. N. Am. Small Anim. Pract.* **1998**, *28*, 1111–1128. [CrossRef] [PubMed]
2. Wallis, C.; Saito, E.K.; Salt, C.; Holcombe, L.J.; Desforges, N.G. Association of Periodontal Disease with Breed Size, Breed, Weight, and Age in Pure-Bred Client-Owned Dogs in the United States. *Vet. J.* **2021**, *275*, 105717. [CrossRef] [PubMed]
3. Niemiec, B.A. Periodontal Disease. *Top. Companion Anim. Med.* **2008**, *23*, 72–80. [CrossRef] [PubMed]
4. Kyllar, M.; Witter, K. Prevalence of Dental Disorders in Pet Dogs. *Vet. Med.* **2005**, *50*, 496–505. [CrossRef]
5. Wallis, C.; Holcombe, L.J. A Review of the Frequency and Impact of Periodontal Disease in Dogs. *J. Small Anim. Pract.* **2020**, *61*, 529–540. [CrossRef]
6. Harvey, C.E.; Shofer, F.S.; Laster, L. Association of Age and Body Weight with Periodontal Disease in North American Dogs. *J. Vet. Dent.* **1994**, *11*, 94–105. [CrossRef]
7. Van Dyke, T.E.; Sheilesh, D. Risk Factors for Periodontitis. *J. Int. Acad. Periodontol.* **2005**, *7*, 3–7. [CrossRef]
8. Whyte, A.; Whyte, J.; Monteagudo, L.V.; García-Barrios, A.; Teresa Tejedor, M. Periodontal and Dental Status in Packs of Spanish Dogs. *Animals* **2021**, *11*, 1082. [CrossRef]
9. Harvey, C.E. Management of Periodontal Disease: Understanding the Options. *Vet. Clin. N. Am. Small Anim. Pract.* **2005**, *35*, 819–836. [CrossRef]
10. Enlund, K.B.; Jönsson, B.; Abrahamsson, K.H.; Pettersson, A. Long-Term Effects of Motivational Interviewing vs. Traditional Counseling on Dog Owners' Adherence to Veterinary Dental Home Care: A Three-Year Follow-up Study. *Front. Vet. Sci.* **2024**, *11*, 1296618. [CrossRef]
11. Silva, C.; Requicha, J.; Dias, I.; Bastos, E.; Viegas, C. Genomic Medicine in Canine Periodontal Disease: A Systematic Review. *Animals* **2023**, *13*, 2463. [CrossRef] [PubMed]

12. Albuquerque, C.; Morinha, F.; Requicha, J.; Martins, T.; Dias, I.; Guedes-Pinto, H.; Bastos, E.; Viegas, C. Canine Periodontitis: The Dog as an Important Model for Periodontal Studies. *Vet. J.* **2012**, *191*, 299–305. [CrossRef]
13. Whyte, A.; Bonastre, C.; Monteagudo, L.V.; Les, F.; Obon, J.; Whyte, J.; Tejedor, M.T. Canine Stage 1 Periodontal Disease: A Latent Pathology. *Vet. J.* **2014**, *201*, 118–120. [CrossRef] [PubMed]
14. Kouki, M.I.; Papadimitriou, S.A.; Kazakos, G.M.; Savas, I.; Bitchava, D. Periodontal Disease as a Potential Factor for Systemic Inflammatory Response in the Dog. *J. Vet. Dent.* **2013**, *30*, 26–29. [CrossRef] [PubMed]
15. Loos, B.G. Systemic Markers of Inflammation in Periodontitis. *J. Periodontol.* **2005**, *76*, 2106–2115. [CrossRef] [PubMed]
16. Rejec, A.; Butinar, J.; Gawor, J.; Petelin, M. Evaluation of Complete Blood Count Indices (NLR, PLR, MPV/PLT, and PLCRi) in Healthy Dogs, Dogs with Periodontitis, and Dogs with Oropharyngeal Tumors as Potential Biomarkers of Systemic Inflammatory Response. *J. Vet. Dent.* **2017**, *34*, 231–240. [CrossRef]
17. Glickman, L.T.; Glickman, N.W.; Moore, G.E.; Goldstein, G.S.; Lewis, H.B. Evaluation of the Risk of Endocarditis and Other Cardiovascular Events on the Basis of the Severity of Periodontal Disease in Dogs. *J. Am. Vet. Med. Assoc.* **2009**, *234*, 486–494. [CrossRef]
18. Pereira dos Santos, J.D.; Cunha, E.; Nunes, T.; Tavares, L.; Oliveira, M. Relation between Periodontal Disease and Systemic Diseases in Dogs. *Res. Vet. Sci.* **2019**, *125*, 136–140. [CrossRef]
19. Cray, C. Acute Phase Proteins in Animals. *Prog. Mol. Biol. Transl. Sci.* **2012**, *105*, 113–150. [CrossRef]
20. D'Aiuto, F.; Parkar, M.; Andreou, G.; Suvan, J.; Brett, P.M.; Ready, D.; Tonetti, M.S. Periodontitis and Systemic Inflammation: Control of the Local Infection Is Associated with a Reduction in Serum Inflammatory Markers. *J. Dent. Res.* **2004**, *83*, 156–160. [CrossRef]
21. Passoja, A.; Puijola, I.; Knuuttila, M.; Niemelä, O.; Karttunen, R.; Raunio, T.; Tervonen, T. Serum Levels of Interleukin-10 and Tumour Necrosis Factor-α in Chronic Periodontitis. *J. Clin. Periodontol.* **2010**, *37*, 881–887. [CrossRef] [PubMed]
22. Rawlinson, J.E.; Goldstein, R.E.; Reiter, A.M.; Attwater, D.Z.; Harvey, C.E. Association of Periodontal Disease with Systemic Health Indices in Dogs and the Systemic Response to Treatment of Periodontal Disease. *J. Am. Vet. Med. Assoc.* **2011**, *238*, 601–609. [CrossRef] [PubMed]
23. Liu, Z.; Li, Y.; Wang, Y.; Zhang, H.; Lian, Y.; Cheng, X. The Neutrophil-to-Lymphocyte and Monocyte-to-Lymphocyte Ratios Are Independently Associated With the Severity of Autoimmune Encephalitis. *Front. Immunol.* **2022**, *13*, 911779. [CrossRef]
24. Balta, S.; Ozturk, C. The Platelet-Lymphocyte Ratio: A Simple, Inexpensive and Rapid Prognostic Marker for Cardiovascular Events. *Platelets* **2015**, *26*, 680–681. [CrossRef] [PubMed]
25. Angkananard, T.; Anothaisintawee, T.; McEvoy, M.; Attia, J.; Thakkinstian, A. Neutrophil Lymphocyte Ratio and Cardiovascular Disease Risk: A Systematic Review and Meta-Analysis. *Biomed. Res. Int.* **2018**, *2018*, 2703518. [CrossRef] [PubMed]
26. Jung, M.J.; Kim, J.H. Prognostic Efficacy of Complete Blood Count Indices for Assessing the Presence and the Progression of Myxomatous Mitral Valve Disease in Dogs. *Animals* **2023**, *13*, 2821. [CrossRef]
27. Summers, C.; Rankin, S.M.; Condliffe, A.M.; Singh, N.; Peters, A.M.; Chilvers, E.R. Neutrophil Kinetics in Health and Disease. *Trends Immunol.* **2010**, *31*, 318–324. [CrossRef]
28. Hernberg, M.; Mattila, P.S.; Rissanen, M.; Hansson, J.; Aamdal, S.; Bastholt, L.; Von Der Maase, H.; Schmidt, H.; Stierner, U.; Tarkkanenz, J. The Prognostic Role of Blood Lymphocyte Subset Distribution in Patients With Resected High-Risk Primary or Regionally Metastatic Melanoma. *J. Immunother.* **2007**, *30*, 773–779. [CrossRef]
29. George-Gay, B.; Parker, K. Understanding the Complete Blood Count with Differential. *J. Perianesth Nurs.* **2003**, *18*, 96–117. [CrossRef]
30. González-Domínguez, A.; Cristobal-Verdejo, J.I.; López-Espinar, C.; Fontela-González, S.; Vázquez, S.; Justo-Domínguez, J.; González-Caramazana, J.; Bragado-Cuesta, M.; Álvarez-Punzano, A.; Herrería-Bustillo, V.J. Retrospective Evaluation of Hematological Ratios in Canine Parvovirosis: 401 Cases. *J. Vet. Intern. Med.* **2024**, *38*, 161–166. [CrossRef]
31. Djordjevic, D.; Rondovic, G.; Surbatovic, M.; Stanojevic, I.; Udovicic, I.; Andjelic, T.; Zeba, S.; Milosavljevic, S.; Stankovic, N.; Abazovic, D.; et al. Neutrophil-to-Lymphocyte Ratio, Monocyte-to-Lymphocyte Ratio, Platelet-to-Lymphocyte Ratio, and Mean Platelet Volume-to-Platelet Count Ratio as Biomarkers in Critically Ill and Injured Patients: Which Ratio to Choose to Predict Outcome and Nature of Bacteremia? *Mediat. Inflamm.* **2018**, *2018*, 3758068. [CrossRef]
32. Li, T.; Dong, G.; Zhang, M.; Xu, Z.; Hu, Y.; Xie, B.; Wang, Y.; Xu, B. Association of Neutrophil-Lymphocyte Ratio and the Presence of Neonatal Sepsis. *J. Immunol. Res.* **2020**, *2020*, 7650713. [CrossRef] [PubMed]
33. Hou, S.K.; Lin, H.A.; Chen, S.C.; Lin, C.F.; Lin, S.F. Monocyte Distribution Width, Neutrophil-to-Lymphocyte Ratio, and Platelet-to-Lymphocyte Ratio Improves Early Prediction for Sepsis at the Emergency. *J. Pers. Med.* **2021**, *11*, 732. [CrossRef] [PubMed]
34. Zahorec, R. Neutrophil-to-Lymphocyte Ratio, Past, Present and Future Perspectives. *Bratisl. Med. J.* **2021**, *122*, 474–488. [CrossRef]
35. Conway, E.A.; Pizarro Del Valle, C.; Waugh, E.M.; French, A.; Ridyard, A.E. Retrospective Investigation of the Neutrophil-to-Lymphocyte Ratio in Dogs with Pneumonia: 49 Cases (2011–2016). *J. Vet. Emerg. Crit. Care* **2021**, *31*, 490–497. [CrossRef]
36. Hodgson, N.; Llewellyn, E.A.; Schaeffer, D.J. Utility and Prognostic Significance of Neutrophil-to-Lymphocyte Ratio in Dogs with Septic Peritonitis. *J. Am. Anim. Hosp. Assoc.* **2018**, *54*, 351–359. [CrossRef]
37. Mutz, M.; Boudreaux, B.; Kearney, M.; Stroda, K.; Gaunt, S.; Shiomitsu, K. Prognostic Value of Baseline Absolute Lymphocyte Concentration and Neutrophil/Lymphocyte Ratio in Dogs with Newly Diagnosed Multi-Centric Lymphoma. *Vet. Comp. Oncol.* **2015**, *13*, 337–347. [CrossRef]

38. MacFarlane, M.J.; MacFarlane, L.L.; Scase, T.; Parkin, T.; Morris, J.S. Use of Neutrophil to Lymphocyte Ratio for Predicting Histopathological Grade of Canine Mast Cell Tumours. *Vet. Rec.* **2016**, *179*, 491. [CrossRef]
39. Macfarlane, L.; Morris, J.; Pratschke, K.; Mellor, D.; Scase, T.; Macfarlane, M.; Mclauchlan, G. Diagnostic Value of Neutrophil-Lymphocyte and Albumin-Globulin Ratios in Canine Soft Tissue Sarcoma. *J. Small Anim. Pract.* **2016**, *57*, 135–141. [CrossRef]
40. Neumann, S. Neutrophil-to-Lymphocyte and Platelet-to-Lymphocyte Ratios in Dogs and Cats with Acute Pancreatitis. *Vet. Clin. Pathol.* **2021**, *50*, 45–51. [CrossRef]
41. Park, J.; Lee, D.; Yun, T.; Koo, Y.; Chae, Y.; Kim, H.; Yang, M.P.; Kang, B.T. Evaluation of the Blood Neutrophil-to-Lymphocyte Ratio as a Biomarker for Meningoencephalitis of Unknown Etiology in Dogs. *J. Vet. Intern. Med.* **2022**, *36*, 1719–1725. [CrossRef] [PubMed]
42. Klinger, M.H.F.; Jelkmann, W. Role of Blood Platelets in Infection and Inflammation. *J. Interferon Cytokine Res.* **2002**, *22*, 913–922. [CrossRef] [PubMed]
43. Anitua, E.; Andia, I.; Ardanza, B.; Nurden, P.; Nurden, A.T. Autologous Platelets as a Source of Proteins for Healing and Tissue Regeneration. *Thromb. Haemost.* **2004**, *91*, 4–15. [CrossRef]
44. Woolcock, A.D.; Keenan, A.; Cheung, C.; Christian, J.A.; Moore, G.E. Thrombocytosis in 715 Dogs (2011–2015). *J. Vet. Intern. Med.* **2017**, *31*, 1691–1699. [CrossRef] [PubMed]
45. Hu, G.; Liu, Q.; Ma, J.Y.; Liu, C.Y. Prognostic Significance of Platelet-to-Lymphocyte Ratio in Cholangiocarcinoma: A Meta-Analysis. *Biomed. Res. Int.* **2018**, *2018*, 7375169. [CrossRef] [PubMed]
46. Pierini, A.; Gori, E.; Lippi, I.; Lubas, G.; Marchetti, V. Are Leukocyte and Platelet Abnormalities and Complete Blood Count Ratios Potential Prognostic Markers in Canine Sepsis? *Front. Vet. Sci.* **2020**, *7*, 578846. [CrossRef]
47. Durán-Galea, A.; Cristóbal-Verdejo, J.I.; Barrera-Chacón, R.; Macías-García, B.; González-Solís, M.A.; Nicolás-Barceló, P.; García-Ibáñez, A.B.; Ruíz-Tapia, P.; Duque-Carrasco, F.J. Clinical Importance of Neutrophil-to-Lymphocyte Ratio, Platelet-to-Lymphocyte Ratio and Systemic Immune-Inflammation Index in Dogs with Leishmaniasis. *Comp. Immunol. Microbiol. Infect. Dis.* **2024**, *107*, 102148. [CrossRef]
48. Becher, A.; Suchodolski, J.S.; Steiner, J.M.; Heilmann, R.M. Blood Neutrophil-to-Lymphocyte Ratio (NLR) as a Diagnostic Marker in Dogs with Chronic Enteropathy. *J. Vet. Diagn. Investig.* **2021**, *33*, 516–527. [CrossRef]
49. Benvenuti, E.; Pierini, A.; Gori, E.; Lucarelli, C.; Lubas, G.; Marchetti, V. Neutrophil-to-Lymphocyte Ratio (NLR) in Canine Inflammatory Bowel Disease (IBD). *Vet. Sci.* **2020**, *7*, 141. [CrossRef]
50. Cristóbal, J.I.; Duque, F.J.; Usón-Casaús, J.; Barrera, R.; López, E.; Pérez-Merino, E.M. Complete Blood Count-Derived Inflammatory Markers Changes in Dogs with Chronic Inflammatory Enteropathy Treated with Adipose-Derived Mesenchymal Stem Cells. *Animals* **2022**, *12*, 2798. [CrossRef]
51. Kurt, M.; Onal, I.K.; Sayilir, A.; Beyazit, Y.; Oztas, E.; Kekilli, M.; Turhan, N.; Karaman, K.; Akdogan, M. The Role of Mean Platelet Volume in the Diagnosis of Hepatocellular Carcinoma in Patients with Chronic Liver Disease. *Hepatogastroenterology* **2012**, *59*, 1580–1582. [PubMed]
52. Karaman, K.; Bostanci, E.B.; Aksoy, E.; Kurt, M.; Celep, B.; Ulas, M.; Dalgic, T.; Surmelioglu, A.; Hayran, M.; Akoglu, M. The Predictive Value of Mean Platelet Volume in Differential Diagnosis of Non-Functional Pancreatic Neuroendocrine Tumors from Pancreatic Adenocarcinomas. *Eur. J. Intern. Med.* **2011**, *22*, 95–98. [CrossRef]
53. Afari, M.E.; Bhat, T. Neutrophil to Lymphocyte Ratio (NLR) and Cardiovascular Diseases: An Update. *Expert. Rev. Cardiovasc. Ther.* **2016**, *14*, 573–577. [CrossRef] [PubMed]
54. Chen, H.; Li, M.; Liu, L.; Dang, X.; Zhu, D.; Tian, G. Monocyte/Lymphocyte Ratio Is Related to the Severity of Coronary Artery Disease and Clinical Outcome in Patients with Non-ST-Elevation Myocardial Infarction. *Medicine* **2019**, *98*, e16267. [CrossRef] [PubMed]
55. Henriques, J.; Felisberto, R.; Constantino-Casas, F.; Cabeçadas, J.; Dobson, J. Peripheral Blood Cell Ratios as Prognostic Factors in Canine Diffuse Large B-Cell Lymphoma Treated with CHOP Protocol. *Vet. Comp. Oncol.* **2021**, *19*, 242–252. [CrossRef]
56. Marôco, J. Regressão Categorial. In *Análise Estatística com o SPSS Statistics 25*; Report Number; Análise e Gestão de Informação, LDA: Pêro Pinheiro, Portugal, 2018; Volume 18, pp. 807–841.
57. Yamauchi, A.; Yoshimoto, S.; Kudo, A.; Takagi, S. Negative Influence of Aging on Differentiation and Proliferation of CD8+ T-Cells in Dogs. *Vet. Sci.* **2023**, *10*, 541. [CrossRef]
58. Oh, B.S.; Jang, J.W.; Kwon, J.H.; You, C.R.; Chung, K.W.; Kay, C.S.; Jung, H.S.; Lee, S. Prognostic Value of C-Reactive Protein and Neutrophil-to-Lymphocyte Ratio in Patients with Hepatocellular Carcinoma. *BMC Cancer* **2013**, *13*, 78. [CrossRef]
59. Lu, R.; Li, W.; Wang, X.; Shi, D.; Meng, H. Elevated Neutrophil-to-Lymphocyte Ratio but Not Platelet-to-Lymphocyte Ratio Is Associated with Generalized Aggressive Periodontitis in a Chinese Population. *J. Periodontol.* **2021**, *92*, 507–513. [CrossRef]
60. Turkmen, K.; Guney, I.; Yerlikaya, F.H.; Tonbul, H.Z. The Relationship Between Neutrophil-to-Lymphocyte Ratio and Inflammation in End-Stage Renal Disease Patients. *Ren. Fail.* **2012**, *34*, 155–159. [CrossRef]
61. Liu, J.H.; Zhang, Y.J.; Ma, Q.H.; Sun, H.P.; Xu, Y.; Pan, C.W. Elevated Blood Neutrophil to Lymphocyte Ratio in Older Adults with Cognitive Impairment. *Arch. Gerontol. Geriatr.* **2020**, *88*, 104041. [CrossRef]
62. Bassani, B.; Cucchiara, M.; Butera, A.; Kayali, O.; Chiesa, A.; Palano, M.T.; Olmeo, F.; Gallazzi, M.; Dellavia, C.P.B.; Mortara, L.; et al. Neutrophils' Contribution to Periodontitis and Periodontitis-Associated Cardiovascular Diseases. *Int. J. Mol. Sci.* **2023**, *24*, 15370. [CrossRef] [PubMed]

63. Kumar, V.; Sharma, A. Neutrophils: Cinderella of Innate Immune System. *Int. Immunopharmacol.* **2010**, *10*, 1325–1334. [CrossRef] [PubMed]
64. Bhansali, R.S.; Yeltiwar, R.K.; Bhat, K.G. Assessment of Peripheral Neutrophil Functions in Patients with Localized Aggressive Periodontitis in the Indian Population. *J. Indian Soc. Periodontol.* **2013**, *17*, 731–736. [CrossRef] [PubMed]
65. Gonzales, J.R. T-and B-Cell Subsets in Periodontitis. *Periodontology 2000* **2015**, *69*, 181–200. [CrossRef]
66. Kinane, D.F.; Lappin, D.F.; Culshaw, S. The Role of Acquired Host Immunity in Periodontal Diseases. *Periodontology 2000* **2024**, 1–15. [CrossRef]
67. Cavalla, F.; Hernández, M. Polarization Profiles of T Lymphocytes and Macrophages Responses in Periodontitis. *Adv. Exp. Med. Biol.* **2022**, *1373*, 195–208. [CrossRef]
68. Han, N.; Liu, Y.; Du, J.; Xu, J.; Guo, L.; Liu, Y. Regulation of the Host Immune Microenvironment in Periodontitis and Periodontal Bone Remodeling. *Int. J. Mol. Sci.* **2023**, *24*, 3158. [CrossRef]
69. Bittner-Eddy, P.D.; Fischer, L.A.; Costalonga, M. Transient Expression of IL-17A in Foxp3 Fate-Tracked Cells in Porphyromonas Gingivalis-Mediated Oral Dysbiosis. *Front. Immunol.* **2020**, *11*, 677. [CrossRef]
70. Yang, B.; Pang, X.; Li, Z.; Chen, Z.; Wang, Y. Immunomodulation in the Treatment of Periodontitis: Progress and Perspectives. *Front. Immunol.* **2021**, *12*, 781378. [CrossRef]
71. Figueredo, C.M.; Lira-Junior, R.; Love, R.M. T and B Cells in Periodontal Disease: New Functions in A Complex Scenario. *Int. J. Mol. Sci.* **2019**, *20*, 3949. [CrossRef]
72. Díaz-Zúñiga, J.; Melgar-Rodríguez, S.; Monasterio, G.; Pujol, M.; Rojas, L.; Alvarez, C.; Carvajal, P.; Vernal, R. Differential Human Th22-Lymphocyte Response Triggered by Aggregatibacter Actinomycetemcomitans Serotypes. *Arch. Oral Biol.* **2017**, *78*, 26–33. [CrossRef] [PubMed]
73. Azab, B.; Bhatt, V.R.; Phookan, J.; Murukutla, S.; Kohn, N.; Terjanian, T.; Widmann, W.D. Usefulness of the Neutrophil-to-Lymphocyte Ratio in Predicting Short- and Long-Term Mortality in Breast Cancer Patients. *Ann. Surg. Oncol.* **2012**, *19*, 217–224. [CrossRef]
74. DeBowes, L.J. The Effects of Dental Disease on Systemic Disease. *Vet. Clin. N. Am. Small Anim. Pract.* **1998**, *28*, 1057–1062. [CrossRef]
75. Preshaw, P.M.; Taylor, J.J. How Has Research into Cytokine Interactions and Their Role in Driving Immune Responses Impacted Our Understanding of Periodontitis? *J. Clin. Periodontol.* **2011**, *38*, 60–84. [CrossRef]
76. Kinane, D.F.; Preshaw, P.M.; Loos, B.G. Host-Response: Understanding the Cellular and Molecular Mechanisms of Host-Microbial Interactions—Consensus of the Seventh European Workshop on Periodontology. *J. Clin. Periodontol.* **2011**, *38*, 44–48. [CrossRef] [PubMed]
77. Iqbal, P.S.; Khan, S.N.; Haris, M.; Narayanan, M.; Laju, S.; Kumar, S.S. Assessment of Systemic Inflammatory Markers in Patients with Aggressive Periodontitis. *J. Int. Oral Health* **2015**, *7*, 48–51.
78. Kamath, S.; Blann, A.D.; Lip, G.Y.H. Platelet Activation: Assessment and Quantification. *Eur. Heart J.* **2001**, *22*, 1561–1571. [CrossRef] [PubMed]
79. Taylor, B.A.; Tofler, G.H.; Carey, H.M.R.; Morel-Kopp, M.-C.; Philcox, S.; Carter, T.R.; Elliott, M.J.; Kull, A.D.; Ward, C.; Schenck, K. Full-Mouth Tooth Extraction Lowers Systemic Inflammatory and Thrombotic Markers of Cardiovascular Risk. *J. Dent. Res.* **2005**, *85*, 74–78. [CrossRef]
80. Mutthineni, R.; Ramishetty, A.; Gojja, P.; Muralidaran, G.; Burle, V. Platelet Indices Be a New Biomarker for Periodontal Disease. *Contemp. Clin. Dent.* **2021**, *12*, 289–293. [CrossRef]
81. Papapanagiotou, D.; Nicu, E.A.; Bizzarro, S.; Gerdes, V.E.A.; Meijers, J.C.; Nieuwland, R.; van der Velden, U.; Loos, B.G. Periodontitis Is Associated with Platelet Activation. *Atherosclerosis* **2009**, *202*, 605–611. [CrossRef]
82. Mishra, S.; Gazala, M.P.; Rahman, W. Clinical and Diagnostic Significance of Blood Leukocyte Ratios in Young Patients with Stage III Grade C Periodontitis. *Acta Odontol. Scand.* **2022**, *80*, 1969035. [CrossRef] [PubMed]
83. Albuquerque, C.; Morinha, F.; Magalhães, J.; Requicha, J.; Dias, I.; Guedes-Pinto, H.; Bastos, E.; Viegas, C. Variants in the Interleukin-1 Alpha and Beta Genes, and the Risk for Periodontal Disease in Dogs. *J. Genet.* **2015**, *94*, 651–659. [CrossRef]
84. Shiroddria, S.; Smith, J.; McKay, I.J.; Kennett, C.N.; Hughes, F.J. Polymorphisms in the IL-1A Gene Are Correlated with Levels of Interleukin-1α Protein in Gingival Crevicular Fluid of Teeth with Severe Periodontal Disease. *J. Dent. Res.* **2000**, *79*, 1864–1869. [CrossRef]
85. Graves, D.T.; Cochran, D. The Contribution of Interleukin-1 and Tumor Necrosis Factor to Periodontal Tissue Destruction. *J. Periodontol.* **2003**, *74*, 391–401. [CrossRef] [PubMed]
86. Buhlin, K.; Gustafsson, A.; Pockley, A.G.; Frostegård, J.; Klinge, B. Risk Factors for Cardiovascular Disease in Patients with Periodontitis. *Eur. Heart J.* **2003**, *24*, 2099–2107. [CrossRef]
87. Marchesi, M.C.; Maggi, G.; Cremonini, V.; Miglio, A.; Contiero, B.; Guglielmini, C.; Antognoni, M.T. Monocytes Count, NLR, MLR and PLR in Canine Inflammatory Bowel Disease. *Animals* **2024**, *14*, 837. [CrossRef]
88. Hutter, J.W.; Van Der Velden, U.; Varoufaki, A.; Huffels, R.A.M.; Hoek, F.J.; Loos, B.G. Lower Numbers of Erythrocytes and Lower Levels of Hemoglob in Periodontitis Patients Compared to Control Subjects. *J. Clin. Periodontol.* **2001**, *28*, 930–936. [CrossRef] [PubMed]
89. Yamamoto, T.; Tsuneishi, M.; Furuta, M.; Ekuni, D.; Morita, M.; Hirata, Y. Relationship Between Decrease of Erythrocyte Count and Progression of Periodontal Disease in a Rural Japanese Population. *J. Periodontol.* **2011**, *82*, 106–113. [CrossRef] [PubMed]

90. Anand, P.S.; Sagar, D.K.; Ashok, S.; Kamath, K.P. Association of Aggressive Periodontitis with Reduced Erythrocyte Counts and Reduced Hemoglobin Levels. *J. Periodontal. Res.* **2014**, *49*, 719–728. [CrossRef]
91. Aljohani, H.A. Association between Hemoglobin Level and Severity of Chronic Periodontitis. *JKAU Med. Sci.* **2010**, *17*, 53–64. [CrossRef]
92. Thongsahuan, S.; Fonghoi, L.; Kaewfai, S.; Srinoun, K. Precision and Accuracy of the Mindray BC-5000Vet Hematology Analyzer for Canine and Feline Blood. *Vet. Clin. Pathol.* **2020**, *49*, 207–216. [CrossRef] [PubMed]

veterinary sciences

MDPI

Article

Extensive Epidural Hemorrhage Associated with Thoracolumbar Disc Extrusion in French Bulldogs

Julius Underberg [1,*], Arianna Maiolini [2], Maja Waschk [1] and Daniela Schweizer [1]

[1] Division of Clinical Radiology, Department of Clinical Veterinary Science, Vetsuisse Faculty, University of Bern, 3012 Bern, Switzerland; maja.waschk@unibe.ch (M.W.); daniela.schweizer@unibe.ch (D.S.)

[2] Division of Clinical Neurology, Department of Clinical Veterinary Science, Vetsuisse Faculty, University of Bern, 3012 Bern, Switzerland; arianna.maiolini@unibe.ch

* Correspondence: julius.underberg@unibe.ch

Simple Summary: French Bulldogs often suffer from disc herniations leading to paralysis of the hind limbs. In this breed especially, a high number of dogs show associated bleeding in the vertebral canal that, additionally to the disc herniation, can lead to compression of the spinal cord. We analyzed magnetic resonance images of 148 French Bulldogs to see if bleeding occurred more often at certain locations of disc herniation. We found that the most common location of disc herniation in French Bulldogs was L4/L5, which is different compared to other breeds, and that in 70% of these dogs, disc herniation was accompanied by extensive bleeding. The highest percentage of disc herniations with extensive bleeding was seen with herniation between the first and second lumbar vertebra.

Abstract: (1) Background: Intervertebral disc extrusion may be accompanied by extensive epidural hemorrhage (EEH) and result in spinal cord compression. EEH is more commonly seen in French Bulldogs compared to other breeds. The majority of French Bulldogs have lumbar intervertebral disc extrusion, but it is unclear if there is an association between the site of thoracolumbar disc extrusion and EEH. This retrospective study's aim was to investigate the association and prevalence between the site of thoracolumbar intervertebral disc extrusion and EEH in French Bulldogs. (2) Methods: Thoracolumbar MRI of French Bulldogs was examined for the site of intervertebral disc extrusion and the presence, distribution, and extent of EEH. (3) Results: In 148 French Bulldogs with thoracolumbar intervertebral disc extrusion, the most common location was L4/L5 (22%), followed by L3/L4 (20%) and T13/L1 (15%), and 70.3% had EEH. The highest incidence of EEH across all thoracolumbar sites was seen at L1/L2 (84.2%), followed by L5/L6 (81.3%) and L2/L3 (72.2%). The incidence of EEH was 70.5% across all lumbar sites and 66.9% for all thoracic sites. (4) Conclusion: In this population of French Bulldogs, the prevalence of lumbar disc extrusion was high, and 70.3% had EEH. There was no significant difference between the site of thoracolumbar intervertebral disc extrusion and EEH.

Keywords: intervertebral disc extrusion; epidural hemorrhage; dorsal epidural space; intramedullary bleeding

1. Introduction

Intervertebral disc extrusion can result in the traumatic rupture of the ventral venous plexus, leading to epidural hemorrhage. The extent of epidural hemorrhage is variable and may span multiple spinal cord segments, potentially necessitating decompressive surgery at more than one site [1]. This phenomenon is described as disc extrusion with extensive epidural hemorrhage, without further specifying the length and compression of the epidural hemorrhage [2]. Epidural hemorrhage is characterized by the presence of erythrocytes and macrophages containing blood degradation products, which are believed to trigger epidural inflammation. Epidural inflammation is important to consider in cases of intervertebral disc herniation, as paraplegic medium- to large-breed dogs with extensive epidural hemorrhage have less favorable outcomes [3]. In this context, it is important to emphasize that the contact area between the epidural hemorrhage and the spinal cord seems to be more critical for prognosis than the extent of spinal cord compression, and dogs with minimal epidural inflammation have a better prognosis for regaining ambulation [4]. Domestic dog breeds with the chondrodystrophic phenotype are strongly associated with the expression of the FGF4 retrogene (fibroblast growth factor 4) [5]. Specifically, the FGF4 retrogene on chromosome 12 appears to be associated with intervertebral disc disease [6].

French Bulldogs appear to develop disc-associated epidural hemorrhage more frequently than Dachshunds, other chondrodystrophic breeds, and non-chondrodystrophic dogs [7,8].

Across all dog breeds, the most common site for intervertebral disc extrusion is at T12/T13, but epidural hemorrhage seems significantly more common in intervertebral disc extrusion at L4/5 and less common at the thoracolumbar junction [9]. However, the association between the location of intervertebral disc herniation and epidural hemorrhage is debated and another study found that in French Bulldogs, epidural hemorrhage is associated with caudal thoracic and thoracolumbar intervertebral disc extrusion [8]. Based on our observations, French Bulldogs at our institution frequently present with epidural hemorrhage in cases of lumbar intervertebral disc extrusion.

The aim of this retrospective study was to determine the prevalence of extensive epidural hemorrhage (EEH) at specific sites of intervertebral disc extrusion in the thoracic and lumbar spine in a population of French Bulldogs and whether certain sites are at an increased risk of developing EEH. We hypothesized that EEH has a higher prevalence with intervertebral disc extrusion at specific caudal lumbar sites.

2. Materials and Methods

Inclusion criteria were French Bulldogs that presented with paraparesis or paraplegia localized to T3–L3 or L4–S3 spinal cord segments, with a clinical onset of less than 14 days and an MRI diagnosis of intervertebral disc extrusion in a 10-year time frame from 2012–2022. Dogs were classified according to the presence or absence of deep pain perception. The deep pain positive (DPP) group included dogs showing different degrees of paraparesis and paraplegia with preserved deep pain. The deep pain negative (DPN) group included paraplegic dogs without deep pain perception. Dogs showing unilateral sensorimotor loss were grouped into the DPP group. Further inclusion criteria were an MRI examination protocol including a T2-weighted sequence in the sagittal and transverse plane, a T1-weighted sequence in the transverse or dorsal plane, a T2-weighted fat-suppressed sequence in the dorsal plane, and a T2* gradient echo sequence in the transverse plane. Exclusion criteria were dogs where MRI images did not allow for correct counting of the vertebrae or incomplete sequence acquisition. The vertebra with the last pair of ribs was defined as T13.

The images of this single institution study were acquired with a 1.0 T magnet (Philips HFO Panorama, Philips Medical Systems, PC Best, The Netherlands), a 1.5 T magnet (Avanto, Siemens Healthineers, Erlangen, Germany), or a 3.0 T magnet (Magnetom Vida 3T, Siemens Healthineers, Erlangen, Germany).

Images were evaluated by a senior resident (JU) and a board-certified radiologist (DS) in consensus, blinded to the original report, using a DICOM viewer (DeepUnity Viewer, Dedalus Healthcare GmbH, Bonn, Germany). Images were assessed for the location of intervertebral disc extrusion based on narrowing of the intervertebral disc space, reduced signal intensity and volume of the nucleus pulposus on T2-weighted images, asymmetric protrusion of the annulus, annular tear, and T1-weighted and T2-weighted hypointense material within the vertebral canal in continuation with the intervertebral disc [2]. Intervertebral disc extrusion was classified as thoracic if located between T1/T2 and T12/T13 and lumbar if between T13/L1 and L6/L7.

Epidural hemorrhage was identified by material within the epidural space with mixed signal intensity on T2-weighted images combined with T1-weighted iso-, hypo-, or hyperintense signals compared to a normal spinal cord and signal void on T2* GRE sequences. Extensive epidural hemorrhage was defined as the extension of epidural hemorrhage over more than two consecutive intervertebral disc spaces and expressed as times vertebral body length of L2.

Epidural material was evaluated for

- location in relation to the affected intervertebral disc space (cranial, caudal, both);
- cranio-caudal extent of the epidural material expressed as times the length of L2;
- location in relation to the spinal cord in the transverse section (dorsal right, dorsal left, ventral left, ventral right);
- location of maximal compression with respect to affected intervertebral disc space;
- maximal spinal cord compression as mild (<25%), moderate (25–50%), or severe (>50%);
- intramedullary signal void in T2* gradient echo sequences;

Recovery of deep pain perception of the DPN dogs was recorded at the time of discharge.

Statistical Analysis

The collected variables were described by means and standard deviations when normally distributed, and median and interquartile ranges were used for non-normally distributed data. Categorical variables were expressed as numbers and percentages. Fisher's exact test and the Chi-squared test were used for comparing the location of intervertebral disc extrusion and the proportion of EEH. NCSS was used for data analysis (NCSS 2023 Statistical Software (2023), NCSS, LLC., Kaysville, UT, USA).

3. Results

3.1. Study Population

A total of 158 French Bulldogs underwent MRI examination of the spine. Eight dogs were excluded due to loss of imaging data or incomplete sequence acquisition. Two dogs were excluded as correct counting of the vertebrae was not possible. This resulted in 148 French bulldogs being included in the image analysis, comprising 87 males and 61 females. The median age of the dogs was 4.02 years (range 0.2–8.77 years, IQR 2.15 years). The mean weight was 13.12 kg (range 7.8–18.8 kg, SD of 2.33 kg). Twelve (8.1%) had previously undergone a hemilaminectomy due to intervertebral disc disease. MRI examination was performed using a 1 T magnet for 132 French Bulldogs, a 1.5 T magnet for 4 dogs, and a 3 T magnet for 12 dogs. Of the 148 French Bulldogs 121 dogs were grouped into the DPP group (82%) and 27 were grouped into the DPN group (18%).

3.2. Extent of Hemorrhage and Material

Epidural hemorrhage was observed in 144 out of 148 French Bulldogs (97%). The longitudinal extent of the hemorrhage ranged from 1 to 6.5 vertebral body lengths (median 2.5). In 104 out of 148 dogs (70.3%), the epidural hemorrhage extended over more than two consecutive intervertebral disc spaces and was therefore classified as EEH. The longitudinal extent in the EEH group ranged from 2 to 6.5 vertebral body lengths (median 3); 44/148 dogs had no extensive epidural hemorrhage (non-EEH group). The extent of epidural hemorrhage in the non-EEH group ranged from 1 to 2 vertebral body lengths (median 1.5). No significant differences between groups were noted for age, sex, or body weight.

The craniocaudal extent of material in relation to the affected intervertebral disc space was seen in the cranial direction in all dogs of both groups; extension in the caudal direction was noted in all but two dogs that belonged to the non-EEH group. The median extent in the EEH group was 2× the length of L2 in the cranial direction and 1× the length of L2 in the caudal direction. For the non-EEH group, the median extent in the cranial and caudal directions was 1× the length of L2. Of the 104 dogs in the EEH group, 18 dogs (17%) were DPN; in the non-EEH group 9 of 44 dogs (20%) were DPN. There was no significant difference between both groups considering deep pain sensation ($p = 0.6$).

3.3. Intervertebral Disc Space

For the entire study population (148 dogs), the most frequent site of intervertebral disc extrusion was L4/L5 (33 dogs, 22%). The second most common location was L3/L4 (30 dogs, 20%), followed by T13/L1 (23 dogs, 16%) and L1/L2 (19 dogs, 13%). The affected intervertebral disc space was classified as lumbar (T13/L1–L6/L7) in 139 out of 148 French Bulldogs (94%) and thoracic (T1/T2–T12/T13) in 9 out of 148 dogs (6%). In the 104 dogs with EEH, the L4/L5 intervertebral disc space was the most affected (22 dogs, 21%), followed by L3/L4 (19 dogs, 18%), L1/L2 (16 dogs, 15%), T13/L1 (15 dogs, 14%), and L2/L3 and L5/L6 (13 dogs, 13%).

Concerning the 44 non-EEH dogs, the most commonly affected intervertebral disc spaces were L3/L4 (11 dogs, 25%) and L4/L5 (11 dogs, 25%), followed by T13/L1 (8 dogs, 18%), and less common locations were L2/L3 (5 dogs, 11%), L5/L6 (3 dogs, 7%), T12/T13 (2 dogs, 5%), and T11/T12 (1 dog, 2%).

The highest prevalence of dogs with EEH across all thoracolumbar sites of disc extrusion was seen at L1/L2 (16 out of 19 dogs, 84%), followed by L5/L6 (13 out of 16 dogs, 81%), L2/L3 (13 out of 18 dogs, 72%), and L4/L5 (22 out of 33 dogs, 66%). At the thoracolumbar junction, 15 out of 23 dogs had EEH (65%). For all lumbar sites (T13/L1–L6/L7), 70.5% exhibited EEH (98 out of 139), while, for all thoracic sites, 6 out of 9 dogs exhibited EEH (66.9%). There was no significant difference in the prevalence of EEH between the specific sites of thoracic or lumbar spine as determined by Fisher's exact test and the Chi-square test. The numbers of dogs with EEH and no EEH for each site of disc extrusion are shown in Figure 1.

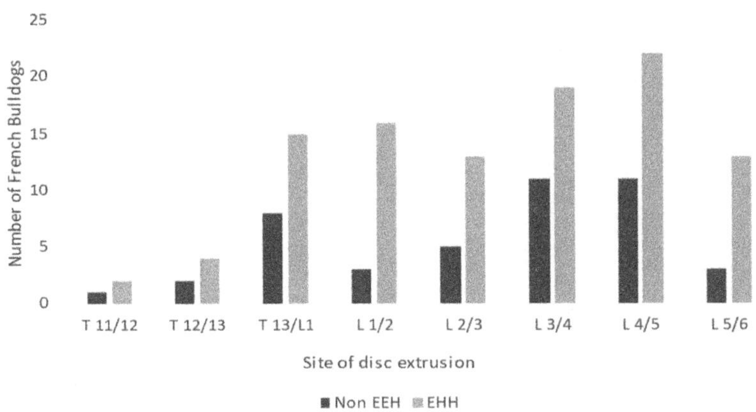

Figure 1. Site of thoracolumbar disc extrusion and numbers of dogs in both groups (EEH and non-EEH). EEH: extensive epidural hemorrhage; non-EEH: no extensive epidural hemorrhage.

3.4. Involvement of Quadrants of the Vertebral Canal

The evaluation of the distribution of material within the vertebral canal based on transverse images revealed that in the study group of 148 French Bulldogs, the material was in the left ventral quadrant in 101 dogs (68%), the right ventral quadrant in 99 dogs (67%), the left dorsal quadrant in 95 dogs (64%), and the right dorsal quadrant in 80 dogs (54%). The involvement of quadrants of the vertebral canal in the EEH versus non-EEH group is shown in Figure 2. Figure 3 shows a dog of the EEH group.

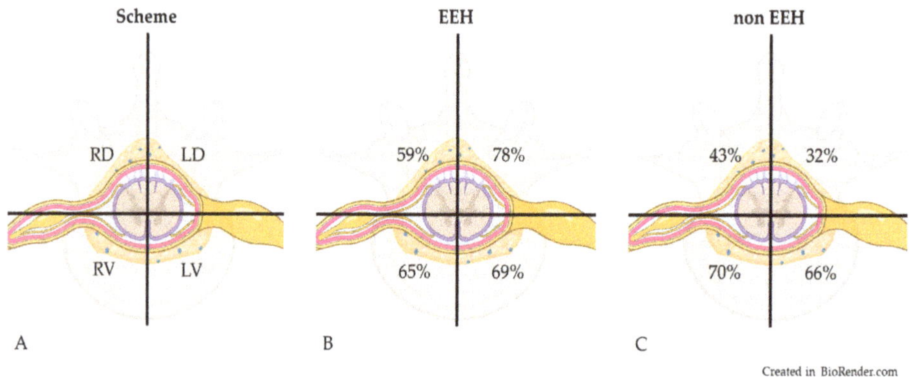

Figure 2. Vertebral canal as seen on transverse images divided in quadrants (**A**). For both groups, EEH (**B**) and non-EEH (**C**), the percentage of dogs with involvement of each quadrant is provided. Note that in most dogs, more than one quadrant was involved. RD: right dorsal; LD: left dorsal; LV: left ventral; RV: right ventral; EEH: extensive epidural hemorrhage; non-EEH: no extensive epidural hemorrhage.

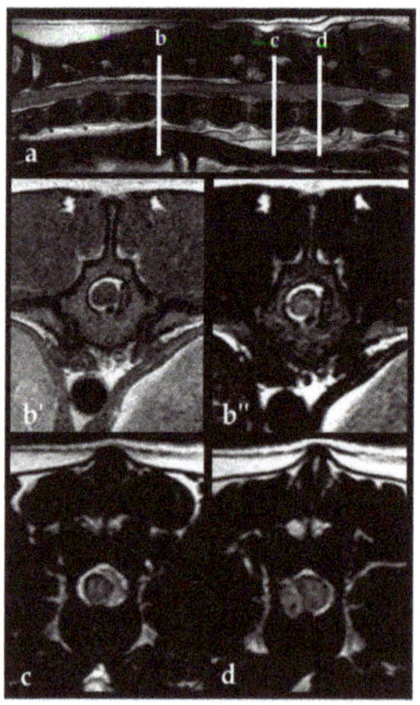

Figure 3. French Bulldog, 5 years, male, entire. T2-weighted sagittal image (**a**) with line b–d indicating the different levels of the transverse images. Epidural material can be seen in the right ventral epidural space at the level of intervertebral disc extrusion L3/4 (**d**). The epidural material extended cranial within the left epidural space at the level L2/3 (**c**) up to the level of mid aspect of T13 (**b′**,**b″**). The epidural material in this case expanded over 5.5× the length of L2 vertebral body lengths. The dog underwent decompressive surgery at two locations.

Material was located in only a single quadrant in four dogs: three non-EEH dogs and one EEH dog. Concerning the EEH dogs, material was present in one or both dorsal quadrants in 102/104 (98%). The involvement of one or two dorsal quadrants was seen in 30 of 44 non-EEH dogs (68%), revealing a significant difference between both groups (p = 0.0001). The involvement of both dorsal quadrants at the same time was observed in 40 dogs with EEH (38%) and only 4 dogs of the non-EEH group (9%). The presence of material in one or two dorsal quadrants without material in a ventral quadrant was found in four dogs with EEH (one with material in a single and three with material in both dorsal quadrants) and was not seen in dogs of the non-EEH group.

3.5. Compression of Spinal Cord

Spinal cord compression was assessed as mild in 23 dogs (16%): 13 dogs with EEH and 10 non-EEH dogs. Moderate spinal cord compression was seen in 76 dogs (51%): 56 dogs of the EEH and 20 dogs of the non-EEH group. Severe spinal cord compression was seen in 49 dogs (33%): 35 dogs with EEH (Figure 4) and 14 dogs without EEH. There was no significant difference between both groups concerning the grade of spinal cord compression ($p = 0.28$).

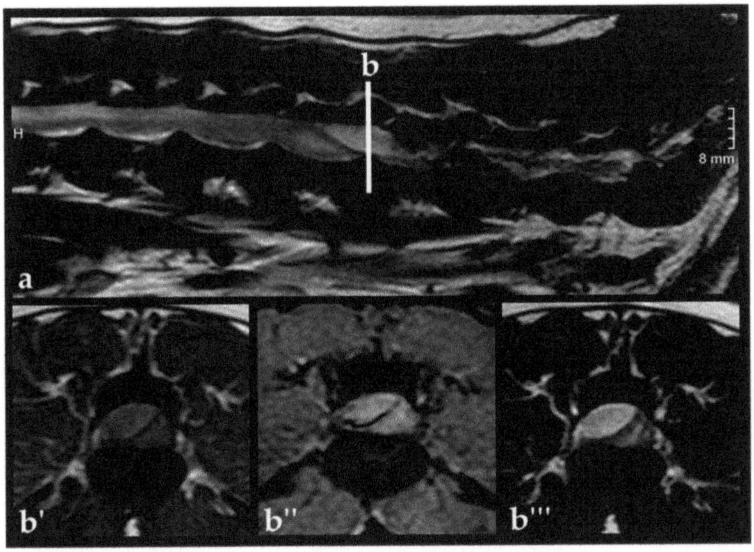

Figure 4. French Bulldog, 5 years, female, intact. T2-weighted sagittal image (**a**), H is indicating the direction towards the head of the dog. The white line (**b**) indicates the level of the transverse images. Transverse T1 (**b′**), T2*-weighted gradient echo (**b″**), and T2-weighted fast spin echo image (**b‴**) at the level L4/5 showing severe compression from left dorsal due to assumed hyperacute epidural hemorrhage. The dog showed intervertebral disc extrusion at the level of L5/6. Decompressive surgery was performed at two locations.

The site of maximal compression was dorsal to the affected intervertebral disc space or a maximum of 0.5 vertebral body lengths away from it in 85 of 104 dogs with EEH (82%) and 41 of 44 non-EEH dogs (93%). Maximal compression at 1–2.5 vertebral body lengths away from the affected intervertebral disc space was seen in 19 of 104 dogs with EEH (18%) and 3 of 44 dogs of the non-EEH group (7%). Maximal compression of 1.5–2.5 vertebral body lengths away from the affected intervertebral disc was only seen in 7 dogs with EEH (7%). Maximal compression was at the intervertebral disc space in 18 of 44 dogs (40%) in the non-EEH group and 46 of 104 dogs (44%) in the EEH group. Maximal compression was cranial to the affected intervertebral disc space in 20 of 44 dogs (45%) in the non-EEH group and 43 of 104 dogs (41%) of the EEH group. Maximal compression was caudal to the intervertebral disc space in 6 of 44 (14%) non-EEH dogs and 15 of 104 (14%) dogs with EEH.

Lateralization of compression was seen in 99 of 104 dogs with EEH (95%) and 38 of 44 dogs of the non-EEH group (86%). Spinal cord compression without lateralization was seen in a total of 11 dogs. Six of these dogs were in the non-EEH group and showed central compression from the ventral. Five of these dogs belonged to the EEH group: four of them had circumferential compression and one dog had central compression from the dorsal.

3.6. Intramedullary Hemorrhage

Intramedullary hemorrhage was identified in seven dogs with EEH (6%) and one dog without EEH (2%). Two of the seven dogs of the EEH group were DPN. Both did not regain deep pain perception after surgery at the time of discharge. The remaining dogs were DPP.

3.7. Decompressive Surgery

Decompressive spinal surgery was performed in 134 of the 148 (90%) French Bulldogs; 14 dogs did not undergo surgery (5 non-EEH, 9 EEH). Decompressive surgery at one location was performed in 85 dogs (64%): 32 non-EEH and 53 EEH dogs. Decompressive surgery at more than one intervertebral disc space was performed in 49 dogs (36%): 42 EEH and 7 non-EEH. EEH was associated with decompressive surgery at more than one location ($p = 0.015$).

3.8. Recovery of Deep Pain Perception

Of the 27 DPN, 4 were euthanized before surgery and the remaining 23 underwent decompressive surgery: 18 dogs belonged to the EEH group, of which 9 did not regain deep pain perception until the time of discharge (range: 1–14 days); 5 dogs belonged to the non-EEH group, of which 4 did not regain deep pain perception. Ten dogs regained deep pain (nine EEH/one non-EEH) until the time of discharge (range: 1–10 days).

4. Discussion

In this retrospective study, 148 French Bulldogs with thoracolumbar intervertebral disc extrusion were assessed for extensive epidural hemorrhage (EEH). We found it especially important to include a large population of French Bulldogs because of the contradicting information in the literature regarding the association of the site of thoracolumbar intervertebral disc extrusion and EEH. A recent study of twelve French Bulldogs reported epidural hemorrhage being more commonly seen in this breed, particularly with intervertebral disc extrusion between T12 and L2 [8], while in other studies with various other dog breeds but including French Bulldogs, an association of epidural hemorrhage with a lumbar or caudal lumbar location of intervertebral disc extrusion has been reported [9]. French Bulldogs are a common dog breed in this country and are overrepresented in the hospital population. The subjective impression at our institution was that extensive epidural hemorrhage is often observed with caudal lumbar intervertebral extrusion, but the results of the present study reveal that this subjective impression may arise from the overall high incidence of lumbar intervertebral disc extrusion. In this study, the most common location of interverbal disc extrusion among all thoracolumbar sites was at L4/L5. This is of clinical interest and appears to be breed-specific as, in the general dog population, disc extrusion at the thoracolumbar junction is the most common [8,10–12].

In a recent study comparing dachshunds and French Bulldogs concerning malformation and disc extrusions, it was shown that intervertebral disc extrusion cranial to the thoracolumbar junction in French Bulldogs is rare, possibly because of the intercapital ligament, which may contribute to preventing intervertebral disc extrusion [13]. Concerning EEH, we found the highest incidence of EEH across all thoracolumbar sites at L1/L2 (84%), but this was closely followed by L5/L6 (82%), with little difference between the two locations. Concerning EEH at the thoracic spine, it needs to be considered that in this study group of 148 French Bulldogs, only 9 dogs had intervertebral disc extrusion in the thoracic spine, 6 of them with EEH. The small number of dogs with thoracic intervertebral disc extrusion does not allow for robust statistical analysis but demonstrates that there is only a small difference in the incidence of EEH between thoracic and lumbar intervertebral disc extrusions. Across all lumbar sites, there was no difference in incidence of EEH and no difference between the cranial and caudal lumbar spine.

The location of intervertebral disc extrusion can be an important risk factor for the development of progressive myelomalacia (PMM), especially in patients with lumbar intumescence intervertebral disc extrusions. Previous studies have shown that French

Bulldogs may be at a higher risk of the development of PMM, and it could be speculated that this may be associated with the increased prevalence of caudal lumbar intervertebral disc extrusions seen in our study and others and the vascular anatomy in this area [14,15].

We found EEH in 104 dogs (70%). This number is marginally higher than expected from the reported incidences of epidural hemorrhage in French Bulldogs, which range from 41–66% [7,8].

Comparison of the incidence of EEH between different studies is difficult as not all studies specify the term epidural hemorrhage or extensive epidural hemorrhage [1,2,7,8]. We observed in almost every French Bulldog in this study some amount of epidural hemorrhage in MRI. Similar observations were made during decompressive surgery of thoracolumbar intervertebral disc extrusion in French Bulldogs, but not for cervical intervertebral disc extrusion [12]. Therefore, it seems important to differentiate between epidural hemorrhage and extensive epidural hemorrhage. In recent publications, extensive epidural hemorrhage is defined as an extension of compressive epidural hemorrhage over more than two consecutive vertebral disc spaces, which takes into account that decompressive surgery might be necessary at more than one site. As the contact area of the hemorrhage to the dura seems to play a role in triggering inflammation, it is therefore considered a possible prognostic factor [3,4]. We used the same definition as Woelfel regarding the extent of epidural hemorrhage, but, in our definition, the epidural did not necessarily need to be compressive. Applying this definition, we found a high incidence of EEH in the present study group of French Bulldogs, which is in line with previous publications, that point out the higher incidence of EEH in French Bulldogs with thoracolumbar intervertebral disc extrusion compared to other breeds. High incidences are also reported for Pitbull Terriers, Cocker Spaniels, Beagles, Labrador Retrievers, and Retriever Mixes, which have similar percentages of intervertebral disc disease with epidural hemorrhage [3,8]. Only in Dachshunds has the percentage of epidural hemorrhage been reported to be low with acute disc disease (11%) [7].

In humans, the location of spontaneous spinal epidural hematoma is described to be posterior (the equivalent to dorsal in veterinary medicine) or posterolateral within the vertebral canal, where anterior and circumferent distribution is much less common [16]. In the present study, we found tendencies of a dorsal involvement of the vertebral canal in epidural hemorrhage. In almost all dogs with EEH, at least one dorsal quadrant was involved. The involvement of both dorsal quadrants was significantly associated with EEH. Also, the finding of no involvement of ventral quadrants was only observed in the EEH group. The involvement of dorsal quadrants could help to identify epidural hemorrhage on imaging, especially if no susceptibility-weighted or CT images are available.

In most dogs, the site of maximal compression was directly dorsal to the affected intervertebral disc space. However, this showed that maximal compression cranial or caudal to the affected intervertebral disc space was more common in dogs with EEH compared to dogs without EEH. Material distributed at a distance of 1.5 to 2.5 vertebral body lengths was only observed in dogs with EEH. It remains unclear why the number of dogs with maximal compression cranial to the disc space was much higher than the number of dogs with maximal compression caudal to the affected intervertebral disc space.

In the present study, intramedullary signal voids indicating intramedullary hemorrhage were only seen in 6% of the dogs. Intramedullary GRE signal void has been associated with a significantly reduced likelihood of regaining nociception in deep pain-negative paraplegic dogs affected by intervertebral disc extrusion [17], underlining the importance of including this sequence in spinal MRI protocols for intervertebral disc extrusion [18]. Although a higher field strength should theoretically increase the free induction decay in T2*-weighted sequences and therefore improve the sensitivity for susceptibility artefacts, we noticed difficulties with T2*-weighted sequences at 3 T due to artifacts from breathing and bone and lung tissue that may limit the visibility of signal void within the vertebral canal. Recent advancements suggest that incorporating susceptibility-weighted imaging (SWI) or 2D spoiled gradient echo multi-echo sequence with magnetization transfer saturation pulse (MEDIC) into standard protocols may offer superior visualization of hemorrhagic lesions and venous structures compared to traditional T2* sequences [19–21].

5. Limitations

Limitations of the present study are the retrospective study design; records of surgical findings or histological confirmation of the extruded material were not part of the inclusion criteria. Possibly, extruded disc material might have been misdiagnosed as hemorrhage based on MRI, possibly overestimating the number of dogs with EEH. Additionally, MRI interpretation might have failed in correctly identifying the affected intervertebral disc space, leading to potentially incorrect results, which may limit the understanding of the overall disease process. Furthermore, while most MRI scans were performed using a 1.0 T scanner, three different magnets were used in this study, potentially introducing variability in imaging quality and the detection of susceptibility artifacts. The retrospective nature of this study did not allow us to examine previous corticosteroid administration and its influence on recovery, nor could a long-term outcome be analyzed for both groups.

6. Conclusions

L4/L5 was found to be the most common site of intervertebral disc extrusion in our large population of 148 French Bulldogs that presented with acute paraparesis/paraplegia and with an MRI diagnosis compatible with thoracolumbar intervertebral disc extrusion. A total of 70% of the dogs had imaging findings consistent with EEH. L1/L2 had the highest prevalence of EEH (84%), but no statistically significant difference was seen between the various thoracic and lumbar intervertebral disc extrusion sites and the prevalence of EEH.

Author Contributions: D.S. initiated the project and was responsible for conceptualization. J.U. and D.S. contributed to all parts of the project and participating in the methodology, data acquisition, and organization. They were involved in data collection, validation, interpretation, and the preparation of figures and graphs. J.U. was responsible for the original draft preparation. A.M., M.W. and D.S. were involved in writing, reviewing, and editing the manuscript. They approved the final version of this article. This work was supported by the Clinical Research Platform, Vetsuisse Faculty Bern, University of Bern, Switzerland. All authors have read and agreed to the published version of the manuscript.

Funding: This research received no external funding.

Institutional Review Board Statement: This research was a retrospective study and was therefore exempt from ethical review according to guidelines from the Canton of Bern, Switzerland.

Informed Consent Statement: No animals or people are identifiable within this publication and, therefore, additional informed consent for publication was not required.

Data Availability Statement: The raw data supporting the conclusions of this article are available from the authors upon request.

Acknowledgments: The authors thank the current and previous team of the Division of Clinical Radiology of the Veterinary Faculty of the University of Bern for acquiring the images.

Conflicts of Interest: No external companies had any role in the writing of, or in the decision to publish, the research presented. All the research was performed by the authors independently.

References

1. Tartarelli, C.L.; Baroni, M.; Borghi, M. Thoracolumbar Disc Extrusion Associated with Extensive Epidural Haemorrhage: A Retrospective Study of 23 Dogs. *J. Small Anim. Pract.* **2005**, *46*, 485–490. [CrossRef] [PubMed]
2. Fenn, J.; Olby, N.J. Classification of Intervertebral Disc Disease. *Front. Vet. Sci.* **2020**, *7*, 579025. [CrossRef] [PubMed]
3. Woelfel, C.W.; Robertson, J.B.; Mariani, C.L.; Muñana, K.R.; Early, P.J.; Olby, N.J. Outcomes and Prognostic Indicators in 59 Paraplegic Medium to Large Breed Dogs with Extensive Epidural Hemorrhage Secondary to Thoracolumbar Disc Extrusion. *Vet. Surg.* **2021**, *50*, 527–536. [CrossRef] [PubMed]
4. Fadda, A.; Oevermann, A.; Vandevelde, M.; Doherr, M.G.; Forterre, F.; Henke, D. Clinical and Pathological Analysis of Epidural Inflammation in Intervertebral Disk Extrusion in Dogs. *J. Vet. Intern. Med.* **2013**, *27*, 924–934. [CrossRef]

5. Parker, H.G.; VonHoldt, B.M.; Quignon, P.; Margulies, E.H.; Shao, S.; Mosher, D.S.; Spady, T.C.; Elkahloun, A.; Cargill, M.; Jones, P.G.; et al. An Expressed Fgf4 Retrogene Is Associated with Breed-Defining Chondrodysplasia in Domestic Dogs. *Science* **2009**, *325*, 995–998. [CrossRef]
6. Brown, E.A.; Dickinson, P.J.; Mansour, T.; Sturges, B.K.; Aguilar, M.; Young, A.E.; Korff, C.; Lind, J.; Ettinger, C.L.; Varon, S.; et al. FGF4 Retrogene on CFA12 Is Responsible for Chondrodystrophy and Intervertebral Disc Disease in Dogs. *Proc. Natl. Acad. Sci. USA* **2017**, *114*, 11476–11481. [CrossRef]
7. Poli, F.; Calistri, M.; Meucci, V.; Di Gennaro, G.; Baroni, M. Prevalence, Clinical Features, and Outcome of Intervertebral Disc Extrusion Associated with Extensive Epidural Hemorrhage in a Population of French Bulldogs Compared to Dachshunds. *J. Vet. Med. Sci.* **2022**, *84*, 1307–1312. [CrossRef]
8. Bridges, J.; Windsor, R.; Stewart, S.D.; Fuerher-Senecal, L.; Khanna, C. Prevalence and Clinical Features of Thoracolumbar Intervertebral Disc-Associated Epidural Hemorrhage in Dogs. *J. Vet. Intern. Med.* **2022**, *36*, 1365–1372. [CrossRef]
9. Mateo, I.; Lorenzo, V.; Foradada, L.; Muñoz, A. Clinical, Pathologic, And Magnetic Resonance Imaging Characteristics Of Canine Disc Extrusion Accompanied By Epidural Hemorrhage Or Inflammation. *Vet. Radiol. Ultrasound* **2011**, *52*, 17–24. [CrossRef]
10. Macias, C.; Mckee, W.M.; May, C.; Innes, J.F. Thoracolumbar Disc Disease in Large Dogs: A Study of 99 Cases. *J. Small Anim. Pract.* **2002**, *43*, 439–446. [CrossRef]
11. Aikawa, T.; Fujita, H.; Kanazono, S.; Shibata, M.; Yoshigae, Y. Long-Term Neurologic Outcome of Hemilaminectomy and Disk Fenestration for Treatment of Dogs with Thoracolumbar Intervertebral Disk Herniation: 831 Cases (2000–2007). *J. Am. Vet. Med. Assoc.* **2012**, *241*, 1617–1626. [CrossRef] [PubMed]
12. Albertini, G.M.; Stabile, F.; Marsh, O.; Uriarte, A. Clinical, Magnetic Resonance Imaging, Surgical Features and Comparison of Surgically Treated Intervertebral Disc Extrusion in French Bulldogs. *Front. Vet. Sci.* **2023**, *10*, 1230280. [CrossRef] [PubMed]
13. Aikawa, T.; Shibata, M.; Asano, M.; Hara, Y.; Tagawa, M.; Orima, H. A Comparison of Thoracolumbar Intervertebral Disc Extrusion in French Bulldogs and Dachshunds and Association with Congenital Vertebral Anomalies. *Vet. Surg.* **2014**, *43*, 301–307. [CrossRef] [PubMed]
14. Olby, N.J.; Moore, S.A.; Brisson, B.; Fenn, J.; Flegel, T.; Kortz, G.; Lewis, M.; Tipold, A. ACVIM Consensus Statement on Diagnosis and Management of Acute Canine Thoracolumbar Intervertebral Disc Extrusion. *J. Vet. Intern. Med.* **2022**, *36*, 1570–1596. [CrossRef] [PubMed]
15. Balducci, F.; Canal, S.; Contiero, B.; Bernardini, M. Prevalence and Risk Factors for Presumptive Ascending/Descending Myelomalacia in Dogs after Thoracolumbar Intervertebral Disk Herniation. *J. Vet. Intern. Med.* **2017**, *31*, 498–504. [CrossRef]
16. Peng, D.; Yan, M.; Liu, T.; Yang, K.; Ma, Y.; Hu, X.; Ying, G.; Zhu, Y. Prognostic Factors and Treatments Efficacy in Spontaneous Spinal Epidural Hematoma: A Multicenter Retrospective Study. *Neurology* **2022**, *99*, E843–E850. [CrossRef]
17. Clark, R.; Ferreira, A.; Behr, S. Significance of Intramedullary T2* Signal Voids in the Magnetic Resonance Imaging of Paraplegic Deep Pain-Negative Dogs Following Intervertebral Disc Extrusion at Short-Term Follow-Up. *Front. Vet. Sci.* **2023**, *10*, 1248024. [CrossRef]
18. Packer, R.A.; Rossmeisl, J.H.; Kent, M.S.; Griffin, J.F.; Mazcko, C.; LeBlanc, A.K. Consensus Recommendations on Standardized Magnetic Resonance Imaging Protocols for Multicenter Canine Brain Tumor Clinical Trials. *Vet. Radiol. Ultrasound* **2018**, *59*, 261–271. [CrossRef]
19. Wang, M.; Dai, Y.; Han, Y.; Haacke, E.M.; Dai, J.; Shi, D. Susceptibility Weighted Imaging in Detecting Hemorrhage in Acute Cervical Spinal Cord Injury. *Magn. Reson. Imaging* **2011**, *29*, 365–373. [CrossRef]
20. Weston, P.; Morales, C.; Dunning, M.; Parry, A.; Carrera, I. Susceptibility Weighted Imaging at 1.5 Tesla Magnetic Resonance Imaging in Dogs: Comparison with T2*-Weighted Gradient Echo Sequence and Its Clinical Indications. *Vet. Radiol. Ultrasound* **2020**, *61*, 566–576. [CrossRef]
21. Held, P.; Dorenbeck, U.; Seitz, J.; Fründ, R.; Albrich, H. MRI of the Abnormal Cervical Spinal Cord Using 2D Spoiled Gradient Echo Multiecho Sequence (MEDIC) with Magnetization Transfer Saturation Pulse. A T2* Weighted Feasibility Study. *J. Neuroradiol.* **2003**, *30*, 83–90. [PubMed]

veterinary sciences

MDPI

Article

Feline Infectious Peritonitis Effusion Index: A Novel Diagnostic Method and Validation of Flow Cytometry-Based Delta Total Nucleated Cells Analysis on the Sysmex XN-1000V®

Ricardo Lopes [1,2,3,*], Filipe Sampaio [3,4], Hugo Lima de Carvalho [3], Andreia Garcês [5,6], Cátia Fernandes [7], Carolina Vitória Neves [3], Alexandre Sardinha de Brito [3], Tiago Marques [8], Carlos Sousa [9], Ana Rita Silva [9], Ângela Martins [6], Luís Cardoso [1,6], Ana Cláudia Coelho [1,6] and Elsa Leclerc Duarte [10,11]

1 Department of Veterinary Sciences, University of Trás-os-Montes e Alto Douro (UTAD), 5000-801 Vila Real, Portugal; lcardoso@utad.pt (L.C.); accoelho@utad.pt (A.C.C.)
2 Department of Veterinary and Animal Sciences, University Institute of Health Sciences (IUCS), CESPU, 4585-116 Gandra, Portugal
3 CEDIVET Veterinary Laboratories, Lionesa Business Hub, R. Lionesa 446 C24, 4465-671 Leça do Balio, Portugal; filipe.sampaio@cedivet.pt (F.S.); hugo.carvalho@cedivet.pt (H.L.d.C.); carolina.neves@cedivet.pt (C.V.N.); alexandresardbrito@gmail.com (A.S.d.B.)
4 Cytology and Hematology Diagnostic Services, Laboratory of Histology and Embryology, Department of Microscopy, ICBAS-School of Medicine and Biomedical Sciences, University of Porto (U.Porto), Rua de Jorge Viterbo Ferreira, 228, 4050-313 Porto, Portugal
5 Wildlife Rehabilitation Centre (CRAS), Veterinary Teaching Hospital, University of Trás-os-Montes e Alto Douro (UTAD), 5000-801 Vila Real, Portugal; andreiamvg@gmail.com
6 Animal and Veterinary Research Centre (CECAV), Associate Laboratory for Animal and Veterinary Sciences (AL4AnimalS), University of Trás-os-Montes e Alto Douro (UTAD), 5000-801 Vila Real, Portugal; angela@utad.pt
7 Anicura Santa Marinha Veterinary Hospital, R. Dom Henrique de Cernache 183, 4400-625 Vila Nova de Gaia, Portugal; catia.fernandes@anicura.pt
8 Infectious Diseases Department, Santa Maria Hospital, Northern Lisbon University Hospital Centre (CHULN), 1649-035 Lisboa, Portugal; tiagomarques@medicina.ulisboa.pt
9 Molecular Diagnostics Laboratory, Unilabs Portugal, Centro Empresarial Lionésa Porto, Rua Lionesa, 4465-671 Leça do Balio, Portugal; carlos.sousa@unilabs.com (C.S.); ana.rita.silva@unilabs.com (A.R.S.)
10 Department of Veterinary Medicine, School of Science and Technology, University of Évora, Polo da Mitra, Apartado 94, 7002-554 Évora, Portugal; emld@uevora.pt
11 Mediterranean Institute for Agriculture, Environment and Development (MED), Global Change and Sustainability Institute (CHANGE), University of Évora, Polo da Mitra, Apartado 94, 7002-554 Évora, Portugal
* Correspondence: lopes.rmv@gmail.com

Simple Summary: Feline infectious peritonitis (FIP) is a severe and even fatal inflammatory disease of cats, caused by feline coronavirus (FCoV), and remains an important diagnostic challenge. This study validates the flow cytometry-based delta total nucleated cells (ΔTNC) on the Sysmex XN-1000V® in effusion samples and introduces the FIP Effusion Index, a novel diagnostic method that integrates cellular and inflammatory markers of FIP in effusions. The FIP Effusion Index demonstrated improved diagnostic accuracy, achieving 96.3% sensitivity and 95.7% specificity for values of 5.06 or higher, and perfect specificity (100%) with 96.3% sensitivity at values of 7.54 or above. This combined approach outperforms traditional methods, providing a superior, rapid, and cost-effective tool for the accurate and timely FIP diagnosis. It offers considerable potential for enhancing the management of suspected FIP cases in clinical practice, with potential applications in both veterinary and human medicine for related coronavirus diseases.

Abstract: The emergence of severe acute respiratory syndrome coronavirus 2 (SARS-CoV-2) has led the medical and scientific community to explore the pathogenesis and clinical manifestations of coronaviruses. In felines, a widespread coronavirus known as feline coronavirus (FCoV) can lead to feline infectious peritonitis (FIP), a highly fatal disease characterised by severe systemic inflammation. Diagnosing FCoV remains challenging due to the limited accuracy of the available methods. The present study introduces the FIP Effusion Index, a novel diagnostic method that combines the albumin-to-globulin (ALB/GLOB) ratio with the delta total nucleated cell (ΔTNC) count obtained via flow cytometry using the Sysmex XN-1000V® analyser in effusions. Samples from cats (n = 50) with suspected FIP were analysed for ΔTNC, with findings showing that a ΔTNC ≥ 2.1 is highly indicative of FIP and a ΔTNC ≥ 4.9 can be considered diagnostic. The FIP Effusion Index enhanced diagnostic precision in our group of samples, achieving 96.3% sensitivity and 95.7% specificity for values ≥ 5.06, and reaching perfect specificity (100%) with 96.3% sensitivity for values ≥ 7.54. This combined approach surpasses the accuracy of individual parameters, establishing the FIP Effusion Index as a superior diagnostic tool for FIP, with potential applications in both veterinary and human medicine for related coronavirus diseases.

Keywords: clinical pathology; coronavirus; delta total nucleated cells (ΔTNC); effusions; feline infectious peritonitis (FIP); flow cytometry; FIP Effusion Index; One Health; Stockholm paradigm; zoonosis

1. Introduction

Coronaviruses (CoVs) constitute a large family of RNA viruses that infect a broad range of mammals including humans avian species, livestock and companion animals. Some cause severe disease, which is significantly challenging to public health, veterinary medicine, and the global economy [1–5]. The high diversity of these viruses facilitates the emergence of new variants with altered target cell preferences or expanded host ranges [6]. Such adaptations can result in cross-species transmission and zoonotic outbreaks. Recent examples of the latter are severe acute respiratory syndrome coronavirus (SARS-CoV), Middle East respiratory syndrome coronavirus (MERS-CoV), and the most recent severe acute respiratory syndrome coronavirus 2 (SARS-CoV-2) causing coronavirus infectious disease 2019 (COVID-19) [7–12].

The COVID-19 pandemic has highlighted the need for effective diagnostic model to better understand coronavirus-related diseases [13]. A notable characteristic of coronaviruses is their ability to alter tissue or cell tropism. The best prototypic example of such evolutionary adaptation is the feline coronavirus (FCoV), which illustrates how a virus can shift its pathogenic profile within a host or between hosts [8,12–14].

Feline infectious peritonitis (FIP), caused by FCoV, is a widespread and considerably fatal, severe systemic inflammatory disease in cats, caused by a mutated FCoV and its interaction with the host's immune response [15]. Although SARS-CoV-2 and FCoV belong to different genera within the *Coronaviridae* family—FCoV to *Alphacoronavirus* and SARS-CoV-2 to *Betacoronavirus*—both exhibit rapid disease progression and diverse extrapulmonary manifestations [16,17]. Cats are susceptible to SARS-CoV-2, and FCoV has also been shown to infect humans [12,18–21]. These similarities suggest that FIP could serve as a valuable model for studying coronavirus-induced pathology, potentially framing diagnostic strategies in both veterinary and human medicine [7,16,21–26].

FCoV strains are subdivided into two distinct biotypes: feline enteric coronavirus (FECV) and feline infectious peritonitis virus (FIPV). The FIPV biotype is more virulent than the FECV biotype and can cause peritonitis with a poor prognosis, while most FECV biotypes do not cause lesions [18,27,28]. Serotype 1 FCoV includes unique feline strains, while serotype 2 appears to have arisen from the recombination between type 1 FCoV and canine coronavirus (CCoV). Although serotype 1 is the most prevalent worldwide, serotypes 1 and 2 can cause FIP. In cats with FIP, the virus replicates to high titres in monocytes and can be found in multiple organs. In subclinical infected cats, on the other hand, FCoV is mainly confined to the intestine. Even pathogenic strains of the FECV biotype can cause only mild enteritis because of their very low virulence [18,29–32].

The pathogenesis of FIP involves complex interactions including two amino acid substitutions (M1058L and S1060A) in the spike protein of the virus, which are associated with the systemic spread of FCoV rather than the direct causation of FIP. These mutations highlight the diagnostic challenge, as current tests are unable to properly differentiate between enteric and pathogenic FCoV strains, emphasising the need to improve diagnostic tools for this purpose, which offer higher sensitivity and specificity [27,29,33–37].

Diagnosing FIP antemortem presents considerable challenges, particularly in the non-effusive ("dry"/parenchymatous) form, due to its variable clinical manifestations and the low specificity of many laboratory tests. The effusive ("wet"/non-parenchymatous) form of FIP is easier to diagnose because it presents more clearly through patient history, clinical signs, and the analysis of effusions, which are typically yellow, turbid, viscous, and often contain fibrin strands [30,38–40].

It is worth mentioning that, although the distinction between effusive and non-effusive forms may provide some assistance in the diagnostic approach, they do not represent two distinct entities of FIP. It is common for cats that initially present with the non-effusive form to eventually develop effusions in the later stages of the disease [30,41].

These effusions typically have a high protein content with a decreased albumin-to-globulin (ALB/GLOB) ratio. Cell counts range from 2–6 $\times$ $10^3/\mu L$, sometimes up to 30 $\times$ $10^3/\mu L$, and the cytologic pattern, which is only highly suggestive but not definitive diagnostic for FIP, mostly consists of nondegenerate neutrophils, macrophages, lymphocytes, and rare plasma cells in a prominent proteinaceous background [42,43].

Current diagnostic methods, such as cytological analysis, real-time quantitative reverse transcription polymerase chain reaction (real-time RT-qPCR), serum protein electrophoresis, alpha-1-acid glycoprotein (AGP) measurement, FCoV antigen detection, and others, especially when combined, can strongly suggest FIP but do not provide a definitive diagnosis [28,38,42,44]. Despite innovations in the field, an ideal diagnosis for FIP still does not exist. Detection of the FCoV antigen by immunohistochemistry (IHC) on histopathological abnormal tissue obtained either post–mortem or via laparotomy/laparoscopy remains the gold standard of diagnosis [38,40].

PCR is a highly sensitive and specific technique for detecting DNA or RNA sequences, making it effective for diagnosing infectious diseases including coronaviruses, where it is commonly used as part of routine testing [45]. RT-qPCR is particularly useful for RNA viruses like FCoV, as it converts RNA to complementary DNA (cDNA) and amplifies specific sequences in real-time, allowing for the rapid detection and quantification of viral load, which is crucial for early FIP diagnosis and management [46–48]. However, RT-qPCR's high sensitivity and specificity come with considerable costs for specialised equipment, reagents, and skilled personnel, limiting its routine use in some veterinary practices. This highlights the need for more affordable and complementary diagnostic methods such as the Rivalta test, ALB/GLOB ratio, and delta total nucleated cell (ΔTNC) count on effusions [30,36].

The Rivalta test, a simple and cost-effective assay, has gained recognition for its high diagnostic value in FIP. The test relies on the precipitation of proteins such as fibrinogen when in contact with acetic acid, a reaction particularly noticeable in FIP-related effusions [36,42,49]. However, this reaction is not exclusive to FIP and can also occur in cases of bacterial peritonitis, pleuritis, or lymphoma. Therefore, it is essential to combine the Rivalta test with cytological analysis to achieve a more accurate diagnosis [42,50].

The ALB/GLOB ratio is a practical and low-cost parameter for evaluating FIP effusions, providing valuable insights into the protein alterations associated with the disease. A decreased ratio suggests increased globulin and decreased albumin levels, indicating an inflammatory response typical of FIP and helping to differentiate it from other conditions [30].

Recent studies have highlighted the utility of combining tests with high positive likelihood ratios (LR+), such as the ΔTNC count on effusions, measured using the Sysmex XN-1000V® analyser. This method offers an innovative approach by quantifying the difference in total nucleated cell counts between specific channels (TNCC-WDF and TNCC-WNR). This accuracy is attributed to the formation of clots in the WNR reagent, which trap cells similarly to the mechanism observed in the Rivalta test, a method also noted for its high diagnostic accuracy for FIP. When used on the Sysmex XN-1000V®, this method has demonstrated promising results comparable to those obtained with the Sysmex XT–2000iV® [36,43].

The FIP Effusion Index integrates two independent variables, each providing unique but complementary insights into the nature of the effusion: the ΔTNC and the ALB/GLOB ratio. By integrating cellular and inflammatory responses with protein changes characteristic of FIP, the FIP Effusion Index aims to enhance diagnostic accuracy, reduce the likelihood of misdiagnosis, and provide a rapid and cost-effective approach for managing suspected FIP cases in clinical practice.

The present study aimed to address three primary objectives in accordance with the Standards for Reporting of Diagnostic Accuracy (STARD) guidelines [51]: (1) to determine the diagnostic accuracy of ΔTNC for FIP effusions using the Sysmex XN-1000V®, with real-time RT-qPCR as the reference method; (2) to compare the diagnostic accuracy of ΔTNC-XN with the Rivalta test and the ALB/GLOB ratio; and (3) to introduce a new diagnostic method, the FIP Effusion Index, which correlates ΔTNC and the ALB/GLOB ratio to minimise false-positive and false-negative outcomes, and evaluate its diagnostic accuracy in diagnosing FIP.

2. Materials and Methods

2.1. Data Collection, Sampling, and Diagnostic Procedures

Effusion samples from suspected cases of FIP were submitted to CEDIVET Veterinary Laboratories (Porto, Portugal). These samples were submitted from veterinary practices including clinics and hospitals from the north region of mainland Portugal. Along with the samples, a laboratory requisition was received, which included the complete clinical information for each cat, particularly their breed, sex, age, vaccination and prophylactic status, clinical suspicion/clinical signs, and requested analyses.

Although a large number of samples were submitted and all were fully analysed in accordance with the methods outlined below, only 50 samples fulfilled all the inclusion criteria. The cases selected for data analysis were based on the following inclusion and exclusion criteria:

Inclusion criteria:

1. Only samples from cats with effusions reported by veterinary practitioners who were clinically suspected of FIP (e.g., lethargy, anorexia, weight loss, fever, jaundice, dyspnoea, abdominal lymphadenopathy, thickened intestinal walls, renomegaly, uveitis, vasculitis/phlebitis, retinal changes, multifocal neurologic signs) were included in this study.
2. Availability of laboratory requisition with the relevant clinical information such as clinical suspicion/clinical signs and requested analyses.
3. Availability of complete documentation of physicochemical analysis of the effusion including specific gravity (SG), protein content, absolute counts of white blood cells and total nucleated cells in body fluid, and total nucleated cells derived from the ratio between specific channel (TNCC-WDF and TNCC-WNR) counts on the Sysmex XN-1000V®.

Exclusion criteria:

1. Absence of complete clinical documentation: Samples lacking a comprehensive laboratory requisition or clinical information including clinical suspicion/clinical signs and requested analyses were excluded.
2. Lack of confirmed FCoV real-time RT-qPCR diagnosis: Cases without a confirmed diagnosis of FCoV through real-time RT-qPCR testing were excluded from the study.
3. Incomplete physicochemical analysis data: Effusion samples without full documentation of physicochemical analysis such as SG, protein content, absolute counts of white blood cells and total nucleated cells in body fluid, and the total nucleated cells derived from the ratio between the TNCC-WDF and TNCC-WNR counts on the Sysmex XN-1000V® were not included.

4. Non-compliance with sampling protocols: Samples collected in non-EDTA tubes or samples submitted beyond the 18–24 h window post-collection were excluded to maintain the integrity of the cell count analysis.
5. Samples with cytological findings that were septic, unclear, or inconclusive were excluded from the study.
6. Previous treatments influencing effusion composition: Samples from cats that had undergone treatments known to affect the composition of effusions, such as recent chemotherapy or immunosuppressive therapy, were excluded to avoid confounding results.
7. Insufficient sample volume: Effusions with insufficient volume to perform all required analyses were excluded to ensure that all parameters could be adequately measured.

Based on these criteria, cats were classified as affected by FIP if the effusion showed elevated protein levels (>2.5 g/dL), the cytological analysis was indicative of FIP, and a positive FCoV real-time RT-qPCR result was obtained [30,52–54].

The age of the animals was categorised into five groups: kitten, <1 year old; young, 1 to <2 years old; adult, 2 to <6 years old; senior, 6 to <11 years old; and old, ≥11 years old.

2.2. Biochemical Analysis

The measurements of total protein (TP) and albumin (ALB) were performed using the Beckman Coulter AU680e Automated Chemistry Analyzer® (Beckman Coulter, Cassina De'Pecchi, Italy). Globulin levels were determined through indirect calculation. The effusion SG was measured using a clinical refractometer (Clinical Refractometer Mod. TS400, Sper Scientific, Scottsdale, AZ, USA).

2.3. Rivalta Test

To conduct the Rivalta test, 7–8 mL of distilled water was dispensed into a 10 mL plastic tube (Cat: 470332-708, VWR®, Radnor, PA, USA). A drop (20–30 µL) of acetic acid (98–100%) was then added using a disposable pipette (Merck, Darmstadt, Germany) and the mixture was thoroughly agitated. Subsequently, a drop (20–30 µL) of effusion fluid was carefully placed on top of the acetic acid solution with a second disposable pipette. The Rivalta test was deemed positive if a precipitate formed, adhered to the surface, retained its shape, or slowly settled to the bottom of the solution as a drop or jellyfish-like (Figure 1). A negative test was indicated by the drop dissipating (disappearing) and the solution remaining clear [49,55,56].

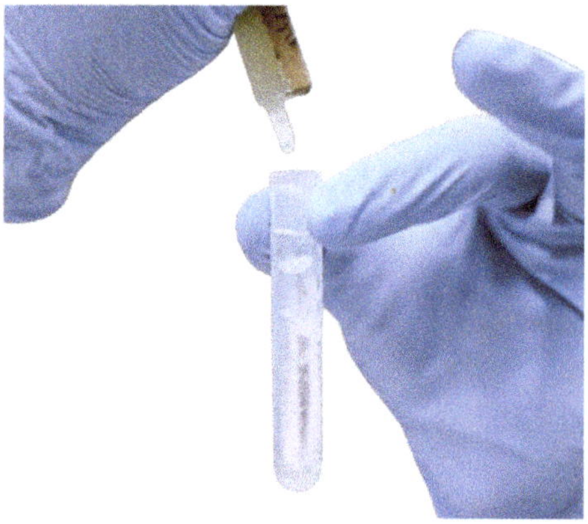

Figure 1. Positive Rivalta test. This is identified by the formation of a precipitate that adheres to the surface, retains its shape, or slowly settles to the bottom of the solution in the test tube. The image shows the precipitate clearly forming within the tube, indicating a positive result for the Rivalta test.

2.4. Flow Cytometry Analysis

The tests for the total white blood cell count in body fluid (WBC-BF), total nucleated cells in body fluid (TC-BF), total nucleated cells in the white blood cell differential channel in whole blood mode (TNCC-WDF), total nucleated cells in the white blood cell nucleated channel in whole blood mode (TNCC-WNR), and the ΔTNC ratio derived from the TNCC-WDF and TNCC-WNR counts were performed using the Sysmex XN-1000V® (Sysmex Europe GmbH, Norderstedt, Germany) (Figures 2–5). Confirmation of the effusion findings and cellular morphology was conducted through cytological evaluation using light microscopy, following staining with May-Grünwald Giemsa, on slides prepared by direct smear, squash preparation, and cytocentrifugation.

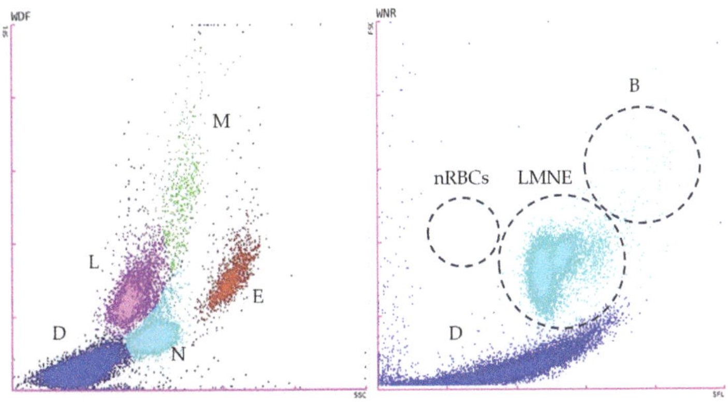

Figure 2. Scattergrams adapted from the Sysmex XN-1000V® in whole blood mode with the EDTA blood feline sample. The figure on the **left** displays the TNCC-WDF (total nucleated cells in the white blood cell differential channel), while the figure on the **right** illustrates the TNCC-WNR (total nucleated cells in the white blood cell nucleated channel). B, basophils; D (blue), debris; E (red), eosinophils; L (purple), lymphocytes; LMNE, all the WBC populations except basophils; M (green), monocytes; N (light blue), neutrophils; nRBCs, nucleated red blood cells.

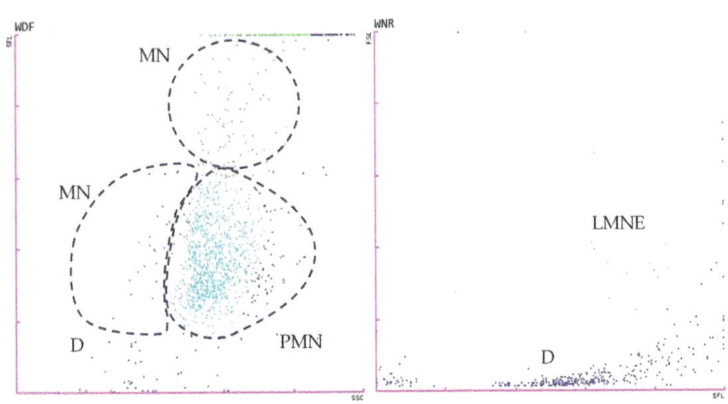

Figure 3. Scattergrams adapted from the Sysmex XN-1000V® in whole blood mode with effusion from a case of feline infectious peritonitis. The figure on the **left** displays the TNCC-WDF (total nucleated cells in the white blood cell differential channel), while the figure on the **right** illustrates the TNCC-WNR (total nucleated cells in the white blood cell nucleated channel). D (blue), debris; LMNE (light blue), all the WBC populations except basophils; MN, mononuclear cells (macrophages, monocytes, lymphocytes); PMN, polymorphonuclear cells (mainly neutrophils). Note that there was a marked reduction in events in the WNR channel.

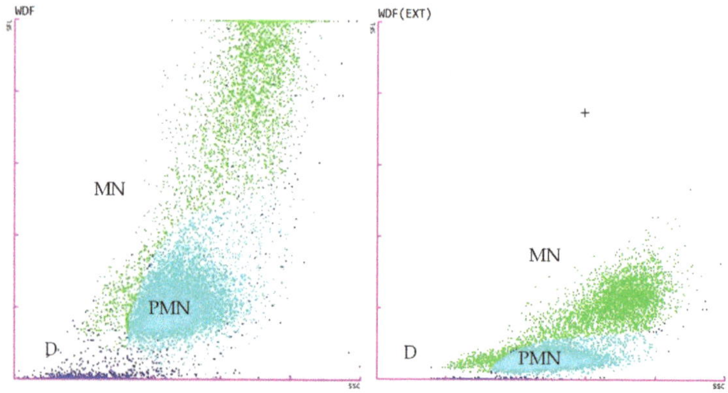

Figure 4. Scattergrams adapted from the Sysmex XN-1000V® in WBC-BF (total white blood cell count in body fluid mode) with effusion from a case of feline infectious peritonitis (FIP). The figure on the **left** displays the WDF-BF, while the figure on the **right** illustrates the WDF(EXT)-BF (extended). D (blue), debris (cell detritus, compound particles, other cells); MN (green), mononuclear cells (macrophages, monocytes, lymphocytes); PMN (light blue), polymorphonuclear cells (mainly neutrophils); +, high fluorescence events (epithelial cells, cell conglomerates).

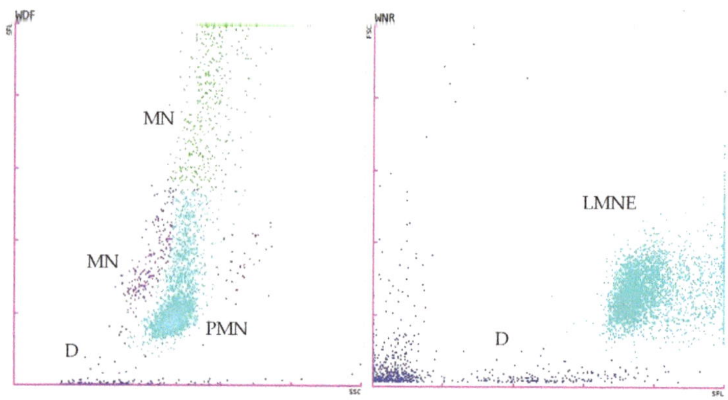

Figure 5. Scattergrams adapted from the Sysmex XN-1000V® in whole blood mode with and effusion from a non-feline infectious peritonitis (FIP) case. The figure on the **left** displays the TNCC-WDF (total nucleated cells in the white blood cell differential channel), while the figure on the **right** illustrates the TNCC-WNR (total nucleated cells in the white blood cell nucleated channel). D (blue), debris; LMNE (light blue), all the WBC populations except basophils; MN (purple and green), mononuclear cells (macrophages, monocytes, lymphocytes); PMN (light blue), polymorphonuclear cells (mainly neutrophils). Note that there was no reduction in events in the WNR channel.

2.5. FIP Effusion Index

The FIP Effusion Index is calculated using Formula (1):

$$\text{FIP Effusion Index} = \frac{\Delta\text{Total Nucleated Cells } (\Delta\text{TNC})}{\text{Albumin/Globulin (ALB/GLOB) ratio}}, \quad (1)$$

The FIP Effusion Index is calculated by dividing the delta total nucleated cell (ΔTNC) count by the albumin-to-globulin (ALB/GLOB) ratio. This formula integrates cellular and biochemical markers to enhance diagnostic accuracy in detecting FIP.

2.6. Effusion Classification

Effusions were classified according to the Traditional Classification Scheme [57,58], based on their protein concentration and total cell count. This scheme categorises effusions into transudate, modified transudates, and exudates, as exemplified in Table 1. The classification criteria include specific thresholds for protein levels and total cell counts, enabling a systematic approach to differentiating between types of effusions.

Table 1. Traditional Classification Scheme using criteria based on protein levels and cell counts.

Type	Total Protein (g/dL)	Total Cell Count (cells/μL)
Transudate	<2.5	<1500
Modified transudate	2.5–7.5	1000–7000
Exudate	>3.0	>7000

2.7. Molecular Analysis

The extraction of viral RNA from the effusion samples was conducted using the Promega AX9760 custom protocol for whole blood, applied through the KingFisher™ Flex system (Thermo Fisher Scientific, Waltham, MA, USA). The protocol was adapted for the extraction of nucleic acids from body fluid samples. All reagents were stable at room temperature and used under the manufacturer's specified conditions, with storage between 15–30 °C. The target RNA needed to be isolated free from contaminants such as proteins, cellular debris, and non-target nucleic acids. To achieve this, the samples were first treated with lysis buffer and Proteinase K. After an incubation period of approximately 27 min, a binding buffer mix containing MagnaCell™ Binding Buffer and paramagnetic resin was added. After the binding step, samples were washed with 80% ethanol to remove contaminants, and RNA was finally eluted using the elution buffer.

For the detection of FCoV in the effusion samples, molecular biology analyses were performed using the NZYTech Feline Coronavirus RT-qPCR Kit® (NZYTech, Lisboa, Portugal), optimised to provide the broadest possible detection profile while maintaining specificity for FCoV, with over 95% homology to a wide range of FCoV genomes based on comprehensive bioinformatic analysis of reference data from the NCBI database. The real-time RT-qPCR reaction was configured according to the manufacturer. For each reaction, an internal extraction control was included. A negative control and a positive control were also included.

The amplification reaction was carried out on a QuantStudio™ 5 real-time PCR System (Thermo Fisher Scientific, Waltham, MA, USA), following the thermal cycling conditions suggested in the manufacturer's protocol (Table 2).

Table 2. Thermal cycling conditions used for the detection of FCoV in the effusion samples.

Cycles	Temperature	Time	Notes
1	50 °C	20 min	Reverse transcription
1	95 °C	2 min	Polymerase activation
40	95 °C	5 s	Denaturation
	60 °C	30 s	Annealing/Extension

During the amplification process, both primers and probes specifically annealed to a designated target region of the FCoV genome. The fluorogenic probe, consisting of a DNA sequence tagged with a fluorescent dye at the 5′ end and a quencher molecule at the 3′ end, was cleaved during the PCR amplification process. This cleavage caused the dye and quencher to separate, leading to an increase in fluorescence intensity. The rise in fluorescence was subsequently measured by the instrument, indicating successful target amplification.

For the final validation of the results, the following criteria had to be met:

- Amplification of the internal extraction control;
- No amplification of the negative control;
- Amplification of the positive control.

Only results meeting these criteria were validated as negative (no amplification of the target), positive (true amplification of the target, indicated by a sigmoidal curve), or invalid (if there was no amplification of the target and no amplification of the internal extraction control, and therefore excluded from the study).

2.8. Statistical Analysis

All the data were available in digital format in Sislab® version 23R12.1.1.1 (Glintt, Global Intelligent Technologies, S.A., Lisboa, Portugal) and transferred to Microsoft Excel® version 2410 (Microsoft, Redmond, WA, USA) sheets. Statistical analysis was conducted using the JMP®, version 14.3 (SAS Institute, Cary, NC, USA, 1989–2023), DATAtab® (DATAtab: Online Statistics Calculator. DATAtab e.U. Graz, Austria, 2024), and MedCalc® Statistical Software version 20.006 (MedCalc Software Ltd., Ostend, Belgium, 2021). Non-parametric tests were employed to study the differences between the observed and expected frequencies of categories within a field including the binomial test, the one-sample Chi-square test, the chi-square test of independence, and Fisher's exact test, depending on the number of categories in the categorical field. For comparisons among three or more independent groups, the Kruskal–Wallis test was used, followed by the Dunn–Bonferroni post hoc test for multiple comparisons when appropriate. The Cochran–Armitage trend test was also used to assess trends across ordered categories. Odds ratios (ORs) were calculated to evaluate the likelihood of a positive Rivalta test in cats with FIP compared to those without, providing a measure of association between the presence of FIP and the test outcomes. To evaluate the diagnostic accuracy of the ALB/GLOB ratio, and the ΔTNC count, derived from the ratio of TNCC-WDF to TNCC-WNR on the Sysmex XN-1000V®, a mixed-model ANOVA was performed on the biochemical and flow cytometry data, and receiver operating characteristic (ROC) curve analysis was utilised to further assess the diagnostic effectiveness of ΔTNC and the FIP Effusion Index in identifying FIP in the effusion samples. The sample parameters were categorised into two groups:

Qualitative variables: These involved the breed, sex and age categories (kitten: <1 year old; young: 1 to <2 years old; adult: 2 to <6 years old; senior: 6 to <11 years old; old: ≥11 years old) as well as the results of the Rivalta test and effusion classification.

Quantitative variables: These involved biochemical parameters such as TP, ALB, GLOB, and ALB/GLOB ratio. The physical parameter, effusion SG, was also included. Flow cytometry data included the WBC-BF, TC-BF, TNCC-WDF, TNCC-WNR, and ΔTNC derived from the TNCC-WDF and TNCC-WNR counts. Additionally, the FIP Effusion Index, which integrates the ALB/GLOB ratio and ΔTNC, and the quantitative molecular analysis results from the Feline Coronavirus RT-qPCR Kit®, classified in accordance with the manufacturer's test specifications, were included.

3. Results

3.1. Qualitative Variables

3.1.1. Descriptive Data

Of the 50 animals included in this study, 23 (46%; 95% CI: 33.0–59.6%) tested negative with the Feline Coronavirus RT-qPCR Kit® while 27 (54%; 95% CI: 40.4–67.0%) tested positive.

3.1.2. Breed

Regarding breed, our study population comprised animals from six different cat breeds that included the following: 42 Domestic Shorthairs (84%), three Scottish Fold (6%), two Persian (4%), one Sphynx (2%), one British Shorthair (2%), and one Siamese (2%).

The Kruskal–Wallis test indicated no significant association between the cat breed and the negative and positive outcomes of the coronavirus molecular biology test in effusions (p =

0.337), suggesting no variability in FIP frequency across breeds. Detailed ranks and pairwise comparisons were performed, revealing specific breed differences, but none reached significance after adjustment for multiple comparisons. The Dunn–Bonferroni post hoc test revealed no significant associations, with all adjusted p-values exceeding the 0.05 significance level.

3.1.3. Sex

From the 50 animals analysed, 32 (64%) were males and 18 (36%) were females. Table 3 represents the percentages of positive and negative tests according to sex. The differences observed between animal sexes were not significant (p = 0.174).

Table 3. Negative and positive molecular biology test by sex in the 50 animals included in this study.

	Feline Coronavirus RT-qPCR		
Sex	**Negative**	**Positive**	**Total**
Female	11 (22%)	7 (14%)	18 (36%)
Male	12 (24%)	20 (40%)	32 (64%)
Total	23 (46%)	27 (54%)	50 (100%)

3.1.4. Age

From the 50 animals analysed, the age distribution ranged from ≤1 year (6 months) to 17 years, with an average age of 4.7 ± 3.9 years. A total of 26% (95% CI: 15.9–39.6%; n = 13) were kittens, 24% (95% CI: 14.3–37.4%; n = 12) were young, 26% (95% CI: 15.9–39.6%; n = 13) were adults, 14% (95% CI: 7.0–26.2%; n = 7) were seniors, and 14% (95% CI: 4.3–21.4%; n = 5) were old.

Table 4 displays the occurrence of coronavirus identification according to age group. The Spearman correlation analysis indicated no significant relationship between age groups and the frequency of FIP, as evidenced by a p-value of 0.083. Therefore, it cannot be concluded that age influences the frequency of feline coronavirus infection among the groups studied.

Table 4. Occurrence of coronavirus identification by age in the 50 animals included in this study.

		Age Group					
		Kitten (<1 Year Old)	**Young (1 to <2 Years Old)**	**Adult (2 to <6 Years Old)**	**Senior (6 to <11 Years Old)**	**Old (≥11 Years Old)**	**Total**
Feline Coronavirus RT-qPCR	Negative	4 (8%)	6 (12%)	5 (10%)	3 (6%)	5 (10%)	23 (46%)
	Positive	9 (18%)	6 (12%)	8 (16%)	4 (8%)	0 (0%)	27 (54%)
	Total	13 (26%)	12 (24%)	13 (26%)	7 (14%)	5 (10%)	50 (100%)

3.1.5. Rivalta Test Results

Regarding the Rivalta test, 33 (66%) were positive and 17 (34%) were negative. The analysis revealed a significant association between the Rivalta test and FCoV real-time RT-qPCR results, confirmed by Fisher's exact test (p < 0.001). This trend is further supported by the Cochran–Armitage test, highlighting a significant association between the two parameters (p < 0.001), suggesting that feline infectious peritonitis effusions are more likely to have a positive Rivalta test. The sensitivity of the Rivalta test, which is its ability to correctly identify cats with FIP, was approximately 96.3%. The specificity, or the test's ability to correctly identify cats without the disease, was approximately 69.6%. The odds ratio (OR) for a cat with FIP having a positive Rivalta test in the effusion was 59.4 (95% CI: 6.68–528.82), indicating that cats with FIP were 59 times more likely to have a positive Rivalta test compared to cats that were negative for FCoV real-time RT-qPCR in the effusion.

Table 5 represents the percentages of positive and negative Rivalta tests according to the FCoV real-time RT-qPCR.

Table 5. Positive and negative Rivalta tests according to the Feline Coronavirus RT-qPCR.

		Rivalta Test				
		Negative		Positive		Total
		n	% Within Rivalta Test	*n*	% Within Rivalta Test	*n*
Feline Coronavirus RT-qPCR	Negative	16 (32%)	69.6% (specificity)	7 (14%)	30.4%	23 (46%)
	Positive	1 (2%)	3.7%	26 (52%)	96.3% (sensitivity)	27 (54%)
	Total	17 (34%)		33 (66%)		50 (100%)

3.1.6. Effusion Classification

Among the effusion samples analysed, 46 (92%) were classified as modified transudates, while 4 (8%) were exudates. No pure transudates were identified. Statistical analysis revealed no correlation between effusion type and the FCoV real-time RT-qPCR results ($p = 0.87$). Table 6 represents the percentages of effusion type according to the FCoV RT-qPCR.

Table 6. Effusion type according to Feline Coronavirus RT-qPCR.

		Effusion Type		
		Modified Transudate	Exudate	Total
Feline Coronavirus RT-qPCR	Negative	21 (42%)	2 (4%)	23 (46%)
	Positive	25 (50%)	2 (4%)	27 (54%)
	Total	46 (92%)	4 (8%)	50 (100%)

The cytological features of effusions associated with FIP observed in the present study included fluid that was often clear and yellow, with a sometimes viscous consistency. The fluid was occasionally mildly haemodiluted. Cellular density varied widely but was generally considered low. The background of these samples was noted to be pale to moderately basophilic and granular, frequently containing numerous protein crescents. The nucleated cells identified were usually a mix of non-degenerate neutrophils, monocytes/macrophages, and small lymphocytes (Figure 6).

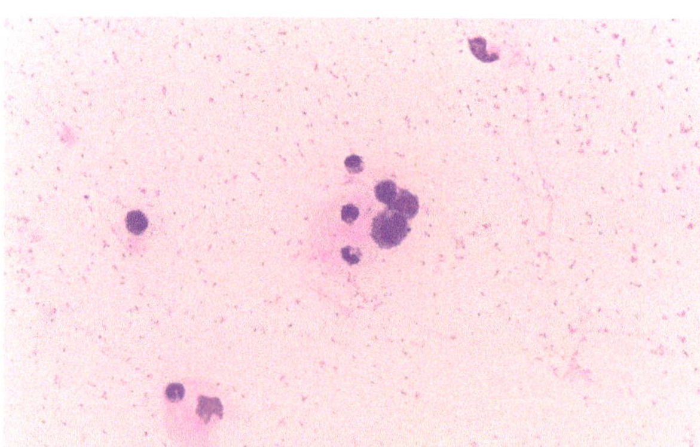

Figure 6. Cytology of a feline infectious peritonitis (FIP) effusion. Mixed cell population (non-degenerate neutrophils and macrophages) dispersed in a granular eosinophilic proteinaceous background commonly observed in FIP effusions. May-Grünwald Giemsa, 400×.

3.2. *Quantitative Variables*

3.2.1. Biochemical and Flow Cytometry Data

Spearman's correlation analysis (Supplementary Materials, Table S1) was performed on the data obtained. Table 7 presents the values, in effusions, of the biochemical and flow cytometry data WBC-BF, TC-BF, TNCC-WDF, TNCC-WNR, and ΔTNC, categorised according to the Feline Coronavirus RT-qPCR Kit® test results, showing comparisons between cats with either negative or positive outcomes.

Table 7. Biochemical and flow cytometry data WBC-BF, TC-BF, TNCC-WDF, TNCC-WNR, and ΔTNC, in effusions, categorised according to the Feline Coronavirus RT-qPCR Kit® test results, showing comparisons between cats with either negative or positive outcomes.

			Feline Coronavirus RT-qPCR in Effusions	
Effusion			**Negative**	**Positive**
Parameter (Units)	**Minimum**	**Maximum**	**Median (IQR)**	**Median (IQR)**
Specific gravity (SG)	1.01	1.05	1.03	1.03
TP (g/dL)	2.80	10.76	4.00	5.64
ALB (g/dL)	1.02	2.47	1.93	1.65
GLOB (g/dL)	1.25	9.11	2.00	4.15
ALB/GLOB ratio	0.18	1.5	0.93	0.35
WBC-BF ($10^3/\mu L$)	0.05	341,612	0.63	1.11
TC-BF ($10^3/\mu L$)	0.05	343,632	0.64	1.12
TNCC-WDF ($10^3/\mu L$)	0.06	335,752	0.88	0.97
TNCC-WNR ($10^3/\mu L$)	0.00	342,445	1.00	0.03
ΔTNC	0.30	583	0.93	27.04

TP (g/dL), absolute levels of total proteins; ALB (g/dL), absolute levels of albumin; GLOB (g/dL), absolute levels of globulins; WBC-BF ($10^3/\mu L$), total white blood cell count in body fluid mode; TC-BF ($10^3/\mu L$), total nucleated cells in body fluid mode; TNCC-WDF ($10^3/\mu L$), total nucleated cells in the white blood cell differential channel in whole blood mode; TNCC-WNR ($10^3/\mu L$), total nucleated cells in the white blood cell nucleated channel in whole blood mode; ΔTNC, total nucleated cell derived from the ratio between the TNCC-WDF and TNCC-WNR counts on the Sysmex XN-1000V®.

To facilitate interpretation, the authors chose to expound the most salient associations. The Spearman correlation analysis showed a significant correlation ($p < 0.001$) between the ALB/GLOB ratio and the FCoV RT-qPCR test results. Similarly, our study identified a strong positive correlation ($r = 0.85$, $p < 0.001$) between the ΔTNC values and FCoV real-time RT-qPCR results, suggesting that higher ΔTNC values are associated with positive test outcomes.

The ΔTNC was significantly higher ($p < 0.001$) in cats with FIP (median: 27.0; min-max: 1.04–583; 95% CI: 14.7–102.2) than in non-FIP cats (median: 0.9; min-max: 0.31–4.90; 95% CI: 0.8–1.8), and the TNCC-WNR and TNCC-WDF counts were significantly higher ($p < 0.001$ and $p < 0.05$, respectively) in cats with FIP (TNCC-WNR = 0.3; 0.0–3.89; TNCC-WDF = 2.32; 0.06–16.75) than in non-FIP cats (TNCC-WNR = 24,570.9; 0.05–342,445; TNCC-WDF = 24,398; 0.1–335,752) (Figure 7). Results from these latter cats were characterised by a high interindividual variability, likely due to the heterogeneity of the diseases responsible for the effusions. All of the cats with FIP had a ΔTNC $\geq$ 2.1, except for one cat that had "atypical" FIP (ΔTNC 1.04). All non-FIP cats had a ΔTNC < 4.9. More specifically, only 2 specimens from all 23 cats without FIP had a ΔTNC $\geq$ 2.1.

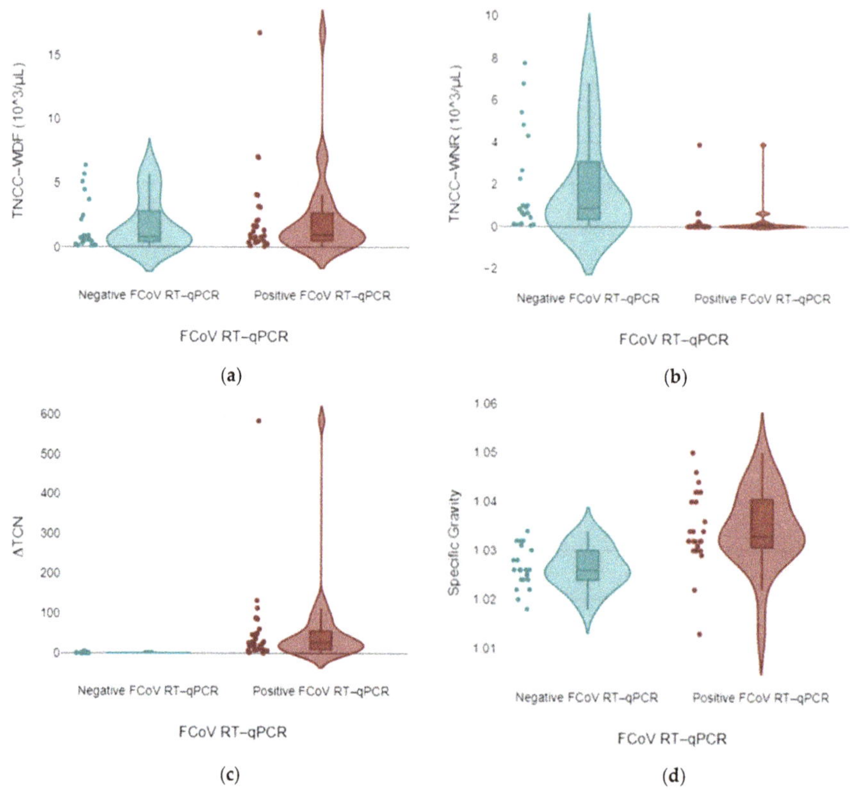

Figure 7. Values of TNCC-WDF (**a**), TNCC-WNR (**b**), ratio between total nucleated cell counts in the WDF and WNR channel of the laser counter Sysmex XN-1000V®, reported as ΔWBC by the instrument but termed ΔTNC for this study (**c**), and values of specific gravity (SG) (**d**) recorded in cats with FIP and in cats with diseases other than FIP (non-FIP). The boxes indicate the I–III interquartile range (IQR), and the horizontal line indicates the median. Dots indicate the values recorded in this study. The TNCC-WDF and the TNCC-WNR graphs do not include the results of two non-FIP specimens that had extremely high TNCC-WDF and TNCC-WNR counts (342.5 and 222.6 cells $\times$ $10^3/\mu$L, respectively).

3.2.2. Diagnostic Accuracy of the ΔTNC

A mixed-model ANOVA was utilised to evaluate the biochemical and flow cytometry data, categorising them based on their responses to the FCoV RT-qPCR Kit® test results as either negative or positive.

The analysis highlighted strong associations of ΔTNC levels in positive cases with statistical significance ($p < 0.001$), suggesting that the total nucleated cells derived from the ratio between the TNCC-WDF and TNCC-WNR counts on the Sysmex XN-1000V® dynamics were notably influenced by the FCoV infection status (Figure 8).

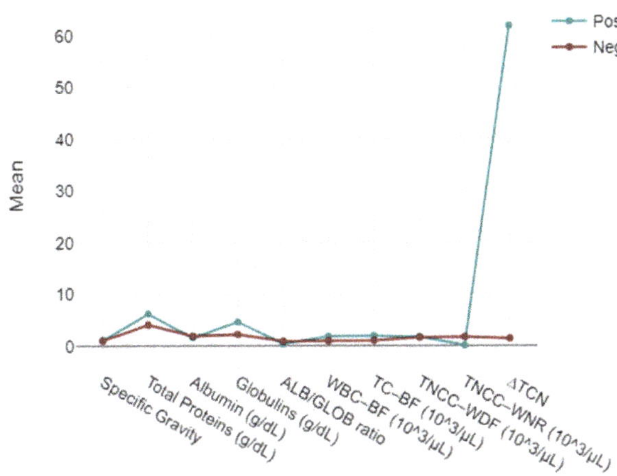

Figure 8. Mixed-model ANOVA findings: biochemical and flow cytometry data, categorising them based on their responses to the Feline Coronavirus RT-qPCR Kit® test as either negative or positive.

3.2.3. Receiver Operating Characteristic (ROC) Curve Analyses

The area under the ROC-AUC for ALB/GLOB ratio in the effusions was 0.968 ($\sigma = 0.024$; 95% CI: 87.4–99.7%), indicating excellent discriminatory power ($p < 0.0001$). The Youden index was 0.9195, with an associated criterion of ALB/GLOB ratio ≤ 0.56, yielding a sensitivity of 96.30% and a specificity of 95.65%.

The analysis of the ΔTNC using the Sysmex XN-1000V® in feline effusions for diagnosing FIP demonstrated significant diagnostic efficacy. Among the 50 samples analysed, 27 (54%) were from cats with a positive FCoV RT-qPCR Kit® test result, and 23 (46%) were from cats with a negative result. The ROC-AUC for ΔTNC was 0.990 ($\sigma = 0.009$; 95% CI: 91.1–100%), indicating excellent discriminatory power ($p < 0.0001$) (Figure 9 and Table 8). The Youden index was 0.9259, with an associated criterion of ΔTNC ≥ 4.903, yielding a sensitivity of 92.59% and a specificity of 100%.

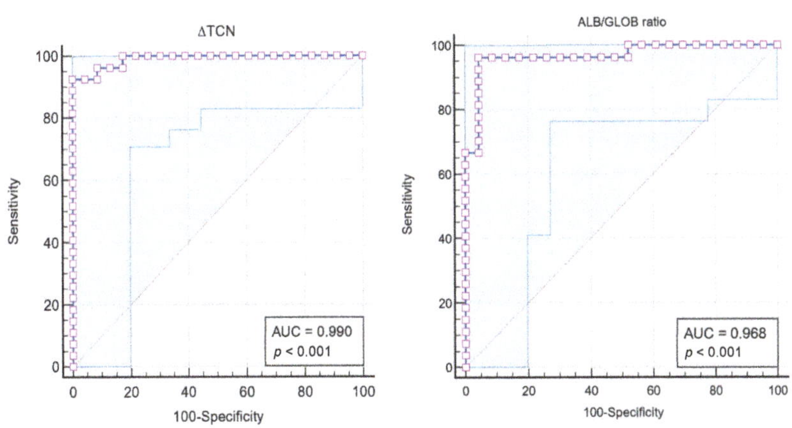

Figure 9. Area under the curve-receiver operating characteristic (AUC-ROC). The figure on the **left** displays the AUC-ROC of ΔTNC in the effusions, total nucleated cell derived from the ratio between TNCC-WDF and TNCC-WNR counts on the Sysmex XN-1000V®, showing strong correlation and statistical significance ($p < 0.001$) with the FCoV real-time RT-qPCR test results. The figure on the **right** illustrates the AUC-ROC of the ALB/GLOB ratio in the effusions, showing a strong correlation and statistical significance ($p < 0.001$) with the FCoV real-time RT-qPCR test results.

Table 8. Criterion values and coordinates of the ROC curve for ΔTNC.

ΔTNC	Sensitivity (%)	95% CI	Specificity (%)	95% CI
≥1.115	96.30	81.0–99.9	86.96	66.4–97.2
≥2.081	96.30	81.0–99.9	91.30	72.0–98.9
≥2.506	92.59	75.7–99.1	91.30	72.0–98.9
≥4.903	92.59	75.7–99.1	100	85.2–100

ΔTNC, total nucleated cells derived from the ratio between the TNCC-WDF and TNCC-WNR counts on the Sysmex XN-1000V®.

3.2.4. FIP Effusion Index

The ROC-AUC for the FIP Effusion Index was 0.995 ($\sigma = 0.006$; 95% CI: 92.0–100%), indicating excellent discriminatory power ($p < 0.0001$). For an FIP Effusion Index ≥ 5.06, the sensitivity was 96.30% and the specificity was 95.65%. Using an optimal criterion determined by the Youden index of 0.9630, a FIP Effusion Index ≥ 7.54 achieved a sensitivity of 96.30% and a specificity of 100% (Figure 10 and Table 9).

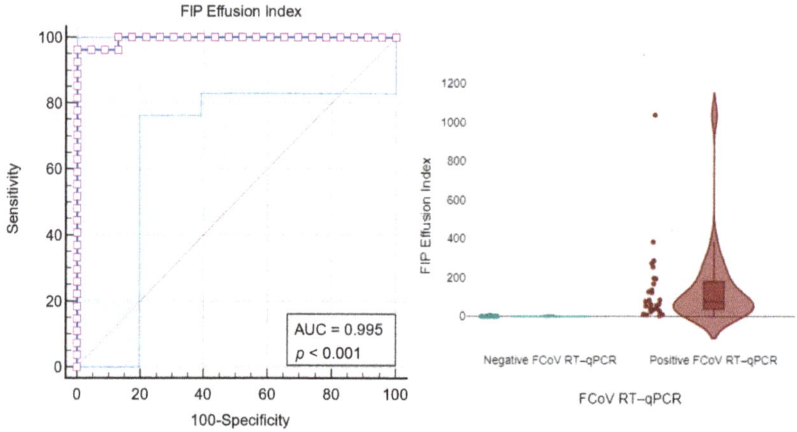

Figure 10. The figure on the **left** displays the area under the curve-receiver operating characteristic (AUC-ROC) of the FIP Effusion Index, showing strong correlation and statistical significance ($p < 0.001$) with the FCoV real-time RT-qPCR test results. The figure on the **right** illustrates the FIP Effusion Index in cats with FIP and cats with diseases other than FIP (non-FIP). The boxes indicate the I–III interquartile range (IQR), and the horizontal line indicates the median. Dots indicate the values recorded in this study.

Table 9. Criterion values and coordinates of the ROC curve for the FIP Effusion Index.

FIP Effusion Index	Sensitivity (%)	95% CI	Specificity (%)	95% CI
≥1.637	100	87.2–100	86.96	66.4–97.2
≥4.968	96.30	81.0–99.9	91.30	72.0–98.9
≥5.059	96.30	81.0–99.9	95.65	78.1–99.9
≥7.542	96.30	81.0–99.9	100	85.2–100

4. Discussion

CoVs are highly adaptable, with significant genetic diversity that allows for the emergence of new variants that are capable of cross-species transmission, leading to outbreaks like SARS-CoV, MERS-CoV, and SARS-CoV-2, which caused the COVID-19 pandemic.

FCoV, responsible for FIP, exemplifies this adaptability, ranging from asymptomatic infections to severe systemic disease. This variability within the same host species makes FCoV an excellent model for studying coronavirus evolution and pathogenesis. FIP is particularly challenging to diagnose due to the lack of pathognomonic signs and its varied presentations within the same individual, highlighting the need for reliable diagnostic tools [3–5,9–11,13,14,16,17,19–21,25,39,59].

The current guidelines [30,34,38,60] for FIP diagnosis highlight the complexities of FIP diagnosis, emphasising the need for understanding the sensitivity, specificity, and predictive values of diagnostic tests, especially when no effusion is present. Guidelines suggest that building an index of suspicion 'brick by brick' through a combination of history, signalment, and targeted diagnostic tests is crucial. Given the fatal nature of untreated FIP, achieving a correct diagnosis is paramount [61–64].

The present study addressed these challenges by both developing the FIP Effusion Index, a novel diagnostic tool that integrates biochemical and cellular markers to enhance diagnostic accuracy and reduce misdiagnosis and validating the flow cytometry-based ΔTNC on the Sysmex XN-1000V® in effusion samples.

The limited sample size of 50 individuals, dictated by stringent inclusion and exclusion criteria, was a constraint. Future studies should include a larger cohort, incorporating IHC FCoV antigen detection and including septic cases to further validate these findings and explore the index's applicability in monitoring patients with persistent effusions.

4.1. Sex, Breed and Age

The findings of this study regarding breed, sex, and age offer a refined perspective on the diagnosis of FIP. Although purebred cats are often regarded as being at a higher risk for FIP, our results indicate that Domestic Shorthairs, the most frequent breed in this study, still represented a considerable proportion of cases, indicating the extensive reach of FIP across all breeds. This observation is consistent with previous studies [33,65,66] and contrary to the expectations set by other published studies [67,68], which often suggest that purebred cats are more likely to develop FIP. However, the lower frequency of FIP in purebred cats within the present study may be attributed to the specific population demographics of cats presented at the participating clinics or even to sampling bias. Concerning sex, while some studies have indicated a marginally higher prevalence of FIP in male cats, potentially due to hormonal influences or behavioural traits, this evidence remains inconclusive [33,69]. Our findings are consistent with this ambiguity, suggesting that sex does not constitute a definitive risk factor for FIP in the studied population. No significant correlation was found between age and the occurrence of FIP. This result contrasts with the expectations set by many published studies [30,33,34,65,70], which suggest that younger cats, particularly those under 2 years of age, are more vulnerable to FIP.

4.2. Effusion Classification

The most common clinical presentation in cats with FIP is the presence of effusion [30]. Typically, FIP effusions contain high protein concentrations and low white blood cell counts and therefore may be classified as either modified transudates or exudates, depending on the total protein concentration and total cell count. However, some cats with FIP can have very high cell counts in the effusion (e.g., secondary bacterial peritonitis) [30,38,71].

Although the vast majority of cats with FIP will present with effusion, diagnostic algorithms for animals with this clinical presentation should consider various pathophysiological mechanisms (or a combination of them) such as altered hydrostatic and oncotic pressures, increased vascular permeability to plasma proteins, leakage of blood from vessels, leakage of lymph from lymphatic vessels, and rupture of hollow organs or tissues to determine the primary diagnosis [53,72].

Another study [65] reviewed the medical records of 231 cats with confirmed FIP and found that 78.1% (175/224) of the cats evaluated had effusion. In contrast, a different study [54] examined a more extensive sample of cats with thoracic and abdominal effusions using cytological criteria and refractometric estimates of the total protein concentration. In this study, only 9.6% (38 out of 396) of the cases were suggestive of FIP, highlighting the range of other potential causes of effusion in felines such as septic, haemorrhagic, neoplastic, and other conditions.

In our study, a large portion of the samples (86%) were classified as modified transudates. Although classifying effusions is essential in clinical and laboratory contexts, this classification alone does not provide a definitive diagnosis. The classification of effusions is primarily used to guide or provide evidence for the type of pathological process that led to the effusion, which helps identify the primary diagnosis [53].

When statistically correlating the type of effusion with the FCoV RT-qPCR test results, no statistical significance was observed. Although FIP is considered a common cause of modified transudates, many of the effusions analysed in this study and classified as modified transudates were not consistent with FIP.

The absence of statistical significance in this context supports the idea that in the presence of an effusion with a clinical suspicion of FIP and with a cytological classification of modified transudate, other differential diagnoses should be also considered. According to some authors [30,53,60], traditional laboratory data, such as protein concentration and cytological evaluation, are often insufficient to diagnose the underlying disorder. In such cases, additional information including patient history, physical examination, imaging findings, or other tests is necessary to accurately interpret the significance of laboratory data and establish the pathogenesis of an effusion. Thus, the absence of statistical significance between modified transudate and FIP further emphasises the need for additional data and techniques to refine the diagnosis, as demonstrated in our study with the FIP Effusion Index or other methods such as immunocytochemistry, PCR, Rivalta, and others.

4.3. ΔTNC-XN, Rivalta Test, and ALB/GLOB Ratio in FIP Diagnostic Approaches

Currently, there are various diagnostic approaches for FIP including cytological analysis, real-time RT-qPCR, serum protein electrophoresis, and the Rivalta test, among others. While a conclusive diagnosis cannot be established, these methodologies, when employed together, may provide strong evidence supporting the presence of FIP. According to the literature review by the authors [15,38], real-time RT-qPCR has a demonstrated high sensitivity and specificity, often reaching 100% in several studies [30,73–81] when compared to the detection of the FCoV antigen by IHC. Therefore, although it is still considered the gold standard, real-time RT-qPCR was selected as the standard test for this study due to its non-invasive nature, speed of results, and suitability for integration into clinical routine.

The Rivalta test is well-known for its simplicity and cost-effectiveness. It demonstrates high sensitivity (96.3%) in identifying FIP-related effusions but has a moderate specificity (69.6%), which results in a higher proportion of false positives [49,50]. This can lead to the overdiagnosis of FIP, particularly in cases of bacterial peritonitis, pleuritis, or lymphoma, where similar effusion characteristics might be present [49,50].

In contrast, the ALB/GLOB ratio in effusions, a biochemical marker, showed excellent diagnostic accuracy with an ROC-AUC of 0.968, sensitivity of 96.3%, and specificity of 95.7% for a criterion of ALB/GLOB ratio ≤ 0.56. This indicates that the ALB/GLOB ratio effectively distinguishes between FIP and non-FIP effusions, outperforming the Rivalta test in specificity and thereby reducing the number of false positives.

The ΔTNC approach, analysed through the Sysmex XN-1000V® flow cytometry, in FIP effusions, demonstrates significant advantages over traditional diagnostic methods. This method outperformed both the Rivalta test and the ALB/GLOB ratio in terms of specificity and presented a near-perfect diagnostic tool. The ability of ΔTNC to effectively differentiate FIP from other effusion-causing conditions, such as inflammation or neoplasms, underlines its utility as a superior diagnostic marker in veterinary practice. In the present study, the ΔTNC method showed high sensitivity (92.6%) and specificity (100%) in detecting FIP-associated effusions, with a ΔTNC value above 2.1 indicating a likely FIP diagnosis and values above 4.9 may be considered diagnostic for FIP. Post hoc Bonferroni tests further supported these results, showing that the total cell count dynamics significantly differed between groups, underscoring the potential of ΔTNC as a sensitive biomarker for diagnosing FCoV infection. These findings align with previous studies [42,43] using similar methodologies, such as the Sysmex XT-2000iV®, reinforcing the robustness and reliability of this innovative approach in clinical practice. Thus, ΔTNC has emerged as a highly dependable marker for distinguishing FIP-associated effusions, with near-perfect accuracy in identifying true positive and true negative cases, underscoring its value in the diagnostic evaluation of suspected FIP in cats [42].

4.4. FIP Effusion Index: A Novel Diagnostic Method

The FIP Effusion Index improves diagnostic accuracy by combining the ALB/GLOB ratio with the ΔTNC for FIP for several reasons. First, the ALB/GLOB ratio and the ΔTNC provide complementary diagnostic information regarding the biochemical and cellular characteristics of effusions. The ALB/GLOB ratio serves as an indirect indicator of specific protein proportions within the fluid, with a decreased ALB/GLOB ratio typically observed in FIP-associated effusions due to increased globulin and decreased albumin levels. Meanwhile, the ΔTNC, obtained via flow cytometry, measures changes in the nucleated cell population, serving as an indirect marker of inflammatory cellular response burden within the effusion. By combining these two measures, the FIP Effusion Index captures both biochemical and cellular alterations, offering a more comprehensive and integrated diagnostic evaluation. This method outperformed the Rivalta test, ALB/GLOB ratio, and ΔTNC individually, underscoring the FIP Effusion Index as a superior method for distinguishing FIP from other conditions with similar effusion characteristics, making it an almost ideal diagnostic tool for FIP.

Furthermore, the use of flow cytometry-based ΔTNC analysis with the FIP Effusion Index offers a less invasive and quicker alternative to molecular diagnostic methods like real-time RT-qPCR. Although PCR has a high diagnostic value, it requires longer processing times and is more resource-intensive. The comparable diagnostic effectiveness of the FIP Effusion Index to real-time RT-qPCR further supports its inclusion in standard diagnostic protocols. With its high sensitivity and specificity, the FIP Effusion Index is a valuable tool in veterinary diagnostics, particularly in differentiating FIP from other causes of effusions in cats.

The study's adherence to the Standards for Reporting of Diagnostic Accuracy (STARD) guidelines ensures that the findings are robust and applicable to larger populations. The optimal cutoff value for ΔTNC and the FIP Effusion Index that minimises false positives and negatives has been clearly defined, providing a practical guideline for clinicians, as demonstrated in the diagnostic algorithm presented in Figure 11. This is particularly important in veterinary practices, where distinguishing FIP from similar conditions is challenging due to overlapping clinical signs and effusion characteristics [51].

The FIP Effusion Index analysis using the Sysmex XN-1000V® represents a significant advancement in diagnosing FIP in cats, offering high accuracy, practicality, and efficiency, which positions it as a superior alternative to traditional methods. This study supports the integration of the FIP Effusion Index into routine veterinary diagnostics for FIP, providing a reliable tool that improves diagnostic accuracy and clinical decision-making.

The use of Sysmex XN-1000V®'s channels, specifically WNR and WDF, introduces an innovative technique for differentiating between FIP-positive and FIP-negative cases. The reagents used in these channels differ from those in the analogous BASO and DIFF channels of the Sysmex XT-2000iV®, offering a more refined and precise assessment of effusion samples. This methodological distinction is crucial as it allows for a more targeted analysis of cellular components involved in FIP, which is not adequately captured by other routine diagnostic tests, which have been shown to provide only limited diagnostic certainty in FIP. This evaluation is essential because the reagents employed for cell counting in the Sysmex XN-1000V® channels (TNCC-WDF and TNCC-WNR) differed from those used in the analogous channels (DIFF and BASO) of the Sysmex XT-2000iV®.

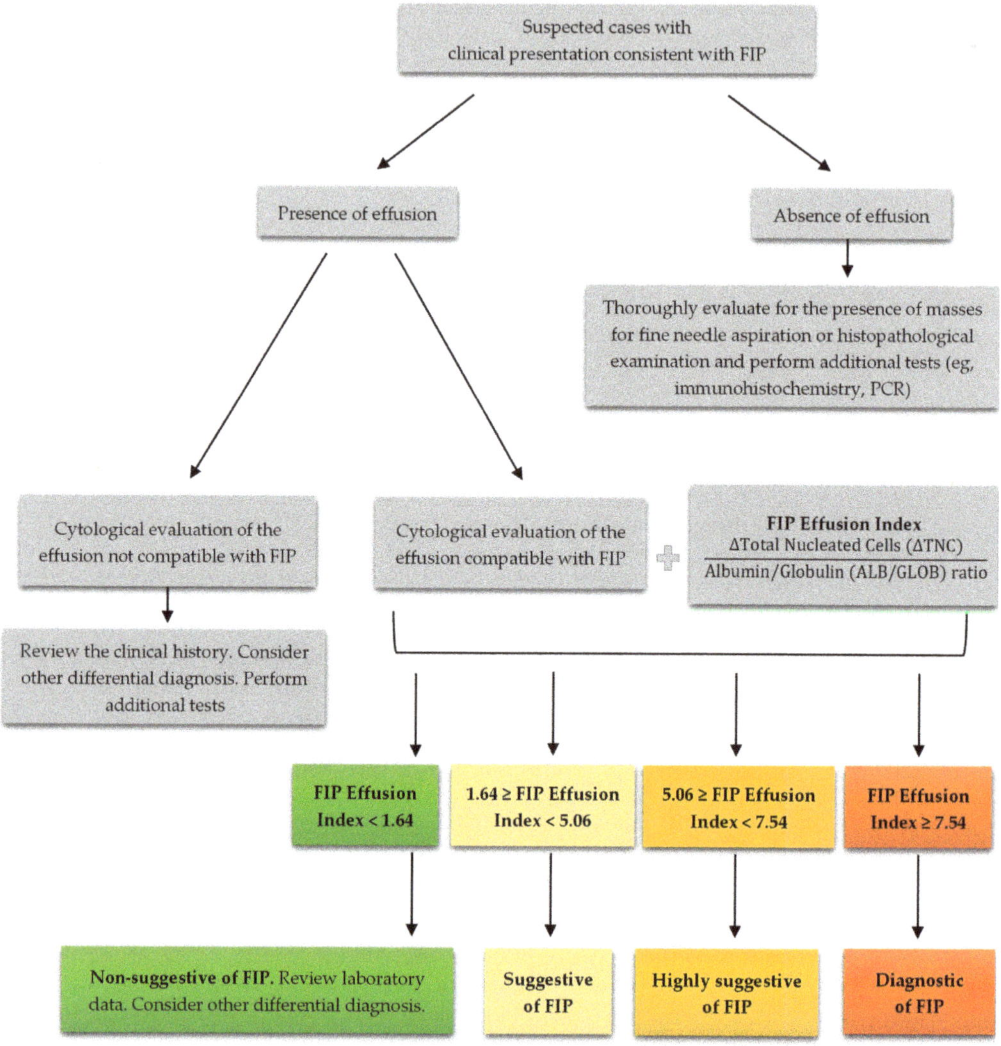

Figure 11. Diagnosis algorithm for evaluating suspected cases of feline infectious peritonitis (FIP) in cats based on their clinical presentation. If effusion is present, the diagnostic process begins with a cytological evaluation of the effusion. If the cytological findings are not compatible with FIP, the clinical history should be reviewed, and other differential diagnoses should be considered. If the effusion is compatible with FIP, the FIP Effusion Index is calculated using the albumin-to-globulin (ALB/GLOB) ratio and the delta total nucleated cell (ΔTNC) count. The interpretation of the FIP Effusion Index is as follows: an index value below 1.64 is not suggestive of FIP, and further review of the laboratory data and consideration of other differential diagnoses are recommended. An index value between 1.64 and less than 5.06 is suggestive of FIP; an index value between 5.06 and less than 7.54 is highly suggestive of FIP, whereas an index value equal to or greater than 7.54 is diagnostic of FIP. If there is an absence of effusion, the algorithm advises a thorough evaluation for the presence of masses through fine needle aspiration or histopathological examination, along with additional tests such as immunohistochemistry or PCR. This structured approach helps clinicians accurately differentiate FIP from other conditions, thereby enhancing diagnostic precision and patient management.

In the present study, the authors presented the first comprehensive analysis of flow cytometry-based ΔTNC using Sysmex XN-1000V® in feline effusions for diagnosing FIP and a novel diagnostic method, the FIP Effusion Index, and its relationship with molecular biology diagnoses.

4.5. Recommendations

Future studies should consider incorporating the detection of the FCoV antigen by IHC as part of the diagnostic process. Immunohistochemistry is currently regarded as the gold standard for FIP diagnosis because it directly visualises viral antigens within macrophages in effusion samples and tissue biopsies. While this study focused on comparing ΔTNC analysis and the FIP Effusion Index with molecular biology techniques in effusions, integrating IHC could provide a more comprehensive diagnostic framework. The combination could help to confirm the presence of FCoV in ambiguous cases, thereby reducing false negatives and improving diagnostic accuracy.

Moreover, future research should explore the use of advanced molecular diagnostics, such as next-generation sequencing (NGS), to detect and characterise mutations in the spike protein of the FCoV. These mutations, particularly M1058L and S1060A, are associated with the systemic spread of the virus and are pivotal in the pathogenesis of FIP. By identifying specific mutations that correlate with disease severity and clinical outcomes, NGS could further enhance the diagnostic process and allow for more targeted therapeutic interventions. However, while NGS shows promise in identifying these mutations, they are not exclusive to the disease, which further underscores the relevance of the present study.

Additionally, studies involving a larger and more diverse sample size including cases with septic effusions are recommended. Larger, multi-centre studies involving cats from various geographical locations and different clinical settings would help to validate the robustness and applicability of ΔTNC and the FIP Effusion Index analysis as diagnostic tools. These studies should also aim to standardise the protocols for sample collection, handling, and analysis to minimise variability and improve the reproducibility of results.

Another key area for future investigation is the longitudinal monitoring of ΔTNC and FIP Effusion Index values in monitoring patients with persistent effusions. Tracking these values over time could provide valuable insights into disease progression and response to treatment, potentially identifying new biomarkers that predict clinical outcomes. This approach could also facilitate the development of dynamic diagnostic models that adapt to the changing clinical picture of FIP, enhancing early detection and improving patient management strategies.

In the non-effusive form, given the complexity and diagnostic challenges, there is a pressing need for future research to improve diagnostic accuracy through the integration of advanced methodologies including the combined use of laboratory biomarkers and imaging techniques. Particular emphasis should be placed on developing diagnostic indices that incorporate imaging modalities, such as ultrasonography, alongside biochemical blood markers and/or molecular techniques, such as RT-qPCR on fine-needle aspirates, to establish robust diagnostic correlations.

Furthermore, given the parallels between FIP and coronavirus-induced diseases in humans, such as COVID-19, there is an opportunity to investigate the translational potential of these findings. Studies should consider how the diagnostic tools and methods used for FIP in veterinary medicine could be adapted or inspire novel approaches for diagnosing and managing coronavirus-related diseases in humans. This cross-species research could lead to a better understanding of coronavirus pathogenesis, zoonotic transmission, and potential therapeutic targets.

Finally, it is essential for future research to focus on the development of rapid, point-of-care diagnostic tests for FIP that combine the sensitivity and specificity of laboratory-based assays like ΔTNC and the FIP Effusion Index analysis with the convenience and speed required for routine clinical practice. Such advancements would significantly enhance the ability of veterinarians to diagnose FIP quickly and accurately, facilitating timely treatment and improving the overall prognosis for affected cats.

By addressing these recommendations, future studies can build upon the current findings and contribute to more effective diagnostic strategies for FIP, ultimately improving the health and welfare of cats and potentially offering broader insights into coronavirus infections across species.

5. Conclusions

The present study demonstrates that the Sysmex XN-1000V® generated ΔTNC provides exceptionally high diagnostic accuracy for FIP. This accuracy is attributed to the formation of clots in the WNR channel after mixing with the reagent, which trap cells similarly to the mechanism observed in the Rivalta test, a method also noted for its high diagnostic accuracy for FIP. This clotting reaction leads to a low TNCC-WNR, even when the TNCC-WDF counts are high. Therefore, in routine practice, it is not recommended to use the default TNCC counts generated by the WNR channel, but to directly use the TNCC-WDF and especially the ΔTNC, particularly when FIP is suspected. In these cases, a $\Delta TNC \geq 2.1$ is highly suggestive of FIP, and a $\Delta TNC \geq 4.9$ may be considered as diagnostic for FIP.

The FIP Effusion Index, which integrates the ALB/GLOB ratio with the ΔTNC, has emerged as an exceptionally robust diagnostic tool. The ALB/GLOB ratio itself provided valuable information about the protein composition of the effusion, with a ratio ≤ 0.56 yielding a sensitivity of 96.3% and a specificity of 95.7% for FIP diagnosis. When combined in the FIP Effusion Index, these parameters enhance diagnostic precision. A FIP Effusion Index value equal to or greater than 5.06 resulted in a sensitivity of 96.3% and specificity of 95.7%, while an optimal cutoff of ≥ 7.54 achieved perfect specificity (100%) and a sensitivity of 96.3%. This makes the FIP Effusion Index a highly effective and reliable tool for accurately identifying FIP, providing superior diagnostic accuracy compared to the individual use of the ALB/GLOB ratio or ΔTNC. The use of these combined measures should be considered a standard approach in veterinary medicine practice for the diagnosis of FIP, enabling more precise and confident clinical decision-making and ultimately improving patient outcomes.

This research provides valuable insights into the broader context of coronavirus-related diseases, offering potential guidance for diagnostic approaches in other species with effusive forms of coronavirus infections, as suggested by the Stockholm paradigm. This cross-species relevance underscores the importance of this research in enhancing our understanding of coronavirus pathogenesis and improving diagnostic and therapeutic strategies in both veterinary and human medicine.

Supplementary Materials: The following supporting information can be downloaded at https://www.mdpi.com/article/10.3390/vetsci11110563/s1, Table S1: Spearman correlation analysis: biochemical and flow cytometry data in effusions, categorised according to the feline coronavirus RT-qPCR test.

Author Contributions: Conceptualisation, R.L.; Methodology, R.L.; Software, R.L.; Validation, R.L., F.S., H.L.d.C., A.G., C.F., C.V.N., A.S.d.B., T.M., C.S., A.R.S., Â.M., L.C., A.C.C. and E.L.D.; Formal analysis, R.L. and Â.M.; Investigation, R.L.; Resources, R.L., F.S., C.F., C.V.N., A.S.d.B., H.L.d.C., C.S. and A.R.S.; Data curation, R.L., F.S. and H.L.d.C.; Writing—original draft preparation, R.L.; Writing—review and editing, R.L., F.S., A.S.d.B., A.G., L.C., A.C.C. and E.L.D.; Visualisation, R.L.; Supervision, L.C., A.C.C. and E.L.D.; Project administration, R.L., H.L.d.C., L.C., A.C.C. and E.L.D.; Funding acquisition, R.L., H.L.d.C., A.G., L.C., A.C.C. and E.L.D. All authors have read and agreed to the published version of the manuscript.

Funding: This study was supported by projects UIDB/00772/2020 (Doi:10.54499/UIDB/00772/2020), UIDP/00681/2020 (Doi:10.54499/UIDP/00681/2020), LA/P/0059/2020, UIDB/05183/2020 (Doi:10.54499/UIDB/05183/2020), UIDP/05183/2020 (Doi:10.54499/UIDP/05183/2020), and LA/P/0121/2020 (Doi:10.54499/LA/P/0121/2020), funded by the Portuguese Foundation for Science and Technology (FCT).

Institutional Review Board Statement: All procedures complied with the Portuguese legislation for the protection of animals used for scientific purposes (i.e., Decree-Law no. 113/2013 of 7 August 2013), which transposes European legislation (i.e., Directive 2010/63/EU of the European Parliament and of the Council, of 22 September 2010). This study project was approved by the Institutional Review Board of CEDIVET Veterinary Laboratories, which ensures that the analysed samples of veterinary medical centres can be used anonymously in studies and scientific research works related with this project.

Informed Consent Statement: Not applicable.

Data Availability Statement: The data presented in this study are available upon request from the corresponding authors.

Acknowledgments: The authors extend their sincere gratitude to the CEDIVET Veterinary Laboratories (Porto, Portugal) for their generous provision of the results that facilitated the research conducted in this study.

Conflicts of Interest: R.L., F.S., H.L.d.C., C.V.N. and A.S.d.B. are employed by the company CEDIVET Veterinary Laboratories. The remaining authors declare that the research was conducted in the absence of any commercial or financial relationships that could be construed as potential conflicts of interest.

References

1. Naseer, S.; Khalid, S.; Parveen, S.; Abbass, K.; Song, H.; Achim, M.V. COVID-19 outbreak: Impact on global economy. *Front. Public Health* **2023**, *10*, 1009393. [CrossRef] [PubMed]
2. Clemente-Suárez, V.J.; Navarro-Jiménez, E.; Moreno-Luna, L.; Saavedra-Serrano, M.C.; Jimenez, M.; Simón, J.A.; Tornero-Aguilera, J.F. The impact of the COVID-19 pandemic on social, health, and economy. *Sustainability* **2021**, *13*, 6314. [CrossRef]
3. Hiscott, J.; Alexandridi, M.; Muscolini, M.; Tassone, E.; Palermo, E.; Soultsioti, M.; Zevini, A. The global impact of the coronavirus pandemic. *Cytokine Growth Factor Rev.* **2020**, *53*, 1–9. [CrossRef] [PubMed]
4. Di Gennaro, F.; Pizzol, D.; Marotta, C.; Antunes, M.; Racalbuto, V.; Veronese, N.; Smith, L. Coronavirus diseases (COVID-19) Current status and future perspectives: A narrative review. *Int. J. Environ. Res. Public Health* **2020**, *17*, 2690. [CrossRef]
5. Arshad Ali, S.; Baloch, M.; Ahmed, N.; Arshad Ali, A.; Iqbal, A. The outbreak of coronavirus disease 2019 (COVID-19)—An emerging global health threat. *J. Infect. Public Health* **2020**, *13*, 644–646. [CrossRef]
6. Amoutzias, G.D.; Nikolaidis, M.; Tryfonopoulou, E.; Chlichlia, K.; Markoulatos, P.; Oliver, S.G. The remarkable evolutionary plasticity of coronaviruses by mutation and recombination: Insights for the COVID-19 pandemic and the future evolutionary paths of SARS-CoV-2. *Viruses* **2022**, *14*, 78. [CrossRef]
7. V'kovski, P.; Kratzel, A.; Steiner, S.; Stalder, H.; Thiel, V. Coronavirus biology and replication: Implications for SARS-CoV-2. *Nat. Rev. Microbiol.* **2021**, *19*, 155–170. [CrossRef]
8. Krentz, D.; Zenger, K.; Alberer, M.; Felten, S.; Bergmann, M.; Dorsch, R.; Matiasek, K.; Kolberg, L.; Hofmann-Lehmann, R.; Meli, M.L.; et al. Curing cats with feline infectious peritonitis with an oral multi-component drug containing GS-441524. *Viruses* **2021**, *13*, 2228. [CrossRef]
9. Guo, Y.-R.; Cao, Q.-D.; Hong, Z.-S.; Tan, Y.-Y.; Chen, S.-D.; Jin, H.-J.; Tan, K.-S.; Wang, D.-Y.; Yan, Y. The origin, transmission and clinical therapies on coronavirus disease 2019 (COVID-19) outbreak—An update on the status. *Mil. Med. Res.* **2020**, *7*, 11. [CrossRef]
10. Li, X.; Geng, M.; Peng, Y.; Meng, L.; Lu, S. Molecular immune pathogenesis and diagnosis of COVID-19. *J. Pharm. Anal.* **2020**, *10*, 102–108. [CrossRef]
11. Prompetchara, E.; Ketloy, C.; Palaga, T. Immune responses in COVID-19 and potential vaccines: Lessons learned from SARS and MERS epidemic. *Asian Pac. J. Allergy Immunol.* **2020**, *38*, 1–9. [CrossRef] [PubMed]
12. Pavlin, D.; Grubelnik, G.; Kogoj, R.; Suljič, A.; Tozon, N.; Tatjana, A.Ž.; Korva, M. Investigation of possible recombination between feline coronavirus and SARS-CoV-2 in cats with feline infectious peritonitis. In Proceedings of the European Congress of Veterinary Internal Medicine—Companion Animals (ECVIM-CA), Lyon, France, 5–7 September 2024; p. 319.
13. Wynants, L.; Van Calster, B.; Collins, G.S.; Riley, R.D.; Heinze, G.; Schuit, E.; Albu, E.; Arshi, B.; Bellou, V.; Bonten, M.M.J.; et al. Prediction models for diagnosis and prognosis of COVID-19: Systematic review and critical appraisal. *BMJ* **2020**, *369*, m1328. [CrossRef] [PubMed]
14. Xia, H.; Li, X.; Zhao, W.; Jia, S.; Zhang, X.; Irwin, D.M.; Zhang, S. Adaptive evolution of feline coronavirus genes based on selection analysis. *BioMed Res. Int.* **2020**, *2020*, 9089768. [CrossRef] [PubMed]
15. Zuzzi-Krebitz, A.-M.; Buchta, K.; Bergmann, M.; Krentz, D.; Zwicklbauer, K.; Dorsch, R.; Wess, G.; Fischer, A.; Matiasek, K.; Hönl, A.; et al. Short treatment of 42 days with oral GS-441524 results in equal efficacy as the recommended 84-day treatment in cats suffering from feline infectious peritonitis with effusion—A prospective randomized controlled study. *Viruses* **2024**, *16*, 1144. [CrossRef]
16. Sweet, A.N.; André, N.M.; Stout, A.E.; Licitra, B.N.; Whittaker, G.R. Clinical and molecular relationships between COVID-19 and feline infectious peritonitis (FIP). *Viruses* **2022**, *14*, 481. [CrossRef]

17. Chu, H.; Chan, J.F.-W.; Wang, Y.; Yuen, T.T.-T.; Chai, Y.; Hou, Y.; Shuai, H.; Yang, D.; Hu, B.; Huang, X.; et al. Comparative replication and immune activation profiles of SARS-CoV-2 and SARS-CoV in human lungs: An ex vivo study with implications for the pathogenesis of COVID-19. *Clin. Infect. Dis.* **2020**, *71*, 1400–1409. [CrossRef]
18. Gao, Y.-Y.; Wang, Q.; Liang, X.-Y.; Zhang, S.; Bao, D.; Zhao, H.; Li, S.-B.; Wang, K.; Hu, G.-X.; Gao, F.-S. An updated review of feline coronavirus: Mind the two biotypes. *Virus Res.* **2023**, *326*, 199059. [CrossRef]
19. Hosie, M.J.; Hofmann-Lehmann, R.; Hartmann, K.; Egberink, H.; Truyen, U.; Addie, D.D.; Belák, S.; Boucraut-Baralon, C.; Frymus, T.; Lloret, A.; et al. Anthropogenic infection of cats during the 2020 COVID-19 pandemic. *Viruses* **2021**, *13*, 185. [CrossRef]
20. Michelitsch, A.; Hoffmann, D.; Wernike, K.; Beer, M. Occurrence of antibodies against SARS-CoV-2 in the domestic cat population of Germany. *Vaccines* **2020**, *8*, 772. [CrossRef]
21. Bosco-Lauth, A.M.; Hartwig, A.E.; Porter, S.M.; Gordy, P.W.; Nehring, M.; Byas, A.D.; VandeWoude, S.; Ragan, I.K.; Maison, R.M.; Bowen, R.A. Experimental infection of domestic dogs and cats with SARS-CoV-2: Pathogenesis, transmission, and response to reexposure in cats. *Proc. Natl. Acad. Sci. USA* **2020**, *117*, 26382–26388. [CrossRef]
22. Moyadee, W.; Sunpongsri, S.; Choowongkomon, K.; Roytrakul, S.; Rattanasrisomporn, A.; Tansakul, N.; Rattanasrisomporn, J. Feline infectious peritonitis: A comprehensive evaluation of clinical manifestations, laboratory diagnosis, and therapeutic approaches. *J. Adv. Vet. Anim. Res.* **2024**, *11*, 19–26. [CrossRef] [PubMed]
23. Liu, Y.; Zhu, Z.; Du, J.; Zhu, X.; Pan, C.; Yin, C.; Sun, W. Development of multiplex real-time PCR for simultaneous detection of SARS-CoV-2, CCoV, and FIPV. *Front. Vet. Sci.* **2024**, *11*, 1337690. [CrossRef] [PubMed]
24. Deng, K.; Nemser, S.M.; Frost, K.; Goodman, L.B.; Ip, H.S.; Killian, M.L.; Ulaszek, J.; Kiener, S.; Kmet, M.; Uhlig, S.; et al. Successful detection of delta and omicron variants of SARS-CoV-2 by veterinary diagnostic laboratory participants in an interlaboratory comparison exercise. *J. Appl. Lab. Med.* **2023**, *8*, 726–741. [CrossRef] [PubMed]
25. Shi, J.; Wen, Z.; Zhong, G.; Yang, H.; Wang, C.; Huang, B.; Liu, R.; He, X.; Shuai, L.; Sun, Z.; et al. Susceptibility of ferrets, cats, dogs, and other domesticated animals to SARS–Coronavirus 2. *Science (1979)* **2020**, *368*, 1016–1020. [CrossRef]
26. Alexander, M.R.; Schoeder, C.T.; Brown, J.A.; Smart, C.D.; Moth, C.; Wikswo, J.P.; Capra, J.A.; Meiler, J.; Chen, W.; Madhur, M.S. Predicting susceptibility to SARS-CoV-2 infection based on structural differences in ACE2 across species. *FASEB J.* **2020**, *34*, 15946–15960. [CrossRef]
27. Decaro, N.; Mari, V.; Lanave, G.; Lorusso, E.; Lucente, M.S.; Desario, C.; Colaianni, M.L.; Elia, G.; Ferringo, F.; Alfano, F.; et al. Mutation analysis of the spike protein in italian feline infectious peritonitis virus and feline enteric coronavirus sequences. *Res. Vet. Sci.* **2021**, *135*, 15–19. [CrossRef]
28. Meli, M.; Helfer-Hungerbuehler, A.K.; Spiri, A.; Meili, T.; Riond, B.; Krentz, D.; Zwicklbauer, K.; Bucht, K.; Zuzzi-Krebitz, A.M.; Hartmann, K.; et al. Alpha-1-acid glycoprotein quantification via spatial proximity analyte reagent capture luminescence assay—Validation and application as diagnostic and prognostic marker in serum and effusions of cats with feline infectious peritonitis during GS-441524 treatment. In Proceedings of the European Congress of Veterinary Internal Medicine—Companion Animals (ECVIM-CA), Lyon, France, 5–7 September 2024; p. 305.
29. Farsijani, F.; Safi, S.; Shirazi Beheshtiha, S.H. Comparison of the performance of bioresonance, electrophoresis and RT-PCR in the diagnosis of feline infectious peritonitis. *Arch. Razi Inst.* **2023**, *78*, 1077–1085.
30. Thayer, V.; Gogolski, S.; Felten, S.; Hartmann, K.; Kennedy, M.; Olah, G.A. 2022 AAFP/EveryCat feline infectious peritonitis diagnosis guidelines. *J. Feline Med. Surg.* **2022**, *24*, 905–933, Erratum in *J. Feline Med. Surg.* **2022**, *24*, e676. [CrossRef]
31. Jaimes, J.A.; Millet, J.K.; Stout, A.E.; André, N.M.; Whittaker, G.R. A tale of two viruses: The distinct spike glycoproteins of feline coronaviruses. *Viruses* **2020**, *12*, 83. [CrossRef]
32. Cook, S.; Castillo, D.; Williams, S.; Haake, C.; Murphy, B. Serotype I and II feline coronavirus replication and gene expression patterns of feline cells—Building a better understanding of serotype I FIPV biology. *Viruses* **2022**, *14*, 1356. [CrossRef]
33. Kennedy, M.A. Feline infectious peritonitis: Update on pathogenesis, diagnostics, and treatment. *Vet. Clin. N. Am. Small Anim. Pract.* **2020**, *50*, 1001–1011. [CrossRef] [PubMed]
34. Felten, S.; Hartmann, K. Diagnosis of feline infectious peritonitis: A review of the current literature. *Viruses* **2019**, *11*, 1068. [CrossRef] [PubMed]
35. Licitra, B.N.; Millet, J.K.; Regan, A.D.; Hamilton, B.S.; Rinaldi, V.D.; Duhamel, G.E.; Whittaker, G.R. Mutation in spike protein cleavage site and pathogenesis of feline coronavirus. *Emerg. Infect. Dis.* **2013**, *19*, 1066–1073. [CrossRef]
36. Stranieri, A.; Giordano, A.; Paltrinieri, S.; Giudice, C.; Cannito, V.; Lauzi, S. Comparison of the performance of laboratory tests in the diagnosis of feline infectious peritonitis. *J. Vet. Diagn. Investig.* **2018**, *30*, 459–463. [CrossRef]
37. Geks, A.K.; Walter, H.; Steffensen, N.; Volk, H.; Rieder, J.C.; Busse, C. Detection of S gene mutation (M 1058L and feline coronavirus ribonucleic acid in the aqueous humour of three living cats with uveitis and three negative controls. In Proceedings of the European Congress of Veterinary Internal Medicine—Companion Animals (ECVIM-CA), Lyon, France, 5–7 September 2024; p. 328.
38. Tasker, S.; Addie, D.D.; Egberink, H.; Hofmann-Lehmann, R.; Hosie, M.J.; Truyen, U.; Belák, S.; Boucraut-Baralon, C.; Frymus, T.; Lloret, A.; et al. Feline infectious peritonitis: European advisory board on cat diseases guidelines. *Viruses* **2023**, *15*, 1847. [CrossRef]
39. Addie, D.; Belák, S.; Boucraut-Baralon, C.; Egberink, H.; Frymus, T.; Gruffydd-Jones, T.; Hartmann, K.; Hosie, M.J.; Lloret, A.; Lutz, H.; et al. Feline infectious peritonitis: ABCD guidelines on prevention and management. *J. Feline Med. Surg.* **2009**, *11*, 594–604. [CrossRef] [PubMed]

40. Fonte-Oliveira, L.; Lopes, C.; Malhão, F.; Canadas-Sousa, A.; Gregório, H.; Marcos, R.; Santos, M. The utility of the cell tube block technique in the diagnosis of effusive feline infectious peritonitis. In Proceedings of the Congress of European Society of Veterinary Clinical Pathology (ESVCP/ECVCP), University of Veterinary Medicine, Budapest, Hungary, 28–31 August 2024; p. 148.
41. Kipar, A.; Meli, M.L. Feline infectious peritonitis. *Vet. Pathol.* **2014**, *51*, 505–526. [CrossRef]
42. Giordano, A.; Stranieri, A.; Rossi, G.; Paltrinieri, S. High diagnostic accuracy of the Sysmex XT-2000iV delta total nucleated cells on effusions for feline infectious peritonitis. *Vet. Clin. Pathol.* **2015**, *44*, 295–302. [CrossRef]
43. Zanetti, M.; Paltrinieri, S.; Meazzi, S.; Novellini, S.; Martini, V.; Lauzi, S.; Rossi, S.; Mangiagalli, G.; Giordano, A. Delta total nucleated cells assessed via Sysmex XT-2000iV and Sysmex XN-1000V on feline effusions: Comparable diagnostic accuracy for feline infectious peritonitis. In Proceedings of the European Congress of Veterinary Pathology & Clinical Pathology (ESVP/ECVP/ESVCP/ECVCP), Lisbon, Portugal, 30 August–2 September 2023.
44. Wenk, J.M.; Meli, M.L.; De Witt, C.C.; Bouzon, A.B.; Pineroli, B.; Felten, S.; Meunier, S.M.; Zwicklbauer, K.; Hartmann, K. Feline coronavirus loads and presence of co-infections in cats diagnosed with feline infectious peritonitis. In Proceedings of the European Congress of Veterinary Internal Medicine—Companion Animals (ECVIM-CA), Lyon, France, 5–7 September 2024; p. 306.
45. Yu, B.; Xu, C.; Huang, S.; Ni, J.; Zhou, J.; Zhang, Y.; Wu, M.; Zhang, J.; Fang, L. Development of a universal real-time RT-PCR assay for detection of pan-SARS-coronaviruses with an RNA-based internal control. *Front. Microbiol.* **2023**, *14*, 1181097. [CrossRef]
46. Phongroop, K.; Rattanasrisomporn, J.; Tangtrongsup, S.; Rungsipipat, A.; Piewbang, C.; Techangamsuwan, S. High-resolution melting analysis for simultaneous detection and discrimination between wild-type and vaccine strains of feline calicivirus. *Vet. Q.* **2023**, *43*, 1–12. [CrossRef]
47. Jähne, S.; Felten, S.; Bergmann, M.; Erber, K.; Matiasek, K.; Meli, M.L.; Hofmann-Lehmann, R.; Hartmann, K. Detection of feline coronavirus variants in cats without feline infectious peritonitis. *Viruses* **2022**, *14*, 1671. [CrossRef] [PubMed]
48. Felten, S.; Leutenegger, C.M.; Balzer, H.-J.; Pantchev, N.; Matiasek, K.; Wess, G.; Egberink, H.; Hartmann, K. Sensitivity and specificity of a real-time reverse transcriptase polymerase chain reaction detecting feline coronavirus mutations in effusion and serum/plasma of cats to diagnose feline infectious peritonitis. *BMC Vet. Res.* **2017**, *13*, 228. [CrossRef] [PubMed]
49. Fischer, Y.; Sauter-Louis, C.; Hartmann, K. Diagnostic accuracy of the rivalta test for feline infectious peritonitis. *Vet. Clin. Pathol.* **2012**, *41*, 558–567. [CrossRef] [PubMed]
50. Sakai, N.; Iijima, S.; Shiba, K. Reinvestigation of clinical value of rivalta reaction of puncture fluid. *Rinsho Byori* **2004**, *52*, 877–882.
51. Bossuyt, P.M.; Reitsma, J.B.; Bruns, D.E.; Gatsonis, C.A.; Glasziou, P.P.; Irwig, L.; Lijmer, J.G.; Moher, D.; Rennie, D.; de Vet, H.C.W.; et al. STARD 2015: An updated list of essential items for reporting diagnostic accuracy studies. *Clin. Chem.* **2015**, *61*, 1446–1452. [CrossRef]
52. Washabau, R.J.; Day, M.J. *Canine and Feline Gastroenterology*; Elsevier: Amsterdam, The Netherlands, 2013. [CrossRef]
53. Stockham, S.L.; Scott, M.A. *Fundamentals of Veterinary Clinical Pathology*, 2nd ed.; John Wiley & Sons: Hoboken, NJ, USA, 2013.
54. Gavazza, A.; Turinelli, V.; Lubas, G. Effusion in the cat: Classification of 396 fluids according to a problem-oriented scheme. *Comp. Clin. Path.* **2013**, *22*, 517–521. [CrossRef]
55. Zuraidawati, Z.; Dewi, M.; Syah, N.Y.; Zamzami, R.S. Use of vinegar as a candidate substitute of acetic acid (CH3COOH) in the rivalta test method in diagnosis feline infectious peritonitis (FIP) domestic cat. *Int. J. Trop. Vet. Biomed. Res.* **2023**, *7*, 11–14. [CrossRef]
56. Fischer, Y.; Weber, K.; Sauter-Louis, C.; Hartmann, K. The rivalta's test as a diagnostic variable in feline effusions—Evaluation of optimum reaction and storage conditions. *Tierärztl. Prax. Ausg. K Kleintiere/Heimtiere* **2013**, *41*, 297–303. [CrossRef]
57. Valenciano, A.C.; Rizzi, T.E. Abdominal, thoracic, and pericardial effusions. In *Cowell and Tyler's Diagnostic Cytology and Hematology of the Dog and Cat*; Elsevier: Amsterdam, The Netherlands, 2020; pp. 229–246. [CrossRef]
58. Hazuchova, K.; Held, S.; Klemm, I.; Bauer, N. Simplified Light's criteria and acute phase proteins reflect aetiology of feline body cavity effusions better than the traditional classification scheme. *Animals* **2023**, *13*, 1918. [CrossRef]
59. Lythgoe, K.A.; Hall, M.; Ferretti, L.; de Cesare, M.; MacIntyre-Cockett, G.; Trebes, A.; Andersson, M.; Otecko, N.; Wise, E.L.; Moore, N.; et al. SARS-CoV-2 within-host diversity and transmission. *Science (1979)* **2021**, *372*, eabg0821. [CrossRef]
60. Tasker, S. Diagnosis of feline infectious peritonitis: Update on evidence supporting available tests. *J. Feline Med. Surg.* **2018**, *20*, 228–243. [CrossRef]
61. Hartmann, K. New treatment options for cats with feline infectious peritonitis. In Proceedings of the European Congress of Veterinary Internal Medicine—Companion Animals (ECVIM-CA), Lyon, France, 5–7 September 2024; p. 340.
62. Buchta, K.; Zuzzi-Krebitz, A.M.; Bergmann, M.; Krentz, D.; Zwicklbauer, K.; Dorsch, R.; Wess, G.; Fischer, A.; Hofmann-Lehmann, R.; Meli, M.L.; et al. Shorter treatment duration of 42 days with oral GS-441524 also leads to complete remission in cats with feline infectious peritonitis. In Proceedings of the European Congress of Veterinary Internal Medicine—Companion Animals (ECVIM-CA), Lyon, France, 5–7 September 2024; p. 309.
63. Schlachet, A.; Schuller, S.; Carranza Valencia, A. Unlicensed antiviral treatment with GS-441524: How are clinicians approaching feline infectious peritonitis in primary care practices? In Proceedings of the European Congress of Veterinary Internal Medicine—Companion Animals (ECVIM-CA), Lyon, France, 5–7 September 2024; p. 308.
64. Cerna, P.; Ross, E.; Willis, M.; Visser, L.; Hawley, J.; Lappin, M. Assessment of cardiac changes in cats with feline infectious peritonitis. In Proceedings of the European Congress of Veterinary Internal Medicine—Companion Animals (ECVIM-CA), Lyon, France, 5–7 September 2024; p. 307.

65. Riemer, F.; Kuehner, K.A.; Ritz, S.; Sauter-Louis, C.; Hartmann, K. Clinical and laboratory features of cats with feline infectious peritonitis—A retrospective study of 231 confirmed cases (2000–2010). *J. Feline Med. Surg.* **2016**, *18*, 348–356. [CrossRef] [PubMed]
66. Yin, Y.; Li, T.; Wang, C.; Liu, X.; Ouyang, H.; Ji, W.; Liu, J.; Liao, X.; Li, J.; Hu, C. A retrospective study of clinical and laboratory features and treatment on cats highly suspected of feline infectious peritonitis in Wuhan, China. *Sci. Rep.* **2021**, *11*, 5208. [CrossRef] [PubMed]
67. Pesteanu-Somogyi, L.D.; Radzai, C.; Pressler, B.M. Prevalence of feline infectious peritonitis in specific cat breeds. *J. Feline Med. Surg.* **2006**, *8*, 1–5. [CrossRef] [PubMed]
68. Norris, J.; Bosward, K.; White, J.; Baral, R.; Catt, M.; Malik, R. Clinicopathological findings associated with feline infectious peritonitis in Sydney, Australia: 42 cases (1990–2002). *Aust. Vet. J.* **2005**, *83*, 666–673. [CrossRef] [PubMed]
69. Worthing, K.A.; Wigney, D.I.; Dhand, N.K.; Fawcett, A.; McDonagh, P.; Malik, R.; Norris, J.M. Risk factors for feline infectious peritonitis in australian cats. *J. Feline Med. Surg.* **2012**, *14*, 405–412. [CrossRef] [PubMed]
70. Pedersen, N.C.; Liu, H.; Gandolfi, B.; Lyons, L.A. The influence of age and genetics on natural resistance to experimentally induced feline infectious peritonitis. *Vet. Immunol. Immunopathol.* **2014**, *162*, 33–40. [CrossRef]
71. Savary, K.; Sellon, R.; Law, J. Chylous abdominal effusion in a cat with feline infectious peritonitis. *J. Am. Anim. Hosp. Assoc.* **2001**, *37*, 35–40. [CrossRef]
72. Zwicklbauer, K.; von La Roche, D.; Krentz, D.; Kolberg, L.; Alberer, M.; Hartmann, K.; von Both, U.U.; Härtle, S. Characterization of the lymphocyte response in cats with feline infectious peritonitis during antiviral treatment using the smart tube technology for flow cytometry in feline full blood samples. In Proceedings of the European Congress of Veterinary Internal Medicine—Companion Animals (ECVIM-CA), Lyon, France, 5–7 September 2024; p. 304.
73. Herrewegh, A.A.; de Groot, R.J.; Cepica, A.; Egberink, H.F.; Horzinek, M.C.; Rottier, P.J. Detection of feline coronavirus RNA in feces, tissues, and body fluids of naturally infected cats by reverse transcriptase PCR. *J. Clin. Microbiol.* **1995**, *33*, 684–689. [CrossRef]
74. Kennedy, M.A.; Brenneman, K.; Millsaps, R.K.; Black, J.; Potgieter, L.N.D. Correlation of genomic detection of feline coronavirus with various diagnostic assays for feline infectious peritonitis. *J. Vet. Diagn. Investig.* **1998**, *10*, 93–97. [CrossRef]
75. Barker, E.N.; Stranieri, A.; Helps, C.R.; Porter, E.L.; Davidson, A.D.; Day, M.J.; Knowles, T.; Kipar, A.; Tasker, S. Limitations of using feline coronavirus spike protein gene mutations to diagnose feline infectious peritonitis. *Vet. Res.* **2017**, *48*, 60. [CrossRef] [PubMed]
76. Tsai, H.-Y.; Chueh, L.-L.; Lin, C.-N.; Su, B.-L. Clinicopathological findings and disease staging of feline infectious peritonitis: 51 cases from 2003 to 2009 in Taiwan. *J. Feline Med. Surg.* **2011**, *13*, 74–80. [CrossRef] [PubMed]
77. Hartmann, K.; Binder, C.; Hirschberger, J.; Cole, D.; Reinacher, M.; Schroo, S.; Frost, J.; Egberink, H.; Lutz, H.; Hermanns, W. Comparison of different tests to diagnose feline infectious peritonitis. *J. Vet. Intern. Med.* **2003**, *17*, 781–790. [CrossRef] [PubMed]
78. Felten, S.; Weider, K.; Doenges, S.; Gruendl, S.; Matiasek, K.; Hermanns, W.; Mueller, E.; Matiasek, L.; Fischer, A.; Weber, K.; et al. Detection of feline coronavirus spike gene mutations as a tool to diagnose feline infectious peritonitis. *J. Feline Med. Surg.* **2017**, *19*, 321–335. [CrossRef] [PubMed]
79. Felten, S.; Matiasek, K.; Leutenegger, C.M.; Sangl, L.; Herre, S.; Dörfelt, S.; Fischer, A.; Hartmann, K. Diagnostic value of detecting feline coronavirus RNA and spike gene mutations in cerebrospinal fluid to confirm feline infectious peritonitis. *Viruses* **2021**, *13*, 186. [CrossRef]
80. Doenges, S.J.; Weber, K.; Dorsch, R.; Fux, R.; Hartmann, K. Comparison of real-time reverse transcriptase polymerase chain reaction of peripheral blood mononuclear cells, serum and cell-free body cavity effusion for the diagnosis of feline infectious peritonitis. *J. Feline Med. Surg.* **2017**, *19*, 344–350. [CrossRef]
81. Longstaff, L.; Porter, E.; Crossley, V.J.; Hayhow, S.E.; Helps, C.R.; Tasker, S. Feline coronavirus quantitative reverse transcriptase polymerase chain reaction on effusion samples in cats with and without feline infectious peritonitis. *J. Feline Med. Surg.* **2017**, *19*, 240–245. [CrossRef]

veterinary sciences

MDPI

Article

Treatment with Leflunomide in Conjunction with Glucocorticoids for Dogs with Immune-Mediated Polyarthritis Is Not Associated with Improved Outcomes: A Retrospective Cohort Study of 93 Dogs from Australia (2017–2024)

Remon Wilson [1,*], Inar Swift [1], Mikaela Groth-Semple [1], Sabrina Lee [1], Tamara Dann [2], Ahmed Arafa [3,4], Curtis Poyton [5] and Mary Thompson [6]

1 Queensland Veterinary Specialists, 53 Flinders Parade, North Lakes, QLD 4509, Australia
2 Manly Road Veterinary Hospital, 219 Manly Road, Manly West, QLD 4179, Australia
3 Department of Preventive Cardiology, National Cerebral and Cardiovascular Centre, Suita 564-8565, Japan
4 Department of Public Health and Community Medicine, Faculty of Medicine, Beni-Suef University, Beni-Suef 62521, Egypt
5 Poyton Group 6 Ettarre Street, Bracken Ridge, QLD 4017, Australia
6 Sydney School of Veterinary Science, The University of Sydney, Camperdown, NSW 2050, Australia
* Correspondence: remon.wilson@qldvetspecialists.com.au

Simple Summary: Canine immune-mediated polyarthritis is a common disease that affects young dogs. This study aims at assessing whether leflunomide as add-on treatment to steroids would reduce the rate of relapse and improve survival in dogs with the disease. However, the results of this study suggest that there was no difference in outcomes between dogs that received or did not receive the drug as an add-on therapy. This could provide another step in understanding the nature of immune-mediated diseases and assist practicing veterinarians in making informed decisions about the most appropriate treatment for these conditions.

Abstract: Immune-mediated polyarthritis (IMPA) has a relatively high relapse rate compared to other immune-mediated diseases. Leflunomide is frequently used to treat dogs with IMPA in conjunction with prednisolone. This retrospective cohort study aimed to evaluate the therapeutic efficacy of leflunomide as an adjunctive therapy to prednisolone in reducing relapse and mortality rates in dogs diagnosed with IMPA in Australia. The medical records of client-owned dogs diagnosed with IMPA at a specialist referral hospital in Southeast Queensland from 2017 to 2024 were reviewed. A total of 93 dogs were included in this study, divided into two groups based on the treatment received: Group PRED, consisting of 53 dogs treated with prednisolone as the sole immunosuppressive agent, and Group L+PRED, consisting of 40 dogs that received leflunomide as adjunctive therapy alongside prednisolone. Data collected included breed, age, weight, sex, serum C-reactive protein concentration, results of synovial fluid analysis and microbial culture, treatment protocol, relapse rates and time to relapse, and mortality rates. There was no difference in relapse or mortality rates, time to relapse, nor time to discontinue prednisolone between the PRED and L+PRED groups. The L+PRED group had higher body weights and lower prednisolone dose rate at discharge compared to those in the PRED group. This study demonstrated that the use of leflunomide as an adjunctive therapy to prednisolone for the treatment of dogs with IMPA had no improved outcomes, reduced relapse rates, or shortening in the duration of prednisolone therapy when compared to dogs receiving prednisolone monotherapy.

Keywords: canine; immune-mediated; polyarthritis; leflunomide; prednisolone; synovial fluid; immunosuppressive; relapse; adjunctive therapy

1. Introduction

Immune-mediated polyarthritis (IMPA) is a common immune-mediated condition of dogs [1]. Lameness, stiffness, reluctance to walk, joint effusion, and fever are the most common clinical signs [2,3]. Managing IMPA poses significant challenges due to the unpredictable nature of its clinical course [4]. One significant concern is the high relapse rate compared to other immune-mediated diseases [3]. Two recent UK studies reported relapse rates of 48% and 53% [5,6]. These frequent relapses often require prolonged immunosuppressive therapy, which can add to the complexity of treatment and represent substantial difficulties for veterinarians and owners. The same two studies reported high mortality rates of 17% and 19% [5,6]. This combination of high relapse rates and relatively high mortality highlights the need for improved therapeutic strategies in managing dogs with IMPA.

Corticosteroids, particularly prednisolone, are the mainstay of treatment for IMPA. The long-term use of prednisolone is associated with numerous side effects which negatively impact the quality of life for both dogs and their owners, prompting the search for prednisolone-sparing therapies [7,8].

Leflunomide, an immunosuppressive agent licensed in Australia and other countries for the treatment of human rheumatoid arthritis, has shown some potential in managing immune-mediated conditions including IMPA with an acceptable safety profile in dogs [9]. While some studies indicate the drug's efficacy, their small sample sizes, design limitations, and inadequate focus on relapse rates undermine the overall strength of the evidence. A small case series of 14 dogs explored leflunomide as an alternative to corticosteroids, but it lacked follow-up data regarding clinical relapse after an initial improvement in only 8 dogs [10]. Additionally, leflunomide was included as part of the treatment in only some of the cases in one of these recent studies, and the study did not assess the impact of leflunomide on relapse or mortality rates [5].

Given these knowledge gaps, further investigation is necessary to assess the efficacy of leflunomide in improving outcomes in dogs with IMPA. Our study aimed to evaluate whether the use of leflunomide alongside prednisolone in dogs with IMPA was associated with a reduction in relapse and mortality rates compared with prednisolone monotherapy.

2. Materials and Methods

2.1. Study Design

This retrospective cohort study evaluated the relapse and mortality rates in dogs with a clinical diagnosis of IMPA, treated with two different regimens: prednisolone as monotherapy (PRED) and a combination of leflunomide with prednisolone (L+PRED).

2.2. Data Collection

Electronic medical records were searched for all dogs diagnosed with IMPA at Queensland Veterinary Specialists, a small animal referral hospital in Southeast Queensland, Australia, between January 2017 and April 2024. The keywords used were "immune-mediated polyarthritis", "IMPA", "polyarthritis", and "polyarthropathy". Records were reviewed and the following information was recorded for all dogs: breed, age, sex, weight, rectal temperature at presentation, hospital days and concurrent conditions. The treatment data collected were starting doses of prednisolone and leflunomide, duration of leflunomide and prednisolone treatment, total follow-up duration, occurrence of relapse, leflunomide and prednisolone dose at the time of relapse, and death due to the disease. When available, additional data collected included the results of synovial fluid microbial culture, serum C-reactive protein (CRP) concentration, and whether the IMPA was classified as associative or non-associative.

2.3. Case Selection

The dogs were divided into two treatment groups based on the treatment received during the follow-up period of this study. Group PRED consisted of dogs treated with prednisolone as monotherapy and served as the control group for comparison with the alternative treatment group. Group L+PRED included dogs treated with leflunomide as adjunctive therapy to prednisolone. To be included in the L+PRED group, dogs must have received leflunomide in addition to prednisolone, either at discharge or during subsequent rechecks, but prior to any occurrence of relapse. Dogs were categorised as having associative IMPA if they presented with signs of resolvable comorbidities affecting areas distant from the joints, such as urinary tract infections (UTIs), acute diarrhoea, superficial pyoderma, or atopic dermatitis. Conversely, dogs were classified as having non-associative IMPA when no concurrent disease process was identified.

2.4. Eligibility Criteria

All dogs enrolled in this study were required to exhibit one or more clinical signs consistent with IMPA, including fever (rectal temp $\geq$ 39.2 °C), shifting lameness, stilted gait, joint pain, and/or effusion. Dogs were required to have cytologic evidence of neutrophilic inflammation on synovial fluid samples from a minimum of two joints. Synovial fluid analysis was performed by a reference laboratory if there was sufficient sample collected in an EDTA tube. Alternatively, direct smears were made for cytological evaluation by board-certified clinical pathologists. Although detailed cytological evaluation data for each joint were not collected as part of this study, the cytological reports were reviewed by the authors and given scores based on the pathologist report interpretation (1 = mild; 2 = moderate; 3 = severe/extreme; or 4 = mixed but predominantly neutrophilic) and only dogs exhibiting predominantly neutrophilic inflammation of varying degrees in multiple joints were included. Dogs with only mononuclear inflammation were not included. When available, synovial fluid samples were collected for bacterial culture into blood culture bottles or transfer media swabs, either from a single joint, multiple joints, or pooled fluid from multiple joints.

Dogs were excluded from this study if they had insufficient medical records, no synovial fluid cytology was performed, or if they were lost to follow-up within 30 days. Dogs were excluded from both groups if they had received any immunosuppressive medication other than prednisolone in the PRED group or leflunomide and prednisolone in the L+PRED group prior to the end of the study follow-up period. Dogs were excluded from the L+PRED group if leflunomide was commenced following a relapse event.

Dogs with other severe conditions potentially contributing to the development of IMPA were excluded, as these conditions could potentially affect the response to treatment. These conditions included concurrent neoplasia, systemic fungal infections, or dogs later diagnosed with septic arthritis with organisms not considered contaminants. However, dogs with concurrent clinical signs related to other non-fatal and non-severe concurrent conditions such as gastroenteritis with associated diarrhoea, pancreatitis, UTIs, oral ulceration, or dermatological conditions (limited to atopic dermatitis or superficial pyoderma) were not excluded but were categorised as associative IMPA.

A dog was considered clinically relapsed if it showed one or more clinical signs of IMPA after previous resolution. Any escalation of immunosuppressive treatment was considered an indicator of a perceived relapse by the attending veterinarian. Treatment escalation was defined as restarting immunosuppressive therapy, increasing the dosage or frequency of the current medication, or switching to or adding an additional immunosuppressive medication. Confirmation of relapse was also obtained through abnormal synovial fluid cytology and/or elevated CRP levels whenever these diagnostics were employed. However, these diagnostics were not mandatory for defining clinical relapse.

2.5. Statistical Analyses

The differences in the mean values and proportions of characteristics between PRED and L+PRED groups were calculated using the t-test and chi-squared test, respectively. The Cox proportional hazards models were used to calculate the hazard ratios (HRs) and their 95% confidence intervals (CIs) of relapse and mortality. The follow-up period was calculated from the baseline date to the date of outcome or end of the follow-up, whenever came earlier. Statistical Package for Social Science (SPSS) 2013 (IBM SPSS Statistics for Windows, version 22.0, IBM Corporation, Armonk, NY, USA) was used for data analysis.

3. Results

3.1. Study Population

Medical records of 142 dogs diagnosed with IMPA between January 2017 and April 2024 were identified. Records were reviewed for eligibility based on the diagnostic criteria for the disease.

From the PRED group, 34 dogs were excluded. Of these, 19 were excluded due to lack of follow-up, and 8 due to receiving a second immunosuppressive medication. The remaining 7 dogs were excluded despite initial diagnosis and treatment for IMPA, due to subsequent findings that compromised the accuracy of the diagnosis and treatment. These findings included positive synovial fluid cultures with organisms not considered contaminants (*Staphylococcus pseudintermedius* and *Pasteurella canis*) in two of the seven dogs, suspected synovial myxosarcoma in one, suspected endocarditis on echocardiography in one, disseminated fungal infection in one, and a history of total hip replacement in one.

From the L+PRED group, 15 dogs were excluded. Nine dogs were excluded due to lack of follow-up, one for receiving additional immunosuppressive medication, and five had concurrent conditions that could affect the response to IMPA treatment. Among these five, one dog had a cardiac tumour with pericardial effusion, one had a total hip replacement, one died from presumed concurrent immune-mediated haemolytic anaemia (IMHA), one was later diagnosed with steroid-responsive meningitis arteritis (SRMA), and one died of unknown causes one week after diagnosis.

A total of 93 dogs met the inclusion criteria for this study. These dogs were retrospectively assigned to either the L+PRED group (n = 40) or the PRED group (n = 53) based on the treatment protocol received (Figure 1).

3.2. Baseline Characteristics

Among the 93 dogs included in this study, there were 52 (56%) females (43 neutered) and 41 (44%) males (27 neutered). The most frequently presented breeds were Border Collie and English Staffordshire Bull Terrier (n = 8 each), Labrador Retriever (n = 7), English Cocker Spaniel (n = 6), Golden Retriever and German Shepherd dog (n = 5 each), Rottweiler and Kelpie (n = 4 each), and 27 breeds were represented with < 4 dogs. There were no differences in breed distribution between the PRED and L+PRED groups (p-value > 0.05). Breed distribution is provided in Supplementary Table S1.

The sex distribution did not vary across groups: male neutered dogs comprised 26.4% in the PRED group and 32.5% in the L+PRED group, male entire 18.9% and 10%, female spayed 41.5% and 52.5%, and female entire 13.2% and 5%, respectively (p-value = 0.169). L+PRED dogs were 10 months older and 5 kg heavier than the PRED dogs (p-values = 0.183 and 0.033, respectively). A summary of age, sex, and body weight are presented in Table 1.

Table 1. Summary of age, body weight, and sex distribution.

Characteristics	PRED n = 53	L+PRED n = 40	p Value	All Dogs n = 93
Age (years) [1]	4.0 ± 2.9	4.9 ± 3.3	>0.05	4.4 ± 3.1
Weight (kg) [1]	19.9 ± 10.6	25.0 ± 12.1	<0.05	22.1 ± 11.5
Sex [2]			>0.05	
Male	24 (45.3)	17 (42.5)		41 (44.0)
Female	29 (54.7)	23 (44.3)		52 (56.0)
Neutered [2]				
Yes	36 (67.9)	34 (85.0)	>0.05	70 (75.2)
No	17 (32.1)	6 (15.0)		23 (24.7)

[1] Age and weight values calculated as mean and ±standard deviation. [2] Sex and neuter status provided as number of dogs and percentage (%).

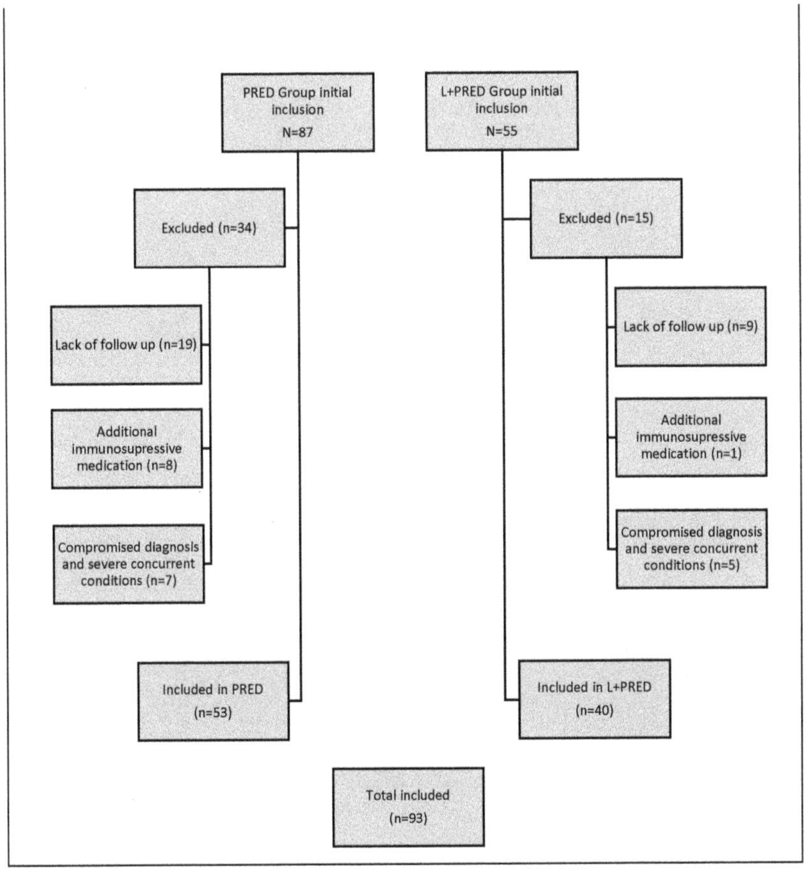

Figure 1. A flowchart summarising the exclusion process for each group.

3.3. Body Temperature and Diagnostic Results

Body temperature (PRED 39.2 ± 0.7 and L+PRED 39.3 ± 0.8), number of joints sampled via arthrocentesis (PRED 2.9 ± 0.8 and L+PRED 3.1 ± 1.0) and the score of the synovial fluid cytology (PRED 1 = 0%, 2 = 30.2%, 3 = 66%, 4 = 3.8%; L+PRED 1 = 2.5%, 2 = 32%, 3 = 60%, 4 = 5%) did not vary between groups (p-value > 0.05).

Synovial fluid microbial cultures were performed in 56/93 (60.2%) dogs (PRED 56.6% and L+PRED 65%). All microbial cultures, except for three in the PRED group and four in the L+PRED group, reported no bacterial growth. In the PRED group, the isolates identified were *Ralstonia insidiosa*, *Serratia marcescens*, and *Achromobacter* spp. In the L+PRED group, *Serratia marcescens* was isolated three times, and *Achromobacter* spp. once. All the positive bacterial cultures reported were regarded by the reference laboratory and treating veterinarian as most likely resulting from bacterial contamination during synovial fluid sample collection.

CRP was measured on presentation in 18/53 (34%) dogs in the PRED group and 8/40 (20%) in the L+PRED group, with median values of 79.6 mg/L (IQR 51.3–95.5) and 78.2 mg/L (IQR 22.2–99.9), respectively. An abnormal CRP concentration was defined as greater than 10 mg/L.

3.4. Concurrent Conditions and Association to IMPA

The presence of concurrent dermatological conditions (PRED 17%, L+PRED 10%), diarrhoea (PRED 49%, L+PRED 30%), classification as associative IMPA (PRED 52.8%, L+PRED 32.5%), and duration of hospitalisation (PRED 2.2 ± 1.5 days, L+PRED 2.0 ± 1.3 days), did not differ between the groups (p-value > 0.05). Concurrent conditions other than diarrhoea and dermatological conditions are summarised in Table 2.

Table 2. Summary of the number of cases with concurrent conditions (other than diarrhoea and dermatological conditions) per treatment group.

Condition	PRED	L+PRED
Pneumonia	2	
Mild anaemia	1	
Pancreatitis	1	1
Urinary tract infection	1	1
Sialadenitis	1	
Hookworm infestation	1	
Oral ulceration	1	1

3.5. Treatment Data

Dogs in the PRED group were more likely to receive intravenous dexamethasone prior to oral prednisolone (PRED 75.5% and L+PRED 55.0%; p-value = 0.038) and received a higher initial dose rate (mg/kg/day) of prednisolone (PRED 1.7 ± 0.5 and L+PRED 1.5 ± 0.4; p-value = 0.010). However, there was no difference in the mean number of days on prednisolone (PRED 176.8 ± 198.9 and L+PRED 159.5 ± 109.9). In the L+PRED group, leflunomide was started at a mean dose of 1.9 ± 0.4 mg/kg/day within 9.9 ± 11.6 days from diagnosis. Immunosuppressive treatments and duration of prednisolone treatment are summarised in Table 3.

Table 3. Immunosuppressive medication dose rates (mean ± standard deviation) and number of prednisolone days for each treatment group.

	PRED (n = 53)	L+PRED (n = 40)	p Value
Dexamethasone *	0.28 ± 0.09	0.44 ± 0.09	0.038
Prednisolone *	1.7 ± 0.5	1.5 ± 0.4	0.010
Leflunomide *		1.9 ± 0.4	
Prednisolone Days †	176.8 ± 198.9	159.5 ± 109.9	0.623

* mg/kg/day. † Number of days (mean ± standard deviation)

3.6. Relapse and Mortality Data

The overall relapse and mortality rates for all 93 dogs in this study were 30.1% and 10.8%, respectively. There was no difference between the relapse rate (PRED 26.4% and L+PRED 35.0%) or the overall mortality rate (PRED 9.4% and L+PRED 12.5%) between the two treatment groups (p-value > 0.05). The median number of days to relapse was 141 (IQR 98–178) in the PRED group and 108 (IQR 54–139) in the L+PRED group, with no difference between the groups (p-value > 0.05). At the time of relapse, 16 (PRED 9 and L+PRED 7) of the total 28 relapsed dogs were receiving prednisolone at a mean dose of 0.5 ± 0.3 mg/kg/day. All of the 14 relapsed L+PRED dogs were receiving leflunomide at a mean dose of 1.7 ± 0.3 mg/kg/day at the time of relapse. Relapse was confirmed in four PRED and three L+PRED cases via repeat arthrocentesis, and elevated CRP confirmed the relapse in three PRED and one L+PRED cases.

In the Cox regression analysis, the unadjusted, age- and sex-adjusted, and multivariable-adjusted models showed no additive benefit of leflunomide in reducing clinical relapse or mortality: HRs for relapse (95% CIs) were 0.88 (0.41–1.89), 0.76 (0.34–1.70), and 0.43 (0.16–1.14), respectively; for mortality: 1.02 (0.29–3.57), 0.66 (0.17–2.53), and 0.34 (0.06–1.86), respectively (Table 4).

Table 4. Association between adding leflunomide to prednisolone and the risk of relapse and mortality.

	Model I	Model II	Model III
Relapse	0.88 (0.41–1.89)	0.76 (0.34–1.70)	0.43 (0.16–1.14)
Mortality	1.02 (0.29–3.57)	0.66 (0.17–2.53)	0.34 (0.06–1.86)

Model I: HR (95% CI) unadjusted. Model II: HR (95% CI) adjusted for age and sex. Model III: HR (95% CI) adjusted for age, sex, weight, receiving dexamethasone, and prednisolone dose at discharge.

3.7. Adverse Effects Related to Leflunomide

Leflunomide-related adverse effects were suspected in 6 of 40 dogs (15.0%) in the L+PRED group. Diarrhoea occurred in 3 dogs; in 2 cases, it was self-limiting, while in the third, it led to discontinuation of the medication after 68 days. One dog experienced a transient decrease in appetite, which resolved without intervention. One dog developed a severe hepatopathy, with serum ALT concentration of 5122 IU/L (reference interval [RI]: 10–125) 57 days after starting leflunomide. This dog was euthanised due to suspected hepatic encephalopathy-induced seizures. Additionally, one dog developed pancytopenia with severe neutropenia (0.05×10^9/L, RI: 2.95–11.64), thrombocytopenia (56 K/μL, RI: 148–484), and anaemia (haematocrit 30%, RI: 37–61) 17 days after initiating leflunomide. This dog recovered after discontinuation of the medication. Notably, none of the dogs that discontinued leflunomide were among the relapsed cases.

3.8. Subsequent Relapses

While the Cox regression analysis included data only up to the first relapse, it is noteworthy that of the total 28 dogs that experienced a relapse, 12 (43%) experienced subsequent relapses ranging from two to five relapses per dog. Additionally, of the 14 relapsed dogs in the PRED group, 8 were later started on leflunomide at a mean dose of 2.2 ± 0.5 mg/kg/day. The median days to relapse and starting leflunomide of those eight dogs was 149 (IQR 112.5–237.5). Of these eight dogs, five experienced further relapses whilst receiving leflunomide. In the L+PRED group, 5 of the 14 relapsed dogs experienced at least one additional relapse while receiving leflunomide during the prednisolone tapering phase.

4. Discussion

We conducted a retrospective evaluation of the use of adjunctive leflunomide in the treatment of dogs with IMPA (L+PRED) and compared the outcomes of relapse and mortality rates to a control group that received prednisolone as monotherapy (PRED). Though focused on a single population, our study provides valuable insight into IMPA in Australia, where data on relapse rates have previously been lacking. While the L+PRED group showed slightly higher relapse and mortality rates, no statistically significant differences in these outcomes were observed. Therefore, the findings of this study do not support the routine addition of leflunomide to prednisolone in the treatment of IMPA.

Although this retrospective study has inherent limitations in fully understanding the factors influencing clinicians' decisions to initiate leflunomide as a second-line treatment, it seems that body weight significantly influenced these decisions. The L+PRED dogs had a higher body weight and lower prednisolone starting dose rate. This likely reflects an attempt by clinicians to reduce prednisolone dosing in large breed dogs when adding leflunomide. Additionally, in our practice, clinicians tend to use a prednisolone dosing regimen of 40 mg/m^2 per day in dogs over 15 kg instead of the commonly recommended immunosuppressive dose of 2 mg/kg/day in an attempt to minimise undesirable prednisolone side effects [9,11]. This could result in the overall lower prednisolone dose rate in the L+PRED group. Clinical impression of severe disease or slow resolution of clinical signs whilst in hospital is another factor that may have influenced the clinician's decision to add leflunomide.

The total relapse rate in our study was 30.1%, which was comparable to one recent and one older UK study [3,4]. The most recent study assessed the relapse rates in 32 dogs with IMPA compared to dogs with IMHA and immune-mediated thrombocytopenia (ITP). The study concluded that IMPA had the highest relapse rate, with 90% of relapses occurring within the first 12 months after diagnosis. This is comparable, albeit with a longer median time to relapse, to our study. Similar to our findings, the authors also found no difference in the use of combination immunosuppressive treatments between dogs that relapsed and those that did not across all disease groups, including IMPA [3].

Another recent UK study reported a relatively higher relapse rate of 48%. This study's aims did not include assessing the impact of leflunomide on relapse rates; however, it did record that 14 out of 23 dogs received leflunomide alongside prednisolone, with 11 of the 14 experienced relapse [5]. A possible explanation for this study's higher relapse rate could be the inclusion of dogs with systemic infections, other immune-mediated diseases, and neoplasia. Although it is difficult to compare outcomes between manuscripts, our study excluded any dog that could have any severe underlying disease. The number of presumed associative IMPA cases in our study appears relatively high; however, most of the associated conditions were non-severe and non-fatal and would not have been expected to impact on the treatment response. The nomenclature system of associative and non-associative was first proposed for IMHA in dogs [12], and it was later adapted for IMPA in a recent study [3], a classification preferred by the authors. However, the term can be broad, potentially leading the readers to assume that the associative IMPA cases in our study are directly related to other more severe diseases.

Assessing the efficacy of individual drugs or dose rates is challenging when medications are used in combination and usually requires prospective randomised controlled trials to evaluate each drug individually. Many prospective studies may have relatively short follow-up periods, limiting the ability to assess long-term relapse rates and outcomes with immunosuppressive drugs. While prospective studies may capture early outcomes, predicting relapse in dogs with IMPA remains difficult. This was shown in a well-designed 90-day prospective trial comparing prednisolone to cyclosporine in 20 dogs (10 per group) and found a 70% response rate to both treatments, highlighting early effectiveness but not long-term relapse risks [13]. The median time to relapse in our study was 141 days in the PRED group (IQR: 98–178) and 108 days in the L+PRED group (IQR: 54–139). This is longer than the 90-day follow-up duration of the aforementioned study, emphasising the need for long-term follow-up for studies to evaluate the relapse rate.

Similarly, the difference in disease duration may explain why leflunomide benefits humans with rheumatoid arthritis (RA) but not dogs with IMPA. The RA disease process in humans is chronic, requiring long-term management to decrease persistent symptoms, compared to canine IMPA, in which most cases achieve remission and do not require indefinite treatment. In dogs, the main concern is acute relapse during corticosteroid tapering, unlike the long-term disease control required in human RA. Many IMPA relapses are mild and can be managed by adjusting the prednisolone dose, while severe or recurrent cases may require addition of other immunomodulating medications [4].

The role of leflunomide in the treatment of more severe chronic and erosive forms of IMPA recognised as having a more similar pathogenesis to human RA has been minimally investigated in the veterinary literature in a small case series of 13 dogs with erosive IMPA, of which 9 were managed with a relatively high leflunomide dose (2.6 to 4.5 mg/kg/day) [14].

Leflunomide as an alternative to prednisolone has been also reported to provide disease control in 8 out of 14 dogs in a previous case series [10]. However, no long-term follow-up was performed to assess the relapse rate, and 6 out of the 14 (43%) did not respond adequately, representing a large portion of the population. Notably, this study was conducted in a different geographical region, and three dogs were seropositive for tick-borne infections and received doxycycline treatment, which may have influenced the outcomes. It remains unclear whether geographical variations in infectious disease prevalence, in particular tick-borne disease, play a role in the outcomes of IMPA treatment [1,15].

It should also be noted that the cases included in our study received a slightly lower initial leflunomide dosage (1.9 ± 0.4 mg/kg/day) than advocated in some previous studies [10,14,16]. The rationale for using a lower leflunomide dosage is not solely anecdotal. The authors observed fewer adverse effects without a difference in outcomes. These findings were also supported by a retrospective study recommending a starting dose of 2 mg/kg/day, which found significant median dose differences between dogs with adverse events (2.9 mg/kg/day) and dogs without adverse events (1.6 mg/kg/day). The same study also found no difference in the median dosages between dogs with immune-mediated diseases that responded (1.9 mg/kg/day) and those that did not respond (1.7 mg/kg/day) [17].

Several limitations arise from the retrospective nature of this study. Firstly, the use of regression analysis does not provide any statistical basis for assessing subsequent relapses. Also, dogs that experienced repeated relapses may have been genetically predisposed to immune-mediated diseases or autoimmunity, re-exposed to an unidentified environmental or infectious trigger that initiated the disease or had an undiagnosed underlying condition. This subset of dogs may have been at higher risk for relapse, independent of the treatment protocol used. Furthermore, variability in treatment protocols such as drug combinations and differences in doses and durations limits a full assessment of leflunomide's role. For instance, the L+PRED group received a lower starting dose of prednisolone compared to the PRED group, likely due to the higher body weight of the dogs in this group. While this could be a potential confounding factor that might obscure the effect of leflunomide, it is unlikely to have had a significant influence, as relapses in both groups occurred much later, when prednisolone doses were either low or discontinued.

Another limitation of this study is our inability to objectively evaluate whether leflunomide provided any clinical benefit in reducing glucocorticoid adverse effects in the L+PRED group compared to the PRED group. The presence of adverse effects based on subjective reports from dog owners was noted in the clinical records of dogs in both groups. A prospective study would be necessary to accurately quantify and compare adverse effects between the two groups, given that both were receiving glucocorticoids. Clinicians commonly initiated leflunomide at diagnosis with the aim of reducing prednisolone-related side effects. However, our data indicated that leflunomide did not significantly reduce the duration of prednisolone treatment, which we believe would be necessary to meaningfully decrease prednisolone-related side effects.

Moreover, while radiography was not part of the eligibility criteria in our study, it may have offered further insights into the potential presence of joint erosion. Nonetheless, the lack of radiography in both groups is not expected to have influenced the results. Finally, in most cases, relapse assessment relied on the recurrence of clinical signs and the response to therapy adjustments, rather than definitive cytological confirmation. Although serum CRP concentration quantification was not available in our practice before 2020, it has now become readily accessible, offering a valuable non-invasive tool to help identify relapse in dogs with IMPA [18].

5. Conclusions

Our study findings indicate that leflunomide did not offer a therapeutic advantage to reduce relapse rate compared to prednisolone as monotherapy within this cohort. When considering the costs and potential adverse effects of leflunomide, one should question using the drug as an adjunctive therapy at the time of diagnosis for non-relapsed canine IMPA cases. A prospective randomised controlled trial involving a higher leflunomide dose may be necessary to thoroughly evaluate the long-term efficacy in dogs with IMPA.

Supplementary Materials: The following supporting information can be downloaded at: https://www.mdpi.com/article/10.3390/vetsci11110537/s1, Table S1: Breed distribution between the treatment groups.

Author Contributions: Conceptualisation, R.W.; methodology, R.W.; software, A.A. and C.P.; validation, I.S.; formal analysis, R.W. and A.A.; investigation, R.W.; data curation, R.W., S.L. and M.G.-S.; writing—original draft preparation, R.W.; writing—review and editing, R.W., I.S., A.A., T.D. and M.T.; visualization, R.W.; supervision, I.S. All authors have read and agreed to the published version of the manuscript.

Funding: This research received no external funding.

Institutional Review Board Statement: Not applicable in this retrospective study.

Informed Consent Statement: Patient consent was waived due to the retrospective nature of this study and the long period of the retrospective analysis.

Data Availability Statement: The data used to generate the results in this manuscript can be made available if requested from the corresponding author. The data are not publicly available due to professional secrecy.

Acknowledgments: The authors thank Anjie Storrer for her guidance and collaboration. The authors extend their appreciation to Lisa Smart for her collaboration and guidance, along with the staff at Queensland Veterinary Specialists who provided assistance in this research paper.

Conflicts of Interest: The authors declare no conflicts of interest.

References

1. Rondeau, M.P.; Walton, R.M.; Bissett, S.; Drobatz, K.J.; Washabau, R.J. Suppurative, Nonseptic Polyarthropathy in Dogs. *J. Vet. Intern. Med.* **2005**, *19*, 654–662. [CrossRef] [PubMed]
2. BENNETT, D. Immune-based Non-erosive Inflammatory Joint Disease of the Dog. 3. Canine Idiopathic Polyarthritis. *J. Small Anim. Pract.* **1987**, *28*, 909–928. [CrossRef]
3. Sparrow, R.; Swann, J.W.; Glanemann, B. Comparison of Timing of Relapse in Dogs with Nonassociative Immune-Mediated Hemolytic Anemia, Thrombocytopenia, or Polyarthritis. *J. Vet. Intern. Med.* **2024**, *38*, 1035–1042. [CrossRef] [PubMed]
4. Clements, D.N.; Gear, R.N.A.; Tattersall, J.; Carmichael, S.; Bennett, D. Type I Immune-Mediated Polyarthritis in Dogs: 39 Cases (1997–2002). *J. Am. Vet. Med. Assoc.* **2004**, *224*, 1323–1327. [CrossRef] [PubMed]
5. Perez, J.C.; Travail, V.; Lamb, V.; Kelly, D. Canine Immune-Mediated Polyarthritis: A Review of 84 Cases in the UK between 2011 and 2021. *Vet. Med. Sci.* **2024**, *10*, e1306. [CrossRef]
6. Ravicini, S.; Kent, A.; Dunning, M.; Baines, S.; Clarke, S.; Allerton, F. Description and Outcome of Dogs with Primary Immune-Mediated Polyarthritis: 73 Cases (2012–2017). *J. Small Anim. Pract.* **2023**, *64*, 142–148. [CrossRef] [PubMed]
7. Johnson, K.C.; Mackin, A. Canine Immune-Mediated Polyarthritis: PART 2: Diagnosis and Treatment. *J. Am. Anim. Hosp. Assoc.* **2012**, *48*, 71–82. [CrossRef] [PubMed]
8. Walker, H. Immune-Mediated Polyarthritis in Dogs: Are Corticosteroids the Best Bet? *Vet. Evid.* **2023**, *8*, 1–9. [CrossRef]
9. Chesne, R.B.; Doornink, M.T.; Sri-Jayantha, L.S.H.; Urie, B.K. Leflunomide with Prednisone or Nonsteroidal Anti-Inflammatory Drug Therapy Is Safe and Tolerated for Long-Term Treatment of Immune-Mediated Polyarthritis in 27 Dogs. *J. Am. Vet. Med. Assoc.* **2024**, *262*, 1188–1192. [CrossRef] [PubMed]
10. Colopy, S.A.; Baker, T.A.; Muir, P. Efficacy of Leflunomide for Treatment of Immune-Mediated Polyarthritis in Dogs: 14 Cases (2006–2008). *J. Am. Vet. Med. Assoc.* **2010**, *236*, 312–318. [CrossRef] [PubMed]
11. Purcell, B.; Wiethoelter, A.; Dandrieux, J. Prednisolone Prescribing Practices for Dogs in Australia. *PLoS ONE* **2023**, *18*, e0282440. [CrossRef] [PubMed]
12. Garden, O.A.; Kidd, L.; Mexas, A.M.; Chang, Y.-M.; Jeffery, U.; Blois, S.L.; Fogle, J.E.; MacNeill, A.L.; Lubas, G.; Birkenheuer, A.; et al. ACVIM Consensus Statement on the Diagnosis of Immune-Mediated Hemolytic Anemia in Dogs and Cats. *J. Vet. Intern. Med.* **2019**, *33*, 313–334. [CrossRef] [PubMed]
13. Rhoades, A.C.; Vernau, W.; Kass, P.H.; Herrera, M.A.; Sykes, J.E. Comparison of the Efficacy of Prednisone and Cyclosporine for Treatment of Dogs with Primary Immune-Mediated Polyarthritis. *J. Am. Vet. Med. Assoc.* **2016**, *248*, 395–404. [CrossRef] [PubMed]
14. Shaughnessy, M.L.; Sample, S.J.; Abicht, C.; Heaton, C.; Muir, P. Clinical Features and Pathological Joint Changes in Dogs with Erosive Immune-Mediated Polyarthritis: 13 Cases (2004–2012). *J. Am. Vet. Med. Assoc.* **2016**, *249*, 1156–1164. [CrossRef] [PubMed]

15. Irwin, P.J.; Robertson, I.D.; Westman, M.E.; Perkins, M.; Straubinger, R.K. Searching for Lyme Borreliosis in Australia: Results of a Canine Sentinel Study. *Parasit. Vectors* **2017**, *10*, 114. [CrossRef] [PubMed]
16. Singer, L.M.; Cohn, L.A.; Reinero, C.R.; Papich, M.G. Leflunomide Pharmacokinetics after Single Oral Administration to Dogs. *J. Vet. Pharmacol. Ther.* **2011**, *34*, 609–611. [CrossRef] [PubMed]
17. Sato, M.; Veir, J.K.; Legare, M.; Lappin, M.R. A Retrospective Study on the Safety and Efficacy of Leflunomide in Dogs. *J. Vet. Intern. Med.* **2017**, *31*, 1502–1507. [CrossRef] [PubMed]
18. Foster, J.D.; Sample, S.; Kohler, R.; Watson, K.; Muir, P.; Trepanier, L.A. Serum Biomarkers of Clinical and Cytologic Response in Dogs with Idiopathic Immune-Mediated Polyarthropathy. *J. Vet. Intern. Med.* **2014**, *28*, 905–911. [CrossRef] [PubMed]

MDPI AG
Grosspeteranlage 5
4052 Basel
Switzerland
Tel.: +41 61 683 77 34

Veterinary Sciences Editorial Office
E-mail: vetsci@mdpi.com
www.mdpi.com/journal/vetsci

www.ingramcontent.com/pod-product-compliance
Lightning Source LLC
Chambersburg PA
CBHW040805250926
51850CB00007BA/1084
* 9 7 8 3 7 2 5 8 4 7 3 9 6 *